Dietrich Werner (Ed.)

Biological Resources and Migration

Springer

Berlin
Heidelberg
New York
Hong Kong
London
Milan
Paris
Tokyo

Dietrich Werner (Ed.)

Biological Resources and Migration

With 43 Figures and 23 Tables

 Springer

Professor Dr. Dietrich Werner
Department of Biology
Institute of Cellbiology and Applied Botany
Philipps-University Marburg
Karl-von-Frisch-Straße
35032 Marburg
Germany

ISBN 3-540-21470-4 Springer-Verlag Berlin Heidelberg New York

Library of Congress Control Number: 2004105119

Springer-Verlag is a part of Springer Science + Business Media
springeronline.com

Springer-Verlag Berlin Heidelberg 2004
Printed in Germany

Cover design: Desgn & Production, Heidelberg
Typesetting: Camera ready by editor
31/3150WI - 5 4 3 2 1 0 - Printed on acid-free paper

Preface

The conference and OECD workshop were organized at the Philipps University in Marburg (Germany) by Dietrich Werner, supported by all other members of the Management Committee "Biological Resources Management for Sustainable Agricultural Systems": Ervin Balazs, Françoise Coudert, Jim Lynch, Kiyotaka Miyashita, Jim Schepers, Paolo Sequi, Jean-Pierre Toutant and Fons Werry and by several national correspondents from the programme: Yvon Martel, Jean Charles Munch, Stefan Martyniuk, Marcello Pagliai and the long-term former chairman of the committee, Heiko van der Borg.

The conference was attended by 102 scientists from 24 countries, including most programme member countries. Equally broad were the disciplines represented, such as ecology, zoology, botany, microbiology, virology, parasitology, molecular biology, plant breeding, soil science, nature conservation, geography, agronomy, socioeconomy and economy. Most of the speakers of the workshop had never met before, indicating the innovative and interdisciplinary character of the conference. Especially the dialogue between natural sciences and socioeconomy was an important goal of the meeting.

When we look at our cultivated land in the OECD member countries as well as in developing countries, we see that a large percentage of the land, but also of the forests, is covered with migrated and introduced plants, such as rice, wheat, maize, soybeans, potatoes, pines and firs, which together cover millions of square kilometers of land. However, also in natural habitats and communities, migration of animals and microbes as well as plants plays a key role in the functioning of the present ecosystems. When we include human activities and people in this view, such central topics as trade, transport and tourism are forms of migration. Therefore, this conference was very much at the center of tasks of several directorates of the OECD.

During the organization of the conference, more than 20 scientists from African countries applied to attend the workshop. The organizer sent a large number of supporting letters to institutions and embassies, especially in Nigeria, Ghana and Cameroon. At the beginning of October, all except four participants from these countries were denied visas, leaving these mostly younger scientists (and the organizer) very frustrated. The dialogue between the OECD member countries and countries with immense problems in human migration is now even more difficult than it was 10 years ago, simply because of the more restricted visa situation.

The organizer of the conference especially thanks Mrs. Lucette Claudet and Eike ter Haseborg for their dedicated work for the conference and for the publication of the proceedings.

Dietrich Werner

The organizer thanks the following institutions for support of this conference/workshop:

- OECD Co-operative Research Programme "Biological Resources Management for Sustainable Agricultural Systems", Paris
- Deutsche Forschungsgemeinschaft, Bonn
- Ursula-Kuhlmann-Fonds, Marburg
- Marburger Universitätsbund

Contents

List of Contributors

Ammann, K., Prof. Dr.
  Botanical Garden, University Bern
  Altenbergrain 21, 3013 Bern
  Switzerland

Bartoš, P., Ing.DrSc.
  Research Institute of Crop Production
  161 06 Praha Ruzyně
  Czech Republic

Becker, S.S., Dr.
  Virology Institute, Faculty of Medicine, Philipps-University Marburg
  Robert-Koch-Strasse 17, 35032 Marburg
  Germany

Bialozyt, R., Dr.
  Working Group of Conservation Biology and Conservation Genetics
  Nature and Conservation Division, Faculty of Biology
  Philipps-University Marburg
  Karl-von-Frisch-Strasse, 35032 Marburg
  Germany

Boyd, I.L., Prof. Dr.
  Sea Mammal Research Unit, Gatty Marine Laboratory
  University of St Andrews,
  St Andrews KY16 8LB
  U.K.

Brecevic, D., Dr.
  Department of Biology, University of Trieste
  34100 - Trieste
  Italy

Carter, Jr., T.E., Prof. Dr.
  Research Geneticist, USDA-ARS
  3127 Ligon Street, Raleigh, NC 27607
  USA

Chen, W.X., Dr.
Dept. of Microbiology, College of Biological Science
China Agricultural University
Beijing 100094
China

Dragan, M., Dr.
Department of Biology, University of Trieste
34100 - Trieste
Italy

Drevon, J.-J., Prof. Dr.
Laboratoire Symbiotes des Racines et Sciences du Sol, INRA
1 Place Viala, 34060 Montpellier-Cedex
France

Feoli, E., Prof. Dr.
Department of Biology, University of Trieste
34100 - Trieste
Italy

Fricke, W., Prof. Dr.
Department of Geography, Ruprecht-Karls-University Heidelberg
Im Neuenheimer Feld 348, 69120 Heidelberg
Germany

Fuchs, J.G., Dr.
Phytopathology Institute of Plant Sciences
Swiss Federal Institute of Technology Zurich (ETH)
8092 Zürich
Switzerland

Graber, W.K.., Dr.
Paul Scherrer Institute
5232 Villigen
Switzerland

Graybosch, R.A., Prof. Dr.
Research Geneticist, USDA-ARS,
344 Keim, UNL, East Campus, Lincoln, Nebraska 68583
USA

Higler, L.W.G., Prof. Dr.
Wageningen UR, ALTERRA
P.O. Box 46, 6700 AA Wageningen
The Netherlands

Holmgren, N., Dr.
  Department of Natural Sciences, University of Skövde
  P.O. Box 408, 541 28 Skövde
  Sweden

Hymowitz, Th., Prof. Dr.
  Plant Genetics, University of Illinois
  Urbana, IL 61801
  USA

Javier, E.L., Dr.
  Plant Breeder and INGER Coordinator
  International Rice Research Institute (IRRI)
  DAPO Box 7777, Metro Manila
  Philippines

Johann, E., Prof. Dr.
  Institute for Socioeconomics
  University of Natural Resources and Applied Life Sciences
  Vienna
  Austria

Kern. J., Dr.
  Institute of Agricultural Engineering (ATB)
  Max-Eyth-Allee 100, 14469 Potsdam
  Germany

Knerr, B., Prof. Dr.
  Institut für Soziokulturelle Studien, FB Landwirtschaft, University Kassel
  Steinstrasse 19, 37213 Witzenhausen
  Germany

Kost, G., Prof. Dr.
  Special Botany and Mycology, Faculty of Biology
  Philipps-University Marburg
  Karl-von-Frischstrasse, 35032 Marburg
  Germany

Kraft, M., Dr.
  General Ecology and Animal Ecology, Faculty of Biology
  Philipps-University Marburg
  Karl-von-Frischstrasse, 35032 Marburg
  Germany

Kreibich, H., Dr.
  Section Engineering Hydrology, GeoForschungsZentrum Potsdam
  Telegrafenberg F325, 14473 Potsdam
  Germany

Kyi, A., Prof. Dr.
  Ministry of Education, Department of Botany, University of Yangon
  Myanmar

Liepelt, S., Dr.
  Working Group of Conservation Biology and Conservation Genetics
  Nature and Conservation Division, Faculty of Biology
  Philipps-University Marburg
  Karl-von-Frisch-Strasse, 35032 Marburg
  Germany

Limpert, E., Dr.
  Lecturer Aerobiology, University Zurich
  Scheuchzerstr. 210, 8057 Zurich
  Switzerland

Loxdale, H.D., Dr.
  Plant and Invertebrate Ecology Division, Rothamsted Research Station
  Harpenden, Hertfordshire, AL5 2JQ
  U.K.

Macdonald, C., Dr.
  Plant and Invertebrate Ecology Division, Rothamsted Research Station
  Harpenden, Hertfordshire, AL5 2JQ
  U.K.

Martyniuk, S., Dr.
  Department of Agricultural Microbiology
  Institute of Soil Science and Plant Cultivation
  Czartoryskich 8, 24-100 Puławy
  Poland

Miyashita, K., Dr.
  Crop Evolutionary Dynamics Team
  National Institute of Agrobiological Sciences (NIAS)
  Kannondai 2-1-2 Tsukuba, Ibaraki 305-8602
  Japan

Miyazaki, S., Dr.
  Crop Evolutionary Dynamics Team
  National Institute of Agrobiological Sciences (NIAS)
  Kannondai 2-1-2 Tsukuba, Ibaraki 305-8602
  Japan

Nelson, R.L., Dr.
  Research Geneticist, USDA-ARS
  Urbana, IL 61801
  USA

Osaki, M., Dr.
  Graduate School of Agriculture, Hokkaido University
  Sapporo, 060-8589
  Japan

Pagliai, M., Dr.
  Istituto Sperimentale per lo Studio e la Difesa del Suolo
  Piazza M. D'Azeglio 30, 50121 - Firenze
  Italy

Plachter, H., Prof. Dr.
  Nature Conservation Division, Faculty of Biology
  Philipps-University Marburg
  Karl-von-Frisch-Strasse, 35032 Marburg
  Germany

Prospero, J.M., Prof. Dr.
  Cooperative Institute for Marine and Atmospheric Studies
  Rosenstiel School of Marine and Atmospheric Science, University of Miami
  4600 Rickenbacker Causeway, Miami FL 33149
  USA

Riede, K., Dr.
  Zoologisches Forschungsinstitut und Museum Alexander Koenig (ZFMK)
  Adenauerallee 160, 53113 Bonn
  Germany

Rinaldi, M., Dr.
  Istituto Sperimentale Agronomico
  Via Celso Ulpiani 5, 70125 – Bari
  Italy

Rodiño, A.P., Dr.
Laboratoire Symbiotes des Racines et Sciences du Sol, INRA
1 Place Viala, 34060 Montpellier-Cedex
France

Rodhouse, P.G.K., Prof. Dr.
British Antarctic Survey
High Cross, Madingley Road, Cambridge CB3 0ET
U.K.

Shinano, T., Dr.
Creative Research Initiative "Sousei" (CRIS), Hokkaido University
Sapporo, 001-0020
Japan

Silva, C., Dr.
Centro de Investigación sobre Fijación de Nitrógeno
Universidad Nacional Autónoma de México
Av. Universidad S/N, Col. Chamilpa, AP 565-A
Cuernavaca, Morelos CP62210
México

Thangaraju, M., Prof. Dr.
Department of Agricultural Microbiology, Tamil Nadu Agricultural University
Coimbatore 641 003, Tamil Nadu
India

Toledo, M.C., Dr.
International Rice Research Institute (IRRI)
DAPO Box 7777, Metro Manila
Philippines

Unno, Y., Dr.
Graduate School of Agriculture, Hokkaido University
Sapporo, 060-8589
Japan

Vaughan, D.A., Prof. Dr.
Crop Evolutionary Dynamics Team
National Institute of Agrobiological Sciences (NIAS)
Kannondai 2-1-2 Tsukuba, Ibaraki 305-8602
Japan

Vogel, K.P., Prof. Dr.
 Research Geneticist, USDA-ARS, University of Nebraska
 344 Keim Hall, Lincoln, NE 68583-0937
 USA

Vinuesa, P., Dr.
 Centro de Investigación sobre Fijación de Nitrógeno
 Universidad Nacional Autónoma de México
 Av. Universidad S/N, Col. Chamilpa, AP 565-A
 Cuernavaca, Morelos CP62210
 México

Wang, E.T., Dr.
 Depto. de Microbiología, Escuela Nacional de Ciencias biológicas
 Instituto Politécnico Nacional
 México D. F.
 México

Wasaki, J., Dr.
 Graduate School of Agriculture, Hokkaido University
 Sapporo, 060-8589
 Japan

Werner, D., Prof. Dr.
 Cell Biology and Applied Botany, Faculty of Biology
 Philipps-University Marburg
 Karl-von-Frisch-Strasse, 35032 Marburg
 Germany

Yan, S., Prof. Dr.
 Institute of Remote Sensing Application
 Chinese Academy of Sciences
 Beijing
 China

Ziegenhagen, B., Prof. Dr.
 Working Group of Conservation Biology and Conservation Genetics
 Nature and Conservation Division, Faculty of Biology
 Philipps-University Marburg
 Karl-von-Frisch-Strasse, 35032 Marburg
 Germany

1 The Rice Genepool and Human Migrations

Duncan A. Vaughan, S. Miyazaki and K. Miyashita

National Institute of Agrobiological Sciences, Tsukuba, Japan, E-mail duncan@affrc.go.jp

1.1 Abstract

The characteristics of the rice grain, particularly its hard, tight husk and nutrient content, make it an ideal grain to accompany migrating people. This paper discusses rice diversity and movement from a historical perspective by providing contrasting examples from the homeland of Asian rice, Madagascar and West Africa. The value of modern analytical techniques to understand the origin, diversification and movement of rice is discussed. The following points emerge:

1. No evidence exists to support the center of origin and the center of diversity of rice being the same area, and molecular studies support multiple domestication events probably over a wide area.

2. Inter-subspecific hybridization and diversification after a genetic bottleneck account for much of the traditional rice diversity of Madagascar.

3. Introduction of Asian rice (*O. sativa*) to West Africa has resulted in partial replacement of indigenous African rice, *O. glaberrima*, introgression has occurred from Asian into African rice, and recently a new interspecific genepool has been developed by researchers with the aim of combining the best characters of both rice species.

Finally, the recent consequences of human migration from rural to urban areas and the impact on the rice ecosystem are discussed with particular reference to Malaysia. The common trend of change from transplanting to direct seeding rice and accompanying ecological changes in rice fields are discussed.

1.2 The Significance of a Grain of Rice

It is necessary to start this paper with a consideration of the rice grain since this is central to the emergence of rice as a major staple food for humankind and also helps explain its spread around the world.

The rice grain evolved in tropical climates from wild rice that grows in still or flowing water, sometimes at a considerable depth. High seed-producing wild rice, from which rice was domesticated, shed their seeds as water recedes and seeds must survive a dry season(s) before germination. Rice grains evolved that can remain viable when subjected to the stresses of tropical climate and (seasonal) aquatic environments.

The hull of the rice grain consists of tightly interlocked palea and lemma. The palea and lemma form a strong structure due to the layer of fibrous schlerenchyma, which protects the embryo and endosperm (Hoshikawa 1989). The high silicon content of rice hulls enhances protection against certain insects and diseases (Chandler 1979). Compared to other cereals rice grains are far better protected and in the hulled state do not readily spoil. In addition, tropical landraces of rice retained the dormancy of wild rice, and this combined with a protective hull enables seed viability to be maintained for a long time in appropriate conditions (Chang 1988).

As for nutrient composition, brown rice has a higher calorie content than wheat and is a good source of protein, particularly containing the most important limiting amino acids lysine, threonine and methionine (Chang 1985; Chandler 1979). Lysine constitutes about 4% of the protein of polished rice, twice the amount found in wheat flour and hulled maize. This combination of highly nutritious food that is well packaged means that rice is an ideal grain to accompany humans on long journeys. From historic documents we know that hulled rice was taken on slave ships from West Africa to the New World and that women slaves dehulled rice during the journey (Carney 2003). It is also known that rice carried on a ship from Madagascar (though it is not clear exactly where the rice came from) was transported over many months to Charleston in the US in 1685. The rice grains were still viable and they marked the beginning of the rice *industry* in the New World (Dethloff 2003). Thus the anatomical and physiological characteristics of the rice grain furnishes it with the ability to remain viable from one planting season to the next in the humid tropics. Again, it is also a nutritious, if not complete, food to accompany humans on long journeys.

1.3 Origin and Diversity and Spread of Rice in Asia

1.3.1 Origins of Rice

Current opinion now suggests that the first domestication of rice occurred at more than one place in Asia, widely separated (Glaszmann 1988; Sato 2000; Cheng et al. 2003). Particularly important in supporting this is a variation in the non-nuclear (cytoplasmic) genomes that shows diversity in *indica* landraces and between *japonica* and *indica* varietal groups (Sato 2000). The earliest evidence of domesticated rice at an archaeological site, dated about 6000 B.C., is in the middle Yangtze valley, from where rice spread in all directions (Fig. 1). Rice was also, probably some millennia later, domesticated in South Asia and possibly also mainland Southeast Asia.

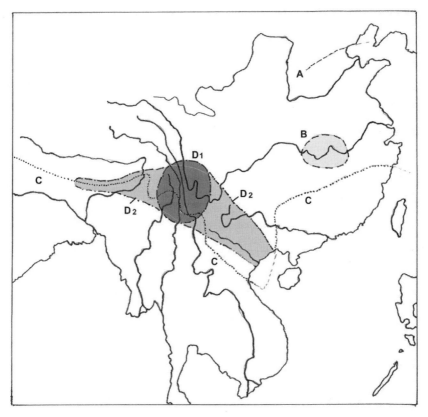

Fig. 1. The home of Asian rice. *A* Northern limit of wild rice reported for ancient times (Chang 1983). *B* Zone of oldest known domesticated rice (Belwood 1997). *C* River arcs of distribution of wild relatives of rice today. D_1 Core zone of rice diversity in the recent past (Watabe 1980). D_2 Asian arc of rice diversity beyond the core zone (Chang 1985)

1.3.2 Center of Diversity of Rice

In the last century collecting of landrace rice germplasm enabled the patterns of rice diversity to be analyzed. This showed that a high level of rice diversity was to be found in a belt from Nepal to the Red river delta of northern Vietnam (Chang 1976). Within this broad area is Yunnan province, China, the core area for rice diversity (Nakagahra 1977; Watabe 1980).

The center of diversity of rice reflects three factors that acted interactively as selection pressures on rice - geography (including land form and climate), ecology and human cultural diversity. Yunnan province and neighboring Assam, India, represents a region through which the great rivers from the eastern Tibetan Plateau flow. The result is a series of deep river valleys on the sides of which tremendous geographic and ecological diversity is found where rice is grown. No other region of the world has a comparable complexity of rivers and mountain systems and consequently environmental diversity as Yunnan province and surrounding areas. Within 300-400km the Brahamaputra (west), Irrawaddy (south), Salween (south), Mekong (southeast), Yangtze (east) rivers flow down from the Tibetan plateau. This area also has a very high level of cultural diversity where peoples of the three sub-continents of Asia meet. For example, in Vietnam's Lai Chau province, adjacent to Yunnan province, there are 13 ethnic groups that speak with languages from 4 linguistic groups. Each of these ethnic groups has its own preference for rice and have selected rice varieties to suit their taste and village environment.

Thus geographical, ecological and cultural diversity explain the location of the center of diversity for rice and also one reason why rice genetic diversity exceeds that of other crops. The genetic diversity and genetic potential to diversify is another reason, along with rice grain structure, to help explain the spread of rice widely with man in early times. There is no archaeological evidence that the current center of rice diversity is where rice was first domesticated. This is supported by the fact that today the greatest variation inthe wild relatives of rice is found in the middle and lower reaches of the river valleys emanating from where cultivated rice is most diverse.

1.3.3 Spread of Asian Rice from its Homeland

Traditional rice varieties are generally photoperiod sensitive. Thus how did rice spread north and south from where it was domesticated and adapt to changes in photoperiod? Wild rice introgression must have played an important role in diversification of early rice. This would have been particularly true on mainland tropical Asia where wild rice was (and is) common.

Spread south across insular Asia may have resulted from different factors (e.g. mutation) since wild rice is not common in many areas. Consequently, the spread of rice culture by Austronesians across the Philippines and some parts of the Malay Archipelago would likely to have been slower than in mainland Asia. However, in Japan, where wild rice has never been reported, wet-rice cultivation spread rapidly over 1200 km from Kyushu to the north of Honshu, 7° latitude, over a period of 300 years (Imamura 1996).

The migration and colonization by the Austronesian peoples has been characterized by Bellwood as "perhaps the most rapid, successful and widespread in the history of humanity prior to the recent dispersals from Europe" (Bellwood 1997). It is not clear how important agricultural capability and/or the crops themselves were in the extraordinary spread of Austronesians that resulted in them emerging apparently from Taiwan (or southern China) and subsequently colonizing a region from Madagascar in the west to New Zealand and the Pacific islands beyond. However, the rapid phase in this expansion between 2000-1000 B.C. occurred after the domestication of rice from a region where domesticated rice was known to have been present. In addition, as discussed in section one of this paper rice has characteristics that lend it to being a useful food source on long sea journeys.

Rice, we can assume, accompanied Austronesians as they traveled but this did not always result in rice cultivation where it was taken. Migration by Austronesians to the east did not result in rice culture west of the Molluca Islands. One reason that rice did not become established as a crop east of the Mollucas by Austronesians may be that during the main period of rapid Austronesian migration east (2000-1000B.C.), rice agriculture and traditions were not embedded in the culture of Austronesians, or at least in that of the migrating peoples. Austronesian migration that led to rice cultivation in the Mariana Islands in the Pacific and westward in Madagascar were later events. The people involved in these later migrations belonged to groups with strong rice culture, as discussed for Madagascar below (Yawata 1961; Bellwood 1997). The presence of traditional *bulu* (tropical *japonica*) varieties in Madagascar and the Marianas points to Austronesian expansion as the source of rice culture in both widely separated regions.

Points
1. Current evidence suggests that domestication of Asian rice occurred more than once and outside what is now considered the core center of rice diversity.
2. Gene flow between rice and its wild relatives probably played an important role in the adaptation of rice to new habitats across mainland Asia.
3. The spread and establishment of rice in new areas reflects the degree to which rice is an essential staple for migrating people and to which their technological ability to grow rice has advanced.

1.4 Migration of Rice and Rice Cultures: The Case of Madagascar

Since the early inhabitants of Madagascar left no written record, the early human history of the island must be deduced indirectly. Expansion of Austronesians to thier furthest point west is believed to have occurred after 400 A.D. based on linguistic evidence (Fukazawa 1988). An understanding of the monsoon winds was critical for the westward expansion of the seagoing Austronesians (Carpenter 1978). It appears likely that crews from southern Borneo, perhaps captained by Malays or Javanese, were the first settlers of Madagascar. Among the characteristic items they bought with them between 400-1000 A.D. were blowpipes, language, rice culture and rice itself. There is no evidence that Madagascar was settled prior to the Austronesians (Bellwood 1997).

Rice is a staple crop in Madagascar and successive waves of people brought rice varieties and cultural practices that established a clear link with Asia. Today there is a complex culture centered on rice (Maeda and Armand 1988). Analysis of present-day traditional varieties of rice from Madagascar has enabled a pattern of evolution of rice in an isolated situation to be deduced from rice cultural practices and rice morphological and biochemical traits.

Based on studies of rice, cultural practices and analyses of traditional varieties for basic *indica* and *japonica* traits, Tanaka (1988) suggested migration occurred in two major waves. First, from the Malay archipelago people brought rice cultural practices and tropical *japonica* varieties. Primarily, the east coast of Madagascar has varieties and cultural practices that reflect Malay influence that include:

1. Tillage using an oar shaped spade;
2. Harvesting panicles with knives to pick each panicle;
3. Foot treading for threshing;
4. Rice storage barns on stilts;
5. Cattle trampling to puddle wetland rice paddies (Tanaka 1988).

Second, rice and rice culture from South Asia were introduced to Madagascar and infiltrated the Malay rice culture and rice varieties (by hybridization). Primarily in northeastern Madagascar rice cultural practices and varieties currently reflect influences from South Asia.

Using a combination of morphological and isozyme analyses, a large collection (about 180) of traditional Madagascar rice varieties have been analyzed (Ahmadi et al. 1991). These analyses revealed that Madagascarhas four main groups of rice-*indica*, temperate *japonica*, tropical *japonica* and a group specific to Madagascar. The Madagascar-specific group appears to be the result of inter-subspecific (*indica/japonica*) recombination and selection.

The results reveal that some minor Asian rice alleles are not found in Madagascar varieties due to a founder effect. On the other hand, Madagascar rice varieties have a much higher frequency of *Amp*-1^2 (located on chromosome 2) than is found in Asian traditional cultivars and this maybe due to genetic drift, environmental conditions that favor certain recombinant forms (natural selection) or selection of specific genotypes to the particular agro-ecologies of Madagascar (human selection).

Points:

1. While the founder effect can result in loss of gene (allelic) diversity, introductions from different parts of Asia have brought together different varietal groups that by farmer/farmer exchange and sympatric planting resulted in gene exchange among varieties and varietal groups (*indica* and *japonica* - intersubspecific recombination) and selection for new environmental niches. In this, wild rice did not play a role in Madagascar, where for the AA genome (the genome of rice) only *O. longistaminata* grows naturally without the *Amp*-1^2 allele that is now found in the background of both *indica* and *japonica* varieties of Madagascar.

1. The founder effect (genetic bottlenecks) does not necessarily lead to a reduced genetic diversity (Polans and Allard 1989; Carson 1990), but rather to a different genetic diversity that may eventually result in new secondary centers of crop diversity (Pickersgill 1998).

2. Despite the relatively narrow genetic base of cultivated rice, rice has adapted to the complex ecosystems of Madagascar (humid tropical and semitropical climate, sea level and 2200 m) and this has resulted in at least 1200 varieties or ecotypes (Arraudeau 1978). Local varieties have a unique adaptation to high altitude

and pathosystems at high altitudes that is not matched by non-adapted introduced rice varieties (Arraudeau 1978).

1.5 African Rice and the Introduction of Asian Rice into West Africa

That mankind has independently followed similar paths in the domestication and exploitation of plants is well illustrated with rice. In both Asia and West Africa today exploitation of wild rice genetic resources occurs. What is remarkable is that the methods used for harvesting these wild species, in West Africa *O. barthii* and in Asia the annual form of *O. rufipogon* (also called *O. nivara*) are very similar. In some areas, standing plants are bundled together with leaves so that the shattering grains mass in the center of the bundle from where they can be harvested. In a similar way flag leaves are used to make standing bundles of cultivated rice by the Ifugao people of the Philippines to prevent tall traditional varieties from lodging (Conklin 1980). Another approach is to swing a basket, the inside of which is smeared with sticky plant latex, over the panicles of maturing wild rice . The awns of spikelets catch on the inside of the basket and are harvested. Present-day exploitation of wild rice genetic resources gives clues as to how early stages of domestication may have occurred.

In Africa where domestication of animals preceded domestication of plants (van der Veen 1999), it may be that use of wild rice as an animal feed first drew attention to these plants. Today, in Asia, seeds of wild rice are sold at markets for planting to produce forage (DAV, pers. observ. Bangladesh).

Traditional African rice (*O. glaberrima*) has three centers of varietal diversity (Porteres 1976) - the inland Niger delta, the Senegal-Gambian river basins and the Guinean highlands. However, Richards (1986, 1996) presents a different picture of West African rice diversity (Fig. 2) that consists of (1) a core zone where rice is the major staple (the West African Rice Zone), (2) the dry savanna river valleys from the Senegal River to Lake Chad where rice is not the major staple but supplements other crops such as sorghum and millet, and, (3) the zone of old established but discontinuous rice cultivation. Recent archaeological finds of what is believed to be domesticated *O. glaberrima* at Dia, upper Middle Niger Delta, Mali, have been dated to between 800 and 500 B.C. (Shawn Sabrina Murray, Wisconsin University 2003, pers. comm.). The large quantities of seeds recovered from the site suggest that *O. glaberrima* was domesticated before this time. The date suggested for domestication for *O. glaberrima* of about 1500 B.C. by Porteres (1976) might be close to the correct time, although clear archaeological evidence is lacking.

Thus, independently in Asia and West Africa very closely related species were domesticated and became major sources of food around which complex rice cultures evolved. By the time Asian rice (*O. sativa*) was bought to West Africa *O. glaberrima* had evolved into numerous ecotypes to fit the ecosystems of the region that were strikingly similar to those where rice grows in Asia. One of the most complex of the West African rice ecosystems is the tidal rice production system of the Casamance, Senegal, that has been described in detail by Linares (2002) and Carney (2001).

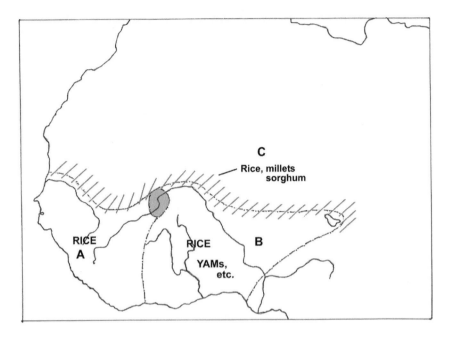

Fig. 2. Rice zones of West Africa. *A* West African rice zone. *B* Area of old established but discontinuous rice cultivation. *C* Transitional zone of rice, millet and sorghum cultivation

Asian cultivated rice is much more diverse than African rice reflecting the longer period since domestication and the larger and more complex area over which it spread (discussed in Section 2 above). After its arrival, Asian rice in West Africa gradually became adopted due to its superior yield even in soils of low clay and low fertility as compared with Asia. While *O. sativa* is the main rice grown in West Africa today, *O. glaberrima* is still grown in some areas because of its:

1. association with religious customs, wild rice being known as "God's rice" in Sierra Leone (Richards 1986, 1996; Linares 2002);

2. superior agronomic traits for stresses such as drought, excess salt and weed competition (Linares 2002; WARDA 2002);

3. taste and quality (WARDA 2003); and

4. self-seeding (shattering), which can reduce labor.

With the introduction of *O. sativa* into the *O. glaberrima* rice growing zone, the opportunity arose for natural hybridization between these two species and wild rice in Africa. Barriers exist to hybridization between *O. glaberrima* and *O. sativa* such as prezygotic isolating barriers, flowering time and hybrids that are male sterile and partially female sterile (Sano 1990). However, natural hybridization from *O. sativa* to *O. glaberrima* of up to 4.5% has been reported (Sano 1989). In addition, gene flow from the rice cultigens to wild rice has been described; hybrids have been called "obake" or ghost rice (Oka 1988). Thus the introduction of *O. sativa* has played a role through hybridization in diversification of the West African rice gene pool.

It is only recently that determined breeding attempts have been made to combine useful traits of *O. glaberrima* and *O. sativa* (Jones et al. 1997). These efforts follow one of the basic principles of plant breeding in that bringing together highly diverged genotypes resulting in new gene combinations can lead to exceptional new plant types. Thus, the FAO sponsored *indica* × *japonica* hybridization project of the 1950s led to one of the most widely grown and widely adapted cultivars of all time, Mashuri. The current program involving crosses between *O. glaberrima* and *O. sativa* has resulted in a series of varieties called NERICA (New Rice for Africa) that may prove to be a major breakthrough in rice varietal development in Africa (WARDA 2003).

During the early part of the slave trade, the peoples of the African rice-growing zone were taken to the New World. Since the slaves left no documentation, the story of the role of African rice (*O. glaberrima*) and African rice technology has remained hidden until recently. It is now becoming clear that early in the slave trade era *O. glaberrima* was introduced many times into the New World (Porteres 1976). The importance of *O. glaberrima* to early migrants to the New World from Africa is hard to discern because Europeans, from whose documents the slave era is recorded, were unfamiliar with the differences between Asian and African rice. What is clear is that the African slaves provided the knowledge of African rice technology developed in the tidal zones of coastal West Africa, and their labor laid the foundation for the development of the South Carolina rice industry. This occurred in 1695 with the export of 438 pounds (one barrel) of rice. By 1770 the figure had reached 83,708,625 pounds (159,445 barrels) (Carney 1996).

Points
Plants

1. Independent domestication of rice occurred in Asia and Africa;

2. Similar processes are used in the exploitation of the wild relatives of rice in Asia and Africa;

3. Asian and African rice diversified into similar ecosystems in the two continents;

4. Plant movement provides opportunities for exploitation of novel variation ("obake" rice by natural means and NERICA rices by modern plant breeding).

Human

1. Movement of people involves movement of their foods and food preferences;

2. Movement of people involves movement of their knowledge and knowledge systems;

3. Documentation in the historic past is biased by both the ability to write and the knowledge of the writer.

1.6 Urban Migration and Loss of Rural Labor and Their Impact on the Rice Ecosystem

Perhaps no human migration in history can match the movement of people from rural to urban areas. This is happening at an accelerated pace. In 1998 the world's population was split about equally between urban and rural areas. However, by 2005 it is expected that urban will exceed rural population. In 1998 Asia's urban population was growing at 3% per year compared to overall growth rate of 1.4%. For Africa the figures were 4and 2.6%,respectively (FAO 1998).

Growth of urban areas has many implications. One of them concerns the cost and availability of labor in rural areas. Labor costs are generally increasing in rural areas. One of the consequences in rice-growing areas has been the rapid replacement of labor-intensive rice transplanting with broadcasting. This change in technology has generally been farmer driven rather than research driven. Hand broadcasting rice has been shown to be 27 times faster than manual transplanting (Yeoh 1972). Broadcasting rice was adopted very quickly in Muda, northwest of Peninsular Malaysia (Table 1). In 1974, the first farmer in the area attempted direct seeding, by 1984 about half of the area was direct seeded. In 1999 more than 90% of the area was direct seeded (Ho et al. 1999).

Table 1. Adoption of direct seeding rice in the first planting season of Muda, Malaysia from 1979-1999. (Adapted from Ho 1991)

	1979	1981	1984	1986	1987	1988	1989	1990
Wet seeding (%)				64.1	0.6	51	85	65
Dry seeding (%)				16.2	59.7	29.8	11	30
Volunteer seeding (%)				19.7	39.7	19.2	4	5
Area under direct seeding				59,402	87,922	83,599	75,581	82,906
Total rice area				90,966	88,956	92,220	92,447	92,587
Direct seeding (%)	0.1	4.6	53,4	65.3	98.8	90.7	81,7	895

Broadcasting rice (or direct seeding) leads to other changes including the emergence of new weeds and increased herbicide use. In the early 1980s in Malaysia *Monochoria vaginalis*, *Ludwigia adscendens*, *Fimbristylis miliacea*, *Leersia hexandra*, *Cyperus haspan* and *Limnocharis flava* were the dominant weeds in Peninsular Malaysia's rice ecosystems. However, by the late 1990s, *Echinochloa crusgalli* and *Leptochloa chinensis* were the ubiquitous and competitive species in rice fields (Azmi et al. 1999). During the same period, a series of factors led to the emergence of weedy rice being a problem in some areas. In the Muda area drought resulted in a lack of irrigation water between 1991 and 1993. As a consequence, farmers were forced to undertake dry seeding and volunteer seeding, i.e., allowing dropped seeds to grow and form the next crop, sometimes supplemented with additional broadcast seeds. Volunteer seeding in particular favored the rapid emergence of shattering weedy rice ecotypes. The use of combine harvesters that travel from one rice growing area to another may have helped spread shattering rice (Vaughan et al. 1998). In many other countries, weedy rice has emerged locally as a result of direct seeding, which has been adopted at least in part due to the labor costs of transplanting rice (such as in Korea; Pyon et al. 2000) and Vietnam (Chin et al. 2000).

Points

1. Human migration to urban areas has profound economic consequences for rural and urban areas;

2. Urban migration can accelerate technological change and innovation in rural areas;

3. Technological change and innovation resulting from migration to urban areas can result in changes in ecosystem structure in rural areas.

Acknowledgement. This paper was prepared as a part of a fellowship to the first author DAV under the OECD Co-operative Research Programme: Biological Resources Management for Sustainable Agricultural Systems.

1.7 References

Ahmadi N, Glaszmann JC, Rabary E (1991) Traditional highland rices originating from intersubspecific recombination in Madagascar. International Rice Research Institute, Rice Genetics II. P.O. Box 933 Manila, Philippines, pp 67-79

Arraudeau M (1978) Rice breeding in Malagasy Republic. In: Buddenhagen I, Persley GJ (eds) Rice in Africa. Academic Press, London, pp 131-135

Azmi M, Pane H, Itoh K (1999) *Echinochloa crus-galli* (L.) Beauv. and *Leptochloa chinensis* (L.) Nees: Comparative biology and ecology in direct-seeded rice (*Oryza sativa* L.). The management of biotic agents in direct seeded rice culture in Malaysia. MARDI/MUDA/JIRCAS Integrated Study Report, pp 52-76

Bellwood P (1997) Prehistory of the Indo-Malaysian Archipelago. University of Hawai'i Press, Hawai'I, 384 pp

Carney J (1996) Rice milling, gender and slave labour in colonial South Carolina. Past and Present 153:108-134

Carney J (2001) Black rice. Harvard University Press, Cambridge, 240 pp

Carney J (2004) Out of Africa: colonial rice history in the Black Atlantic. In:Schiebinger L and Swan C (eds) Colonial Botany: Science, Commerce, Politics. University of Pennsylvania Press, Philadelphia (in press)

Carpenter AJ (1978) The history of rice in Africa. In: Buddenhagen IW, Persley GJ (eds) Rice in Africa. Academic Press, London, pp 1-10

Carson HL (1990) Increased genetic variation after a population bottleneck. TREE 5(7):228-230

Chandler RF (1979) Rice in the tropics: a guide to the development of National Programs. Westview Press, Boulder, 256 pp

Chang TT (1976) The origin, evolution, cultivation, dissemination, and diversification of the Asian and African rices. Euphytica 25:425-441

Chang TT (1983) The origins and early cultures of the cereal grains and food legumes. In: Keightley DN (ed) The origins of Chinese civilization. University of California Press, Berkeley, pp 65-94

Chang TT (1985) Crop history and genetic conservation: rice - a case study. Iowa State J Res 59(4):425-455

Chang TT (1988) Seed processing, storage conditions and seed viability. Rice seed health proceedings of the international workshop on rice seed health, 16-20th March 1987. International Rice Research Institute, Manila, Philippines, pp 343-352

Cheng C, Motohashi R, Tsuchimoto S, Fukuta Y, Ohtsubo H, Ohtsubo E (2003) Polyphyletic origin of cultivated rice: Based on the interspersion pattern of SINEs. Mol Biol Evol 20:67-75

Chin DV, Hien TV, Thiet LV (2000) Weedy rice in Vietnam. In: Baki BB, Chin DV, Mortimer M (eds) Wild and weedy rice in rice ecosystems in Asia - a review. IRRI Limited Proceedings, IRRI, Los Banos Philippines; pp 45-50

Conklin HC (1980) Ethnographic atlas of the Ifugao: a study of environment, culture and society in northern Luzon. Yale University Press, New Haven

Dethloff HC (2003) American rice industry: Historical overview of production and marketing. In: Wayne Smith C, Dilday RH (eds) Rice; origin, history, technology and production. Wiley, Hoboken, New Jersey, pp 67-85

FAO (1998) The state of food and agriculture. FAO, Rome

Fukazawa H (1988) Madagascar: its unity and diversity. In: Takaya Y (ed) Madagascar: perspectives from the Malay world. Center for Southeast Asian Studies, Kyoto University, Japan, pp 3-24

Ho NK (1991) Comparative ecological studies of weed flora in irrigated rice fields in the Muda area. Muda Agricultural Development Authority monograph 44. Percetakan Siaran Sdn. Bhd. Aloe Setar, Malaysia

Glaszmann JC (1988) Geographic pattern of variation among Asian native rice cultivars (*Oryza sativa* L.) based on fifteen isozyme loci. Genome 30:782-792

Ho NK, Dahuli K, Ahmad RAG (1999) Recent advances in direct seeding culture in the Muda area, Malaysia. The management of biotic agents in direct seeded rice culture in Malaysia. MARDI/MUDA/JIRCAS Integrated Study Report, pp 1-6

Hoshikawa K (1989) The growing rice plant: An anatomical monograph. Nobunkyo, Tokyo

Imamura K (1996) Jomon and Yayoi: the transition to agriculture in Japanese prehistory. In: Harris DR (ed) The origins and spread of agriculture and pastoralism in Eurasia. UCL Press, London, pp 442-464

Jones MP, Dingkuhn M, Aluko GK, Semon M (1997) Interspecific *Oryza sativa* L. X *O. glaberrima* Steud. progenies in upland rice improvement. Euphytica 94:237-246

Linares OF (2002) African rice (*Oryza glaberrima*): History and future potential. Proc Natl Acad Sci USA 99(25):16360-16365

Maeda N, Armand R (1988) Vandrozana: a Sihanaka village in the Southeastern Region of Lake Alaotra. In: Takaya Y (ed) Madagascar: perspectives from the Malay world. Center for Southeast Asian Studies, Kyoto University, pp 165-224

Maclean JL, Dawe DC, Hardy B, Hettel GP (eds) (2002) Rice almanac. Los Baños, (Philippines): International Rice Research Institute, Bouaké (Côte d'Ivoire): West African Rice Development Association, Cali (Colombia): International Center for Tropical Agriculture. FAO, Rome, 253 pp

Nakagahra M (1977) Genetic analysis for esterase isozymes in rice cultivars. Jpn J Breed 27(2):141-148

Oka HI (1988) Origin of cultivated rice. Elsevier, Tokyo, 254 pp

Pickersgill B (1998) Crop introductions and the development of secondary areas of diversity. In: Prendergast HDV, Etkin NL, Harris DR, Houghton PJ (eds) Plants for food and medicine. Royal Botanic Gardens, Kew, pp 93-105

Polans NO, Allard RW (1989) An experimental evaluation of the recovery potential of ryegrass population from genetic stress resulting from restriction of population size. Evolution 43(6):1320-1324

Porteres R (1976) African cereals: Eleusine, Fonio, Black Fonio, Teff, *Brachiaria*, paspalum, *Pennisetum*, and African rice. In: Harlan JR, de Wet JM, Stemler ABL (eds) Origins of African plant domestication. Mouton Publishers, The Hague, pp 409-452

Pyon JY, Kwon WY, Guh JO (2000) Distribution, emergence and control of Korean weedy rice. In: Baki BB, Chin DV, Mortimer M (eds) Wild and weedy rice in rice ecosystems in Asia - A review. IRRI Limited proceedings, pp 37-40

Richards P (1986) Coping with hunger. The London research series in Geography no 11. Allen and Unwin, Hemel Hempsted, UK

Richards P (1996) Agrarian creolization: the ethnobiology, history, culture and politics of West African rice. In: Ellen R, Fukui K (eds) Redefining nature: ecology, culture and domestication. Berg, Oxford, pp 291-318

Sano Y (1989) The direction of pollen flow between two co-occurring rice species, *Oryza sativa* and *O. glaberrima*. Heredity 63:353-357

Sano Y (1990) The genic nature of gamete eliminator in rice. Genetics 125:183-191

Sato YI (2000) Origin and evolution of wild, weedy, and cultivated rice. In: Baki BB, Chin BV, Mortimer M (eds) Wild and Weedy rice in rice ecosystems in Asia - a review. Limited proceedings no 2. International Rice Research Institute, Los Banos, Philippines, pp 7-15

Tanaka K (1988) Rice and rice culture in Madagascar. In: Takaya Y(ed) Madagascar: perspectives from the Malay world. Center for Southeast Asian Studies, Kyoto University, pp 25-92

Van der Veen M (1999) The exploitation of plant resources in ancient Africa. Kluwer Academic/Plenum Publishers, New York, 283 pp

Vaughan DA, Watanabe H, HilleRisLambers D, Abdullah Zain Md (1998) Weedy rice complexes in direct seeding rice cultures. World food security and crop production technologies for tomorrow, Kyoto, Japan. Jpn J Crop Sci 67 extra issue 2:277-280

Watabe T(ed) (1980) Some aspects of historical changes of rice. Kyoto University. Kyoto, Japan

WARDA (West African Rice Development Association) (2002) WARDA annual report 2001-2002 (http://www.warda.cgiar.org/publications/KBtext.pdf)

WARDA (West African Rice Development Association) (2003) Bintu and her new rice for Africa (http://www.warda.cgiar.org/publications/KBtext.pdf)

Yeoh KC (1972) Pre-project study on the direct seeding of rice. Malaysian Agriculture Research and Development Institute. Serdang, Malaysia (monograph)

Yawata I (1961) Rice cultivation of the ancient Mariana islanders. In: Plants and the migrations of Pacific peoples. Barrau J (ed) Symposium held at Tenth Pacific Science Congress of the Pacific Science Association. University of Hawaii, Honolulu, 21 Aug -6 Sept 1961, pp 91-92

2 Movement of Rice Germplasm Around the World

Edwin L. Javier[1] and Ma. Concepcion Toledo[2]

[1]Plant Breeder and INGER Coordinator and [2]Assistant Scientist II; International Rice Research Institute, Los Baños, Laguna, Philippines; E-mail e.javier@cgiar.org

2.1 Abstract

The strong partnership between the International Rice Research Institute (IRRI) and the National Agricultural Research Systems (NARS) has contributed substantially to the massive movement of rice germplasm around the world from the 1960s to the present. The three major sources of rice germplasm at IRRI are the Genetic Resources Center (germplasm conservation), plant breeding programs (variety development), and the International Network for Genetic Evaluation of Rice (INGER). From 1989 to 2002, 1.2 million seed packets were distributed by IRRI with 75% going to 38 countries in Asia.

In 1975, INGER was established by IRRI as a global partnership between NARES and International Agricultural Research Centers. It is the major vehicle for worldwide sharing of superior rice genetic resources from various parts of the world. INGER has received around 23,000 unique materials since 1975, at least half of which were from NARS. Over a period of 29 years, INGER has dispatched around 2.7 million seed packets of some 46,000 test entries. Some 62 NARS have released 213, 151, and 272 varieties that could be traced back to IRRI, other IARCs, and 29 NARS, respectively. Direct utilization of introduced varieties has shortened the time involved in variety development, saved a lot of resources for NARS, and hastened the flow of materials from research station to farmers' fields. Thousands of INGER-distributed germplasm have been used by NARS as parents in crosses to improve the performance of their local varieties. New varieties generated have more diverse and complex genealogies. Genetic materials distributed by INGER are accompanied by Material Transfer Agreements. The major challenges facing international exchange of germplasm are emerging intellectual property rights legislations and *sui generis* plant variety protection laws.

2.2 Introduction

Rice is the primary food of more than half of the world's population. More than 90% of rice is grown and consumed in Asia. In many rural areas in Asia, rice is consumed three times a day. In Africa and South America, hundreds of millions depend on rice for their energy source. India and China are the leading countries in rice area and production, while Thailand and Vietnam are the top rice exporters in the world.

Rice is grown in a wide range of water regimes, from no standing water (upland rice) to a few cm of flood water (lowland rice), and from a water depth of 50 cm (deepwater rice) to more than 3 m (floating rice). Upland rice can be rainfed as in Asia and Africa or irrigated as in Brazil. Similarly, lowland rice can also be rain-fall dependent or irrigated. Rainfed rice cultivation is always prone to drought and/or submergence due to the unpredictable occurrence, amount and duration of rainfall. Rice environment can also be a continuum in regard to soil fertility, acidity and temperature. Rice is also planted, harvested and processed in various ways. Genetic variability exists among traditional varieties for various agronomic traits and resistance to different biotic and abiotic stresses.

2.3 International Rice Research Institute as a Germplasm Source for Worldwide Sharing

The International Rice Research Institute (IRRI), which was established in the Philippines in 1960, has a global mandate to improve the well-being of present and future generations of rice farmers and consumers, particularly those with low incomes through rice improvement research and technology transfer. It has developed strong programs on germplasm conservation, varietal development and worldwide germplasm exchange and evaluation.

IRRI's International Rice Genetic Resources Center (IRGC) is responsible for rice germplasm collection, conservation and documentation. It holds the world's largest collection of rice in its genebank. It stores 93,681 accessions of *Oryza sativa* (Asian cultivated rice), 1,543 accessions of *Oryza glabberima* (African cultivated rice) and 4,448 accessions of wild relatives of rice originating from more than 110 countries. Around 82% of these accessions are under the auspices of the Food and Agriculture Organizations (FAO) of the United Nations. A black box storage of duplicate samples of the collection is in Fort Collins, Colorado, USA. Most of the genebank accessions are traditional varieties.

IRRI has several rice breeding programs for different ecosystems like irrigated lowland, rain-fed lowland, favorable (aerobic) and unfavorable upland, and flood-prone environments. Rice breeders work with scientists from various disciplines to come up with varieties having stable and high yields and good gain qualities, and with resistance to major biotic and abiotic stresses. The IRGC germplasm collections are screened by the rice breeding teams for genetic donors of important traits that will improve rice adaptation and productivity in various rice environments. Rice breeding teams have collaborative undertakings with some NARS. IRRI-generated segregating populations and breeding lines, and FAO designated germplasm in the IRRI genebank are freely available to rice scientists around the globe.

Since its inception, IRRI has recognized the importance of germplasm sharing between and among International Agricultural Research Centers (IARCs) and NARS around the globe. Thus, in 1975 IRRI initiated a formal global network of NARS and International Agricultural Research Centers (IARCs) working on rice. The network is called the International Network for Genetic Evaluation of Rice (INGER). Members of the network contribute their elite breeding lines and varieties to INGER for worldwide sharing. All rice genetic resources distributed by IRRI are accompanied by Material Transfer Agreements.

The Consultative Group for International Agricultural Research has 16 members. Three are presently involved in rice improvement research and one previously worked on rice. In addition to IRRI, the other IARCs that are actively involved in rice improvement research are the West Africa Rice Research Development Association (WARDA) in Cote d'Ivoire, constituted in 1970, land Centro Internacional de Agricultura Tropical (CIAT) in Colombia, which was established in 1967. The International Institute for Tropical Agriculture (IITA) in Nigeria was founded in 1967 and worked on rice from 1971 to 1990. IITA's responsibility for rice research was transferred to WARDA in December 1990. All rice research in Africa is now being handled by WARDA. Like IRRI, these CGIAR centers have programs on germplasm conservation, utilization and exchange. All are active members of INGER, with WARDA/IITA looking after the needs of Africa and CIAT concentrating on rice problems of Latin America and the Caribbean.

2.4 Overview of IRRI's Worldwide Rice Germplasm Distribution

IRRI distributed more than 1.2 million seed packets to rice scientists all over the world from 1989 to 2002 (Table 1). The major exporter of rice germplasm is INGER (55.6%), followed by plant breeding programs (30.4%), IRGC (11.2%), and other units (2.8%). The total number of seed packets distributed annually had a general declining trend, with a sharp decrease after 1995, largely because of the declining budget of INGER for seed multiplication and distribution. Of the total seed packets dispatched for the last 14 years, 75% went to Asia (38 countries), 8% each to Africa (36 countries) and North America (2 countries), 5% to South America (20 countries), and 2% each to Europe (22 countries) and Oceania (6 countries). India is the top recipient of rice germplasm, followed by Thailand and South Korea (Table 2). Among the top 20 recipients, 14 countries are in Asia, 2 each in Africa and South America, and 1 each in North America and Oceania.

Table 1. Number of seed packets distributed around the world by IRRI, 1989-2002

Year	INGER	Breeding programs	GRC	Other units	Total
1989	64,960	31,699	16,771	973	114,403
1990	78,159	26,476	8618	972	114,225
1991	89,020	20,079	14,955	1839	125,893
1992	67,945	17,739	6577	2575	94,836
1993	38,528	25,523	19,195	1175	84,421
1994	55,661	43,180	11,783	1881	112,505
1995	61,162	37,373	12,703	1209	112,447
1996	44,546	25,360	4055	972	74,933
1997	29,104	26,204	2117	6783	64,208
1998	27,996	23,265	15,886	450	67,596
1999	26,476	24,861	15,039	800	67,176
2000	22,212	22,376	3653	2353	50,594
2001	37,857	19,570	1874	3145	62,446
2002	26,145	22,147	2005	8534	58,831
Total	669,771	365,852	135,231	33,661	1,204,515

Table 2. Top 20 recipients of rice germplasm from IRRI, 1989-2002

Rank	Country	Total	Rank	Country	Total
1	India	222,462	11	Myanmar	25,709
2	Thailand	143,874	12	Nigeria	22,335
3	Korea, South	109,020	13	Japan	20,781
4	China	86,424	14	Philippines	21,173
5	Vietnam	60,026	15	Nepal	21,047
6	Egypt	44,252	16	Sri Lanka	19,633
7	Bangladesh	45,094	17	Colombia	18,826
8	Indonesia	37,281	18	Iran	18,139
9	USA	31,454	19	Australia	17,793
10	Pakistan	30,240	20	Brazil	14,149

2.5 Movement of Rice Germplasm: The INGER Experience

Over the years, INGER has been funded by the United Nations Development Program (1975-1996), the World Bank (1991-1996), the Swiss Agency for Development and Cooperation (1995), the Federal Ministry of Economic Cooperation/German Agency for Technical cooperation (1995-1997) and IRRI Core Budget (1998 to date). The objectives of INGER are: (1) to ensure safe and free

exchange of rice germplasm and information on its characterization and adaptation; (2) to broaden the genetic diversity and genetic base of rice varieties used by
farmers; (3) to acquire, characterize and evaluate superior rice germplasm; (4) to
assess and validate important traits of superior germplasm including resistance to
stresses and quality characteristics; (5). to characterize and evaluate genotype ×
environment interaction for important traits so that rice improvement programs,
particularly those of NARES, can capitalize on general and specific adaptation; (6)
to enhance the capacity of NARES to utilize and improve germplasm.

IARCs and NARES send small amounts of seeds of their outstanding rice genetic materials, together with their pedigree and salient characteristics, to INGER.
These seeds are multiplied at IRRI until the required amounts for the nurseries are
obtained. A variety or line will take at least 2 years from being nominated to being included in a nursery. The seeds are processed according to the phytosanitary
requirements of importing NARES. INGER always ensures that it distributes
seeds of the highest quality. Each year, cooperators are informed of the available
ecosystem-based and stress-oriented nurseries, and given lists of nursery entries
with details on origin, pedigree and agronomic characteristics. Based on the given
information, they decide what nurseries or specific entries they want to evaluate.
At the end of the season, cooperators share the evaluation data with INGER for
analysis and interpretation. Outstanding test entries are identified, stored and distributed to interested cooperators. Results of analysis across environments are
sent back to the cooperators.

INGER is the major vehicle by which the world's superior rice varieties and
breeding lines are shared. NARS considers INGER as an integral component of
its national breeding program. The total number of varieties contributed by all
NARS was greater than that of all IARCs (Table 3). Among IARCs, IRRI contributed the most number of test entries to INGER.

Table 3. Percent variety contribution of NARS and IARCs to INGER

Period	Number of entries	Contribution (%)		
		NARS	IRRI	Other IARCS
975-1979	8616	62	37	1
1980-1984	7863	63	35	2
1985-1989	6322	56	41	3
1990-1994	6088	63	31	6
1995-1999	4370	60	34	6
2000-2004	3861	50	42	8

Entries are unique by year but some entries are repeated in the following year in different
nurseries

INGER has received more than 23,000 unique breeding lines from 4 IARCs
and 79 members. Many of these entries were evaluated in more than one nursery.
Thus, the total number of test entries distributed is about 46,000. Around 74
countries have received INGER materials at one time or another. Through

INGER, diverse rice genetic resources flow freely within and between continents (Table 4). Asian NARS contributed around 19,000 test entries for 29 years and received more than twice that amount of materials from INGER. The proportion of the number of test entries contributed by Africa, South America and Europe relative to the number of test entries they received ranged from 4-7%. This shows that small contributions can reap substantial profits.

Table 4. Intra- und intercontinental movement of elite rice germplasm through INGER, 1975-2003

Region	No. INGER test entries	No. INGER test entries
	Contributed by the region	Received by the region
Africa	1,576	36,308
Asia	19,126	45,683
Europe	273	10,894
Latin America	2,312	32,675
North America*	26	33
Oceania[a]	51	133

[a]Data from 1996-2003 only

2.6 Direct Utilization of INGER Test Entries

Reports received so far indicated that Africa has the highest number of INGER test entries released as varieties, with 165 and 159 releases tracing back from Asia and Africa, respectively (Table 5). Asian varietal releases are predominantly from Asia. Of the 92 varietal releases in South America, 65 came from South America and the rest from Asia. The regional focus of the IARCs are reflected in the outcome of varietal release. WARDA/IITA and CIAT generate materials which are suited to Africa and South America, respectively. IRRI meets the requirements of Asia and at the same time contributes to the needs of other parts of the world.

Table 5. Number of INGER entries emanating from various NARS and IARCs released in different regions, 1975-2003

Region	Asia		Africa			S. America		Europe	Oceania	Total
	NARS	IRRI	NARS	IITA	WARDA	NARS	CIAT	NARS	NARS	
Asia	74	117	4	2	0	2	1	3	1	204
Africa	83	82	62	44	53	10	2	0	0	336
South America	12	13	2	0	0	9	56	0	0	92
Oceania	0	1	2	0	0	1	0	0	0	4
Total	169	213	70	46	53	22	59	3	1	636

A total of 272 varietal releases could be traced back to 29 NARS. Leading the list is India, the source of 41 releases, followed by Sri Lanka (source of 27 releases), Cote d'Ivoire (25), Taiwan (19), Bangladesh (14), Indonesia (12), Thailand (12) and Philippines (10). Twenty-two countries from Asia, Africa and South America released some of India's variety contributions to INGER. This reflects the strength of India's breeding programs, which are being implemented in different states with very diverse rice environments. For example, India's CR123-23 and OR142-99 from the State of Orissa were released in Nepal and Cambodia, respectively; R22-2-10-1 and PR 106 from the State of Punjab were released in Paraguay and Venezuela, respectively. India also benefited from germplasm exchange in a similar way. India released 30 INGER test entries originating from Bangladesh, Indonesia, IRRI, Philippines and Sri Lanka.

A number of INGER test entries were released in more than one country, indicating their wide adaptation. Some 89 INGER test entries were released as 297 varieties. IRRI contributed 31 of the 89 varieties; IITA, 8 varieties; CIAT, 7 varieties; WARDA, 4 varieties, and 17 NARS, 39. Cote d'Ivoire had the most number of test entries (10) that turned out as multiple releases. The INGER test entry with the highest number of releases was BG 90-2, a variety developed at the Central Rice Breeding Station (now called Rice Research and Development Institute) in Sri Lanka. It was released in four Asian (China, India, Myanmar, Nepal) and ten African (Benin, Ghana, Guinea, Cote d'Ivoire, Kenya, Mali, Nigeria, Senegal, Sierra Leone and Tanzania) countries. IRRI's IR36 was released in five Asian (Bhutan, India, Myanmar, China and Vietnam) and three African (Central African Republic, Gambia and Mozambique,) countries. Its late maturing sister line IR 42 was also released in four Asian and four African countries. Cica 8 of Colombia was released in seven Latin American countries (Belize, Bolivia, Brazil, , Guatemala, Honduras, Panama and Paraguay) and in Cameroon in Africa.

The varietal release discussed above only covers the most outstanding INGER materials introduced in other countries. It should be noted that many of the INGER materials were recommended varieties in their countries of origin before they were tested in other places. Thus, the total number of released of rice varieties around the world is much more than the number of releases associated with INGER.

Evenson and Gollin (1997) analyzed 1,709 released varieties from 1960 to 1991 in various parts of the world, with more complete information for South and Southeast Asian releases. Some 390 released varieties were introduced materials (developed outside the releasing country), of which 75% were developed at IRRI. Many of the released foreign introductions were made available to NARS through INGER. Of the 1709 releases, 45% have at least one IRRI developed line as parent and 36% have at least one parent from other NARS. About 80% of the parents appeared to have come from INGER (Chaudhary et al. 1998). Hossain et al. (2003) studied the derivation of 2,040 released varieties between 1960 and 1999 in 12 South and Southeast Asian countries. About 11% were direct releases of IRRI varieties and about 39% had IRRI developed materials as parents, grandparents or remote ancestors.

2.7 Indirect Utilization of INGER Test Entries

In Asia, the number of released INGER materials from 1975 to 1990 (first period) was substantially more than the number released from 1991 to date (second period). In the first period, Asian NARS were in the stage of building up their capability in varietal improvement research, and direct utilization of advanced breeding lines and varieties was the fastest and the cheapest strategy to meet the demand for varieties with higher yields than local varieties. By the second period, some Asian NARS like those of India, China, Sri Lanka, Thailand, Philippines, to mention a few, had strengthened their research capability. These advanced NARS had started working on breeding problems that are peculiar to their respective countries. Rice scientists were aware that direct utilization of introduced varieties was very successful in an irrigated lowland rice ecosystem. However, released varieties succumbed later to certain new insect pest biotypes or pathogenic races, which could be country-specific. During the first period, yield was the major focus and grain quality of secondary importance. Having released varieties with high yield potential, the next goal of NARS was to develop varieties with high yield and at the same time with good grain qualities, whose definitions vary among and within countries. The adoption rate of released varieties was low for unfavorable environments such as rain-fed lowland, upland and deepwater rice areas. The characteristics of these environments are likely to be different from country to country. Advanced NARS are in the best position to do develop location-specific varieties since their selection process can be done in the target environments. Thus, advanced NARS are using extensively introduced materials as parents in their hybridization programs. Countries with less developed research capability such as Cambodia and East Timor expend more effort on selection among introductions than generating segregating populations with introduced varieties as parents. The same can be said of African NARS who are also in the course of developing their research capabilities.

Since 1975, more than 3,000 INGER test entries have been used as parents for generating more than 16,000 crosses by NARS around the globe, and many lines derived from these crosses have also been released. In the 1960s and early 1970s, IRRI materials were used primarily as a source of semi-dwarfing gene (Hargrove 1978, 1979). Semi-dwarf stature is associated with lodging resistance, a trait that is important for nitrogen responsive varieties. In the mid-1970s, IRRI started sharing materials with many useful traits like high yield potential, early maturity, and multiple resistance to biotic stresses (Hossain et al. 2003). The genealogy of IRRI materials generated over time reflects increasing number of land races. For example, the first green revolution variety IR 8 was derived from a cross between Peta and Dee Gee Woo Gen, where the first parent is an Indonesian variety and the second parent is a traditional variety from Taiwan. The Indonesian variety was a progeny of a cross between Cina, a traditional variety from Taiwan, and Latisail, a traditional variety from India. Thus, IR 8 has three ancestral (ultimate) parents, with Cina serving as a source of it cytoplasm. In contrast, IR 36 released in the seventies has 16 ancestral parents (Fig. 1) originating from China (Cina),

Indonesia (Benong), Japan (unknown variety), Taiwan (Dee Gee Woo Gen, Tsai Yuan Chung, Pa Chiam), Philippines (Tadukan, Sinawpagh, Marong Paroc) and India (Latisail, Kitchili Samba, Eravapandi, Thekkan, Vellaikar, one unknown variety, and *O. Nivara* acc. 101508). National programs utilizing IRRI materials as parents in breeding programs are therefore increasing the genetic diversity of their future varieties (Hossain et al. 2003).

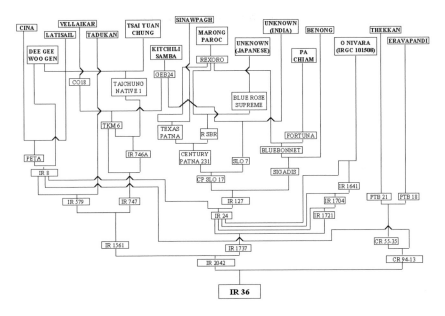

Fig. 1. Genealogy of IR36

2.8 INGER's Major Challenges

The advent of powerful molecular tools that could be used in generating novel plants and the development of new techniques for the production of hybrid rice have triggered the private sectors to invest in plant research. The role of the private sector communities in the production and delivery of new technologies is expected to increase over time. This will be enhanced by the legislation and implementation of intellectual property rights (IPR) and *sui generis* laws for plant variety protection (PVP). These developments are affecting the INGER tradition of free and unrestricted flow of germplasm as reflected in the decline in variety contribution from NARS starting in mid nineties. Many NARS partners think that their materials might be misappropriated. In 2001, the steering committee of INGER required that any seeds received and distributed by INGER should be accompanied by a Material Transfer Agreement (MTA). INGER has three types of

MTAs – MTA for FAO designated germplasm, MTA for IRRI developed seeds, and MTA for Non-IRRI seeds (contributions from NARS and other IARCs). In all MTAs, the recipient must agree not to claim any IPR protection on the material and related information. The MTA for non-IRRI seeds also covers the direct derivatives of the materials received. The IPR awareness campaign of IRRI and the use of MTAs appeared to have a positive effect on germplasm exchange. In 2003, INGER received 139 variety contributions from NARS which is more than INGER had received from 1999 to 2002. It is also important to note that there is a sharing of rice germplasm that is independent of INGER, such as between-country seed exchange. The rapidly changing IPR environment around the globe has also affected the flow of germplasm through this system. The increasing demand for INGER material could not be matched by the declining budget of INGER. Thus, there is need for IARCS and NARS partners of INGER to discuss prioritization of seed requests, formulate a system of equitable distribution of INGER materials among NARES, and devise means of sharing INGER materials within countries, particularly in big countries like India and China.

2.9 Conclusions

INGER has been the major instrument of NARS, IRRI and other IARCs in the worldwide sharing of their superior genetic resources. It has been successful in the safe, free and unrestricted movement of rice germplasm within and among continents for use in rice varietal improvement research, with the rice farmers and rice consumers as the ultimate beneficiaries. All INGER partners are mutually benefited. A member receives substantially more varieties than the member contributes.

Some 636 varieties released around the world were variety introductions derived through INGER. Direct utilization of INGER materials has saved NARS a lot of resources and time involved in developing new varieties, since the time involved in the production of crosses until fixed lines are produced for yield trials has been eliminated. This has also accelerated the flow of new varieties from the research station to the farmers' fields. Farmers' income is improved because these varieties not only have high yields but also have multiple resistance to biotic stresses, and thus save on the use of pesticides and abiotic stresses. INGER entries have very diverse genetic backgrounds and complex genealogy. A number of them have been used as parents in crosses and some of their progenies have been released. INGER materials have therefore contributed in increasing the genetic diversity of varieties in farmers' fields. Countries with limited research capability are the most benefited by direct utilization of INGER materials.

According to Evenson and Golin (1997), the annual value of each INGER material released to the world rice economy was around US$ 2.5 million. Cessation of INGER activity would result in 25% reduction in the number of varieties released in the world. Thus, the continued operation of INGER is economically justified.

2.10 References

Chaudhary RC, Seshu DV, Alluri K, Cuevas-Perez F, Lopez VC, Khush GS (1999) INGER-derived rice varieties directly released in various countries. IRRI, Los Banos, Philippines

Evenson RE, Gollin D (1997) Genetic resources, international organizations and improvement in rice varieties. Econ Dev Cult Change 45:471-500

Hargrove TR (1978) The diffusion of genetic materials and the objectives for their use among rice breeding programs in India. Field Crops Res 1:197-213

Hargrove TR (1979) The diffusion and adoption of semi-dwarf rice cultivars as parents in Asian rice breeding programs. Crop Sci 19:571-574

Hossain M, Gollin D, Cabanilla V, Cabrera E, Johnson N, Khush GS, McLaren G (2003) International research and genetic improvement in rice: evidence from Asia and Latin America. In: Evenson RE, Gollin D (eds) Crop variety improvement and its effect on productivity. The impact of International Agricultural Research, IRRI, Los Baños, Philippines, pp 71-108

3 Potential for Gene Flow from Cultivated Wheat to Weedy Relatives in the Great Plains of North America

Robert A. Graybosch

USDA-ARS, 344 Keim, UNL, East Campus, Lincoln, Nebraska 68583, USA, E-mail rag@unlserve.unl.edu

3.1 Abstract

Common or bread wheat (*Triticum aestivum*), introduced to North America, has become a stable and often dominant component of agricultural systems in the semi-arid western portion of the Great Plains and Intermountain regions of North America. To date, transgenic (GMO) wheats have not been deployed in this region. Concern over potential loss of export markets has been the primary factor in restricting, to date, use of GMO wheat. An additional factor has been concern over possible outcrossing between cultivated wheat and weedy relatives, with potential establishment of transgenes in naturalized populations. Throughout the Great Plains, common wheat grows in close proximity to introduced, and often permanently established, populations of the related species *Triticum cylindricum* (=*Aegilops cylindrica*, jointed goat-grass), *Secale cereale* (rye) and *Elytrigia intermedia* (=*Agropyron intermedium*, intermediate wheat grass). While hybrids can be produced between wheat and both rye and intermediate wheat grass, and both have been used as sources of genes for wheat improvement, naturally occurring hybrids are rare, always sterile, and unlikely to serve as bridges for gene flow. Jointed goat-grass, however, is more closely related to wheat, and natural hybrids are quite common. Common wheat is an allohexaploid containing the A, B and D genomes. Jointed goat-grass is an allotetraploid, and carries the C and D genomes. Thus, in F_1 hybrids between the two species, there are seven common chromosomes, which will pair at meiosis and allow for limited fertility as a female parent. Male (pollen) fertility in such hybrids is, however, low or non-existent. Still, seed may be formed if pollen arrives from either wheat or jointed goat-grass populations. Thus, potential bridges for gene flow exist. A survey of more than 60 naturalized populations of jointed goat-grass has been conducted using genes encoding the wheat seed storage proteins known as high-molecular-weight glutenins and gliadins. To date, no evidence for transfer and permanent establishment of these genes in goat-grass populations has been uncovered. A survey during the 2003 season did reveal the presence of natural hybrids in 10% of observed populations; however, no seed was recovered from these hybrid plants. Jointed goat-grass is naturalized over a wide expanse of agricultural land in western North America. Even if successful, seed set by natural F_1 hybrids is a rare event, the potential for gene flow from wheat remains a possibility, and will require continuous monitoring to adequately assess the risk of such events.

Common or bread wheat (*Triticum aestivum*) is a non-native species presently cultivated over vast expanses of the semi-arid regions of North America. It was introduced along with the first European settlements in the New World, but its cultivation in the region west of the Mississippi began in earnest in the late 1800s. Wheat is a member of a large, complex Tribe (Triticeae) of the grass family (Poaceae). Members of the Triticeae probably evolved fairly recently in geological terms. Chromosomes of species even of different genera are closely related. Within the genus *Triticum*, diploid, tetraploid and hexaploid species exist, produced by a pattern of evolution consisting of radiation of ancestral diploid species, followed by hybridization and chromosome doubling to form tetraploid species. Tetraploid species, in turn, also radiated to form various species, and also hybridized with diploid species to give rise to hexaploid species. Species of *Triticum* still share common groups of chromosomes (genomes). For example, one ancestral species, *T. monoccocum* (2n=2x=14) has the genomic composition AA. Durum wheat and its wild and semi-wild relatives, considered subspecies of *T. turgidum* (2n=4x=28) carry the A and B genomes (genomic composition AABB). Common wheat (*T. aestivum*, 2n=6x=42, AABBDD) is believed to have evolved via hybridization of an AABB tetraploid, and a DD diploid species, *Triticum tauschii.* It should be noted that some authors classify the wild diploid species of *Triticum* with a sister genus, *Aegilops*. In this work, all will be considered species of *Triticum*.

Natural hybrids, often resulting in self-sterile F_1 progeny, are possible between common wheat, and many wild, semi-wild or cultivated relatives, especially those sharing at least one genome in common. In such progeny, male fertility typically is limited; however, partial female fertility often exists, and some F_2 seed might be produced if pollen is provided by natural wind distribution. Wheat geneticists and breeders have made use of this phenomenon to transfer a number of genes of agronomic importance to wheat from both wild and domesticated relatives.   Wild species with at least one common genome are widely distributed in the Middle East and southwest Asia (Kimber and Feldman 1987). These include *T. monoccocum* (AA), occurring in Transcaucasia, Iran, Iraq, Turkey, Syria and Lebanon, *T. timopheevii* (AAGG) known from Transcaucasia, Iran, Iran and Turkey, various subspecies of *T. turgidum* (AABB) native to Iran, Iraq, Turkey, Syria, Lebanon, Jordan and Israel, and *T. ventricosum* (DDUnUn) found in Egypt. *T. crassum* (DDMM) and *T. cylindricum* (CCDD) occur in a similar range as *T. monoccocum*, but in addition are found in Afghanistan and in south Asia, in the region east of the Caspian Sea. *T. juvenale* (DDMuMu) also is found in the Caspian Sea region, as well as in Iran, Iraq and Syria.

Hybridization also may take place between wheat and other species of *Triticum* and even other members of other Triticeae genera, even though no common genomes exist. In addition to the species listed above, natural hybrids have been documented between common wheat and *T. umbellulatum, T. turgidum* subsp. *dicoccoides, T. ovatum, T. triunciale,* and *T. peregrinum* (Kimber and Feldman 1987). In addition, artificial hybrids have been created in greenhouse and laboratory settings between *T. aestivum* and nearly all of the diploid, tetraploid and hexaploid species of *Triticum.* Hybrids also have been produced between common wheat and a number of species of various Triticeae genera, including *Secale* (rye), *Hordeum,* and various genera formerly grouped under the large complex genus *Agropyron.* In general, such hybrids are possible only via man's intervention, and require the use of embryo rescue techniques (in the case of the Agropyron types) or the presence of recessive crossability genes (Sharma and Gill 1983). Wheat cultivars, due to the existence of crossability genes differ in their ability to cross with rye; the landrace 'Chinese Spring' is recessive for the two genes (*Kr1, Kr2*) controlling seed set with rye (and perhaps other species), and has been used to produce a number of artificial hybrids (Riley and Chapman 1967). Via subsequent chromosomal translocations, a number of genes of economic importance have been transferred to wheat via wide crosses (Friebe et al. 1996).

In the western Great Plains of North America, common wheat fields often are bordered by, or contaminated with, naturalized populations of three related species, namely, *T. cylindricum* (=*Aegilops cylindrica,* jointed goat-grass), *S. cereale* (rye) and *Elytrigia intermedia* (=*Agropyron intermedium,* intermediate wheat grass. Rye has escaped cultivation, while intermediate wheat grass has been introduced for use in restoration of over-grazed rangeland or for use in pastures. So called feral rye populations, however, rarely persist in the western Great Plains. Jointed goat-grass migrated, probably as a contaminant of seed shipments, to North America in the early 1900s. The earliest report of *T. cylindricum* west of the Mississippi River Valley is from 1917, when specimens were collected from farmers fields in Edwards County, Kansas (Johnston and Parker 1929). The herbarium of the Smithsonian Institution in Washington, D.C., houses the following collected specimens: Pullman, WA, 1918; Wakita, OK, 1924; Stillwater, OK, 1925; Wray, CO, 1926; Raymond, KS, 1926; State College, NM, 1931; and Ogden, UT, 1938. Thus, within 20 years of its first known collection, jointed goatgrass had spread across a wide range in the western United States.

Natural hybrids between common wheat and rye, or wheat and intermediate wheatgrass are rarely observed in the western Great Plains. Ribeiro-Carvalho et al. (1997) did present evidence, however, suggesting that natural introgression of rye genes occurred during the development of some wheat landraces in Portugal. However, the lack of persistence of feral rye populations in the Great Plains might make natural hybridization a rare event. In contrast, jointed goat-grass populations are well and permanently established. Hybrids between common wheat and jointed goat-grass are easily produced, both via manual cross-pollinations, and in natural settings. In controlled cross-pollination experiments, common wheat × jointed goat-grass crosses are highly fertile. In the same greenhouse environment in the year 2000, the author produced 138 wheat × wheat crosses, with an average seed set of 12 seed per cross. Twenty-five common wheat × jointed goat-grass crosses set an average of nine seeds per cross. The subsequent year, all common wheat × jointed goat-grass F_1 plants were greenhouse grown, but all failed to set seed. Zemetra et al. (1998) also produced hybrids under greenhouse conditions, and found that partial self-fertility was restored after only one back-cross using jointed goat-grass as a male (pollen) parent.

Hybrids are common in natural field situations (Guadagnuolo et al. 2001; Morrison et al. 2002). Hybrids (Fig. 1) were observed in 3 of 30 (10%) natural populations sampled by the author in Nebraska in the summer of 2003. One population, located approximately 16 km east of the town of Sidney, appeared to be composed of approximately 100 hybrid plants.

Fig. 1. Natural F_1 hybrid (*center*) between common wheat (*below*) and jointed goat-grass (*above*). Note the different characteristics of the rachis; in the jointed goat-grass it disarticulates, while in the hybrid it is persistent, as in the wheat parent

Studies have shown that natural F_1 hybrids can produce seed if pollen can reach the stigmas, either via manual cross-pollination (Zemetra et al. 1998), or via wind pollination. Morrison et al. (2002) showed that, in eastern Oregon, USA, natural hybrids were common, with jointed goat-grass being the more frequent female parent. Morrison also demonstrated seed production (BC_1) by hybrid plants, with the majority resulting from pollination by wheat, rather than goat-grass parents. Zemetra et al. (2002) placed hybrids and subsequent BC_1 plants in the field with jointed goat-grass and with wheat, and found seed production both on hybrids and the BC_1 plants. They concluded that the self-sterility observed in greenhouse plantings is overcome if viable pollen reaches the stigma.

Natural hybridization between common wheat and jointed goat-grass raises concern over the possibility of gene transfer between the two species, especially if genes from wheat can become permanently established in naturalized goat-grass populations. Throughout much of the western Great Plains, jointed goat-grass is a problem weed in wheat fields. Both non-transgenic (imidazolinone resistance) and transgenic (glyphosate resistance) genes are being developed, and, in the case of imidazolinone resistance, deployed in wheats to help control jointed goat-grass. As cross-pollination between common wheat and jointed goat-grass is a common event, the possibility exists that goat-grass populations will acquire herbicide resistance traits from cultivated wheat fields. Seefeldt et al. (1998) found what appeared to be natural hybrids between imidazolinone-resistant wheat and jointed goat-grass, and these hybrid plants produced seven BCF_1 seeds. The herbicide resistance trait was found in six of these seven progeny.

Hybridization between common wheat and jointed goat-grass has been substantially documented. The more important question, however, is whether or not hybridization results in permanent gene flow; that is, do genes from common wheat move to goat-grass populations and become permanently established, and inherited in goat-grass progeny. For such an event to occur, F_1 hybrids would have to be pollinated by jointed goat-grass for perhaps one or more generations to restore fertility, and the resultant goat-grass-like progeny would have to be at no selective disadvantage in competition with "true" goat-grass.

To determine whether or not permanent gene flow has occurred, either DNA or phenotypic markers could be used. The seed storage (gluten) proteins of the genus *Triticum* have proven to be especially useful in such investigations. Wheat gluten actually is a complex mixture of perhaps 100 individual protein molecules, belonging to two large classes, the gliadins (soluble in 70% ethanol/water solutions) and the glutenins (solubilized only in the presence of reducing agents). Two classes of glutenins exist. The high-molecular-weight (HMW) glutenin subunits are encoded by complex loci (*Glu-1*) on the long arms of wheat group 1 chromosomes (Payne et al. 1984). Each locus contains two open reading frames encoding single HMW glutenin subunits. However, in nearly all hexaploid common wheats one of the genes on chromosome 1A is inactive; hence, most modern common wheat cultivars produce five HMW glutenin subunits. Low-molecular-weight (LMW) glutenin subunits arise from similar complex loci on the short arms of group 1 chromosomes. A single wheat cultivar might produce as many as 25 different LMW glutenin subunits. Both HMW and LMW glutenin subunits, in their native states, link via intermolecular disulfide bonds to form large polymers. Gliadins, in contrast, exist only as monomeric proteins. They arise from complex loci on the short arms of both group 6 and group 1 chromosomes (Payne et al. 1984). Group 1 produced gliadins arise from the same complex loci as LMW glutenins. Despite the observations that these complex loci are large and contain numerous open reading frames, genetic recombination does not occur within them. Thus, gliadins and glutenin subunits are inherited as members of "blocks."

Individual glutenin subunits and gliadins may be separated by techniques such as SDS-PAGE (sodium dodecyl sulfate polyacrylamide gel electrophoresis), high performance liquid chromatography or capillary electrophoresis. Extensive genetic studies conducted over the past 25 years, especially concentrating on the HMW glutenin subunits (due to their effects on wheat baking quality), have established the chromosomal locations and inheritance of many glutenin and gliadin proteins. Genetic variation for HMW glutenin subunits in hexaploid common wheat is extensive. However, nearly all hard winter wheats ever cultivated in the Great Plains of North America have carried the following alleles, and respective subunits, at the *Glu-1* loci: *Glu-A1a*, subunit 1, or *Glu-A1b*, subunit 2*; *Glu-B1b*, subunits 7+8, *Glu-B1c* (subunits 7+9); *Glu-D1a*, subunits 2+12; *Glu-D1d*, subunits 5+10 (Graybosch 1992). HMW glutenin subunits from closely linked genes in the complex loci always are inherited in these paired combinations, due to the lack of recombination within the *Glu-D1* locus. Jointed goat-grass, in contrast, has been found to be rather uniform, with all populations surveyed to date producing the same pattern of three HMW glutenin subunits (Morrison et al. 2002), and this pattern does not contain proteins that co-migrate with 1, 2* 7+8, 7+9, 2+12 or 5+10.

Gliadins demonstrate more diversity than glutenins, and, while their encoding loci are known, not all observed forms have been assigned to specific alleles. The omega gliadins, arising from complex loci on the short arms of wheat group 1 chromosomes, display the highest level of diversity when resolved via SDS-PAGE. Extensive genetic variation exists in omega gliadin patterns of hexaploid wheat. In this study, the ancestral wheat cultivars Turkey and Scout 66 were used as controls to search for goat-grass populations with wheat-like patterns, which might then indicate transfer of gliadin-containing genes.

Wheat seed storage proteins, therefore, make excellent markers for the analysis of gene flow from wheat to goat-grass populations. They are stably inherited, co-dominant traits. Their analysis is simple, and they can be readily extracted from ground samples of jointed goat-grass heads or joints (groups of spikelets disarticulating together). In addition, seed storage proteins likely are not subjected to selective pressure, and jointed goat-grass individuals carrying such proteins from wheat should not be at a selective disadvantage relative to goat-grasses without such proteins. Morrison et al. (2002) used analyses of HMW glutenin patterns on SDS-PAGE to verify the hybrid nature, and female parentage, of natural F_1 and BCF_1 wheat × jointed goat-grass hybrids.

In the present study, analysis of HMW glutenin subunits and gliadins via SDS-PAGE was used to determine whether wheat genes have become established in naturalized jointed goat-grass populations. Unlike previous studies concentrating on F_1 or BCF_1, the investigations described below have been concentrating on surveying jointed goat-grass populations by sampling non-hybrid plants. Our rationale is that, as jointed goat-grass has grown side by side with cultivated wheats in the Great Plains for nearly 100 years, and all of these cultivated wheats have produced glutenin subunits 1 or 2*, 7+9 or 7+9 and 2+12 or 5+10, the presence of any of these subunits in goat-grass would be indicative of stable and permanent gene flow. Glutenin and gliadin proteins arising from genes on D chromosomes would be most likely encountered as this genome is common to wheat and jointed goat-grass, and D chromosomes are more likely to introgress into jointed goat-grass nuclei, or recombine with jointed goat-grass chromosomes. However, introgressions of genes from other chromosomes also might occur, if chromosomal re-arrangements were to take place.

Jointed goat-grass populations were sampled as follows. At least 24 individual joints from mass collections were analyzed. Samples were pooled into three samples per population, with each sample consisting of eight joints. Samples were ground (any storage proteins present in any of the seed will be detected, even in bulk samples), and glutenin proteins and gliadins were analyzed by SDS-PAGE. Populations (number in parentheses) were sampled from the following Great Plains and western states: Oklahoma (2), Idaho (4), Colorado (8), Washington (3),

Kansas (4), Oregon (2), Idaho (2), Utah (4), Montana (1), South Dakota (1), Wyoming (4) and Nebraska (31).

In agreement with the findings of Morrison et al. (2002), goat-grass populations were found to be uniform with regards to HMW glutenin protein patterns as detected by SDS-PAGE. No polymorphism was detected, and no goat-grass populations were found to contain HMW glutenins derived from common wheat.

Omega gliadin protein patterns of goat-grasses, observed via SDS-PAGE, revealed genetic polymorphisms between populations. At least three different phenotypic patterns were observed amongst sampled goat-grass populations. None of the observed omega gliadin proteins co-migrated with those of the control cultivars Scout 66 or Turkey. However, as extensive variation was observed amongst the goat-grasses, at this point, we cannot unequivocally state that at least some of these proteins did not arise from wheat. A more extensive survey of omega gliadin variation in wheats cultivated in the Great Plains will be necessary before a wheat origin of all the omega gliadins observed in goat-grass populations may be eliminated. Of further interest is the conclusion that North American jointed goatgrass might be more variable than previous studies (Pester et al. 2003) have suggested.

3.2 Conclusions

Throughout much of its range in the Great Plains and Intermountain regions of North America, cultivated bread wheat is grown in close proximity to naturalized populations of jointed goat-grass. Natural hybridization between the two is quite common, resulting in viable F_1 seed. Such plants are self-sterile, but seed set can occur if pollen arrives from an external source. Hence, there is the real possibility of genes moving to jointed goat-grass populations from cultivated wheat. Sampling of 66 naturalized goat-grass populations, using storage proteins as marker genes, failed to establish the presence of permanent gene flow from wheat. However, additional survey work is necessary before the occurrence of such events can be discounted.

3.3 References

Friebe B, Jiang J, Raupp WJ, McIntosh RA, Gill BS (1996) Characterization of wheat-alien translocations conferring resistance to diseases and pests: current status. Euphytica 91:59-87

Graybosch R (1992) The high-molecular-weight glutenin composition of cultivars, germplasm and parents of U.S. red winter wheats. Crop Sci 32:1151-1155

Guadagnuolo R, Savova-Bianchi D, Felber F (2001) Gene flow from wheat (*Triticum aestivum* L.) to jointed goatgrass (*Aegilops cylindrica* Host.), as revealed by RAPD and microsatellite markers. Theor Appl Genet 103:1-8

Johnston CO, Parker JH (1929) *Aegilops cylindrica* Host., A wheat-field weed in Kansas. Trans Kansas Acad Sci 32:80-84

Kimber G, Feldman M (1987) Wild wheat: an introduction. Special report 353. College of Agriculture, University of Missouri, Columbia, USA

Morrison LA, Riera-Lizarazu O, Crémieux L, Mallory-Smith CA (2002) Jointed goatgrass (*Aegilops cylindrica* Host) × wheat (*Triticum aestivum* L.) hybrids: hybridization dynamics in Oregon wheat fields. Crop Sci 42:1862-1873

Payne P, Holt LM, Jackson EA, Law CN (1984) Wheat storage proteins: their genetics and their potential for manipulation by plant breeding. Philos Trans R Soc Lond B 304:359-371

Pester TA, Ward SM, Fenwick AL, Westra P, Nissen SJ (2003) Genetic diversity of jointed goatgrass (*Aegilops cylindrica*). Weed Sci 46:632-634

Ribeiro-Carvalho C, Guedes-Pinto H, Harrison G, Heslop-Harrison JS (1997) Wheat-rye chromosome translocations involving small terminal and intercalary rye chromosome segments in the Portuguese wheat landrace Barbela. Heredity 78:539-546

Riley R, Chapman V (1967) The inheritance in wheat of crossability with rye. Genet Res Cambr 9:259-267

Seefeldt SS, Zemetra R, Young FL, Jones SS (1998) Production of herbicide-resistant jointed goatgrass (*Aegilops cylindrica*) × wheat (*Triticum aestivum*) hybrids in the field by natural hybridization. Weed Sci 46:632-634

Sharma HC, Gill BS (1983) Current status of wide hybridization in wheat. Euphytica 32:17-31

Zemetra RS, Hansen J, Mallory-Smith CA (1998) Potential for gene transfer between wheat (*Triticum aestivum*) and jointed goatgrass (*Aegilops cylindrica*). Weed Sci 46:313-317

Zemetra RS, Mallory-Smith CA, Hansen J, Wang Z, Snyder J, Hang A, Kroiss L, Riera-Lizarazu O, Vales I (2002) The evolution of a biological risk program: gene flow between wheat (*Triticum aestivum* L.) and jointed goatgrass (*Aegilops cylindrica* host). Proceedings of the gene flow workshop. Ohio State University, Columbus, Ohio, pp 178-187

4 Humans, Climate, and Plants: the Migration of Crested Wheatgrass and Smooth Bromegrass to the Great Plains of North America

Kenneth P. Vogel

Research Geneticist, USDA-ARS, 344 Keim Hall, Univ. of Nebraska, Lincoln, Nebraska 68583, USA, E-mail kpv@unlserve.unl.edu

4.1 Abstract

The cultivation practices that were used in Europe and the eastern half of North America were utilized in the initial settlement of the Great Plains. Unfamiliarity with the climate of the Great Plains and Midwest and insufficient knowledge and technology to adapt crop production systems to the soils and climate lead to a major agriculture disaster which resulted in millions of hectares of land that needed to be re-seeded to grasses. Unrestricted grazing on public lands in the intermountain west resulted in severe rangeland degradation. Lack of knowledge and technology for using native plants and some specific characteristics of native plants that made them difficult to use resulted in the use of crested wheatgrasses and smooth bromegrass which had characteristics that met specific revegetation and production requirements. Crested wheatgrass and smooth bromegrass plant materials were from regions that were climatic analogs of the Great Plains and were adapted. These two grasses literally preserved the remaining top soil on millions of hectares of land. In the subsequent half-century, agronomist, geneticists, and rangeland scientists have learned how to establish and manage native grasses such as switchgrass (*Panicum virgatum* L.), big bluestem (*Andropogon gerardii* Vitman), indiangrass [*Sorghastrum nutans* (L.) Nash] and others so they are now available for use in revegetation. Although native grasses are available for use in the Great Plains and the Midwest of North America, crested wheatgrass and smooth bromegrass are now naturalized North American species and will continue to be vital to the economy of the USA and Canada. Their forage production patterns fits gaps in the forage production cycle for ruminant livestock that cannot be adequately met by native species in regions where bromegrass and crested wheatgrasses are well adapted.

4.2 The Plants

Humans throughout history have migrated to new areas and taken their food plants and animals with them (Harlan 1975; Diamond 1997). They also have utilized plants indigenous to the area into which they have migrated. The Great Plains and Midwest of North America were vast grasslands prior to the migration and settlement by people of largely northern European origin. One hundred and fifty years later, significant land areas have been converted to cultivation and are in crop production. Millions of hectares of land also have been planted to crested wheatgrass (*Agropyron* spp.) and smooth bromegrass (*Bromus inermis* Leyss.) which are nonnative forage and pasture species. The purpose of this report is to describe the interactions of humans, plants, and climate that resulted in the migration and utilization of Eurasian grasses on millions of hectares of North American grasslands. The use and status of these species in North America continue to evolve as humans, plants, and their environment interact.

The genus *Agropyron* has contained many of the species of the perennial Triticeae but in current taxonomy it is restricted to species of the crested wheatgrass complex which is a polyploid series based on the 'P' genome (Asay and Jensen 1996). The genus consists of a series of diploid (2n=14), tetraploid (2n=28), and hexaploid (2n=42) species. The cultivated type species are *A. cristatum* or Fairway crested wheatgrass and *A. desertorum* (Fisch. ex Link) Schultes or Standard crested wheatgrass. The Standard type has been the most widely used in the USA; the Fairway type has been used in Canada in addition to the Standard type. Crested wheatgrasses are persistent, long-lived perennial bunchgrasses adapted to temperate sites receiving 200 to 450 mm of annual precipitation. The inflorescence is a spike, and culms range in height from 15 to more than 100 cm. Crested wheatgrasses are native to the steppe region of European Russia and southwestern Siberia (Asay and Jensen 1996). Crested wheatgrasses were apparently first cultivated in the Volga district east of Saratov (Asay and Jensen 1996). Crested wheatgrasses have excellent seed production and the seed is easy to harvest, clean, and plant. They have excellent seedling vigor, drought and cold resistance, and can tolerate heavy grazing. In areas where they are adapted, they typically produce equivalent or greater forage yields and livestock carrying capacity than native rangelands.

Smooth bromegrass is a leafy, tall-growing, sod-forming perennial cool-season grass adapted to temperate regions (Vogel et al. 1996). The commonly grown form of smooth bromegrass is an autoallooctaploid with a chromosome number of 2n=56 while the lesser used tetraploid (2n=28) is an allotetraploid. The flowering culms are 50 to 100 cm in height. The inflorescence is a panicle that is erect and 7 to 20 cm long with whorled branches, and becomes contracted and purplish brown at maturity (Vogel et al. 1996). It is of Eurasian origin and it is a polymorphic species which can be divided into several subspecies or ecotypes that are related to their origin (Tsvelev 1984). Zerebina (1931, 1933, 1938), whose research was subsequently summarized by Knowles and White (1949), recognized two main ecological-geographical groups: the "meadow" group or northern climatype, and the "steppe" group or southern climatype. Descriptions of these groups correspond to the northern and southern types of smooth bromegrass later recognized in North America (Newell and Keim 1943). The meadow types were found from Murmansk in the north to the Caucasus in the south, although south of the central Chernozem region they were found only in valleys and moist habitats. The steppe type was found with the meadow type in the central Chernozem region and was the principal ecotype in the dry steppe areas of the mid and lower Volga districts, Kazakhstan, the northern Caucasus, the eastern Ukraine, and the southern Altai regions of the former USSR. Smooth bromegrass is widely adapted because of the ecotypic variation that exists in the species in its native range. It is similar to the crested wheatgrasses in that it is an excellent seed producer, seed is easy to harvest and clean, and it has excellent seedling vigor and establishes rapidly. It can tolerate mismanagement and heavy grazing pressure and is very productive if well managed. It is not as drought tolerant as crested wheatgrass but has better drought tolerance than other mesic cool-season grasses such as perennial ryegrass (*Lolium perenne* L.), timothy (*Phleum pratense* L.), or cocksfoot (*Dactylis glomerata* L.; Vogel et al. 1996). It produces significantly more forage than crested wheatgrasses in regions where annual precipitation exceeds 500 mm.

4.3 Settlement, Agriculture and War

The Great Plains of the United States were settled by people of largely northern European origin (Morison 1965). The Midwest was settled from about 1820 to 1865 (Morison 1965). The conclusion of the American Civil War in 1865 marked the beginning of the primarily settlement period of the Great Plains which ended about 1900 (Morison 1965). When the prairies and grasslands were first settled, there were abundant grasslands for farmers' cattle and horses. Farmers plowed areas of their farms for grain crop production, primarily maize, oats, and wheat, but retained grasslands for use by their livestock. Although the crops that the settlers brought with them were adequately adapted to the Great Plains, the production systems that the settlers brought with them were not. These systems involved yearly plowing and intensive cultivation for weed control. This left the soil bare and subject to wind and water erosion.

In the prairies and plains of these sections of North America, most of the grasslands contained grasses which had the C_4 photosynthesis system. These warm-season grasses did not have the production patterns of cool-season or C_3 grasses to which the settlers were accustomed (Moser and Vogel 1995). Warm-season grasses start growing 4 to 6 weeks later in the spring than cool-season grasses and go dormant as temperatures drop in the autumn. They require different management than cool-season grasses. As a result, the native pastures were often mismanaged and overgrazed and gradually were reduced in productivity. Farmers wanted new pasture plants that could produce more forage than the native grassland and which would provide grazing in spring and autumn. Plant explorers were employed by the United States Department of Agriculture (USDA) to collect plant material for use in these and other regions of the United States.

The demand for grain crops including wheat before and after World War I resulted in additional millions of hectares of land being plowed during the period 1905 to 1920 and used for grain production (Dillman 1946). This included land in marginal crop production areas that had not been converted to grain crop agriculture in the initial settlement period. Prices for grain crops were high and climatic conditions were favorable for grain crop production which encouraged farmers to convert grasslands to cultivated croplands.

4.4 Disaster - The Drought of the 1930s

In the 1930s or the "Dust Bowl" era, a major drought affected large parts of North America resulting in massive soil erosion problems, particularly on lands that were only marginally suited for crop production. In seven of the ten years between 1928 and 1937, average annual precipitation was below the long-term average (Lorenz 1986). The drought resulted in severe wind erosion and soil drift onto cropland, and loss of plants on overgrazed grasslands. By the end of the 1930s millions of hectares of land in the former prairie and plains states of the USA and Canadian provinces needed to be reseeded to grasslands to preserve the soil and the ecosystems. This drought resulted in the realization that considerable land had been plowed that needed to be rapidly converted into grasslands, and severely damaged grasslands needed to be restored by reseeding.

There were problems with using native species. Native grasslands had been abundant and common and as a result almost no information was available on seed production and establishment of native species. Many of the native species have seed with chaffy appendages which made harvesting, cleaning, and planting difficult and only limited amounts of seed could be obtained from native harvests (Cornelius 1950; Masters et al. 1993). Research was initiated on native grasses beginning in the mid-1930s but a more immediate solution to the revegetation problem was needed. The introduced species that survived the drought, had good seed yields, and that could be readily established were smooth bromegrass and the crested wheatgrasses. Grasses such as cocksfoot and timothy often did not survive the drought in the midwestern USA. After their introduction in the late 1880s, experiment stations in both the USA and Canada had evaluated the available germplasm and made selections. Research had been conducted on seed production, establishment, forage production, and grazing management. By the end of the 1930s, plant materials and the associated technology were available for a rapid increase in use of the species (Dillman 1946; Lorenz 1986; Rogler and Lorenz 1983; Newell 1973; Vogel et al. 1996).

4.5 Crested Wheatgrass Story

Crested wheatgrasses were first introduced into North America in 1892 by N.E. Hansen of the South Dakota Experiment Station (Dillman 1946; Rogler and Lorenz 1983; Lorenz 1986). Hansen was born in Denmark in 1866 and immigrated with his family at the age of seven to the USA (Taylor 1941). His family lived in Iowa and he attended Iowa State College (now Iowa State University) where he received B.S. and M.S. degrees. Even as a student, he was interested in the history and migration of plants and their improvement (Taylor 1941). In 1895, he went to South Dakota State College at Brookings, SD, as Professor of Horticulture. While there, he was contracted by USDA to do plant exploration and collection because of his knowledge of plants and languages (Taylor 1941). He was the first official plant explorer for the USDA. In his first collection trip, which was in 1897-1898, he collected five railroad cars of several hundred kinds of grains, grasses, and plants from Russia, Turkestan, China, Siberia, and Transcaucasus. The material, which included seed of crested wheatgrass from the eastern Volga region and Siberia, was shipped to USDA, Division of Seed and Plant Introduction, Washington, D.C., for distribution to experiment stations. On this trip, he also obtained 12 tons of bromegrass seed from the Volga region of Russia. This seed was distributed to several experiment stations but no permanent plantings or seed increases of crested wheatgrasses were apparently made (Dillman 1946; Lorenz 1986; Asay and Jensen 1996).

N.E. Hansen received a second importation of crested wheatgrass in 1906 from the Valuiki Experiment Station in Russia which consisted of five *A. desertorum* seedlots (PIs 19537-19541) and one *A. cristatum* seedlot (PI 19536). Seed was distributed to 15 experiment stations (Dillman 1946). The first known research planting of these accessions was at the Belle Fouce, SD, research station in 1908. A.C. Dillman was a USDA agronomist at the station and maintained plantings of the crested wheatgrass from 1908 to 1915. He harvested and supplied seed to other stations including what is now the Agricultural Research Service (ARS), USDA Laboratory at Mandan, ND (Dillman 1946; Rogler and Lorenz 1983; Lorenz 1986). Plantings made in 1915 at Mandan were the foundation from which the initial cultivar released in the United States was developed. Over 100 kg of seeded was distributed to experiment stations and farms in the Dakotas, Montana, and Wyoming from 1920 to 1923 (Dillman 1946). Before the release of the first cultivar, over 1800 kg of seed of four of the better accessions was distributed to farmers for testing (Dillman 1946). The first pasture trial was established in 1923 at the Ardmore, SD, field station and performance trials with dairy cattle were successful (Lorenz 1986). The first grazing trial with beef cattle was established at Mandan, ND, in 1932. The pasture was still in production in 1983 (Rogler and Lorenz 1983; Lorenz 1986). The first farm planting was in Montana. Direct increases in the available accessions were used in the initial plantings of the crested wheatgrasses. The first commercial seed field was started by a farmer near Dickenson, ND, in 1926 and first seed was offered for sale in 1928. This farmer had over 80 ha in seed production when the demand for seed came in 1933 as a result of government revegetation programs (Lorenz 1986. In the 25- period between 1906 when N.E. Hansen received his second shipment of crested wheatgrass germplasm and 1933, agronomists at several experiment stations developed the basic agronomic information that was needed for it to become a widely utilized grass. The standard crested wheatgrass cultivar "Nordan" was released by the Mandan, ND, station in 1953 and was the most widely used cultivar in the United States for many years (Asay and Jensen 1996).

In Canada, crested wheatgrass research began in 1916 at Saskatoon, Saskatchewan, by L.E. Kirk who at that time was a graduate assistant (Dillman 1946). Seed was obtained from the USA and included PI 19536 and PI 19540 which were among those received by N.E. Hansen in 1906. Other seed lots were sent to Saskatoon at a later date from Mandan, ND. The first crested wheatgrass cultivar, "Fairway" was a mass selection of fine leafy plants from PI 19536 which were harvested in 1925 (Dillman 1946). Breeding work has continued on crested wheatgrass at several research stations in both the United States and Canada and improved cultivars and management practices are periodically documented and released (Asay and Jensen 1996).

4.6 Bromegrass Story

Smooth bromegrass was introduced into the USA and Canada in the 1880s (Newell 1973; Vogel et al. 1996). The first introduction was by the California Experiment Station in 1884 which distributed seed to other experiment stations (Newell 1973). Major introductions were from central Europe including Hungary, northern Germany, and the Russian collections of N.E. Hansen (Taylor 1941; Newell 1973). By the late 1890s bromegrass was being grown in the Midwest and Great Plains of the USA and Canada. Bromegrass identified as "Hungarian" bromegrass was sown on the University of Nebraska Experiment Station farm at Lincoln, NE, in 1897 in a small field (Lyon 1899). Seed and forage yields were determined. In 1898, a 6.5-ha pasture was established that was used in grazing trials by cattle and horses. In his summary, Lyon (1899) noted that its advantages over the native prairie grasses was that it became green fully a month earlier in the spring and remained green later into the autumn. He indicated that it was not as good for butter and milk production as bluegrass (*Poa pratensis* L.) and white clover (*Trifolium repens* L.) from dairy cattle but that it was safe and widely adaptable. Similar results were being obtained from other experiment stations (Waldron and Porter 1919). The initial spread and distribution of smooth bromegrass in North America were based on the simple increase and distribution of the introduced strains from experiment stations to farmers. The first recorded breeding work on smooth bromegrass was conducted in the early 1900s by two Kansas farmers, the Achenbach brothers, who did mass selection on one of their best bromegrass fields to develop "Achenbach" bromegrass (Vogel et al. 1996). Initial breeding work by several experiment stations was initiated between 1910 and 1920 but did not lead to the development of any cultivars and was not re-initiated until the drought of the 1930s stimulated breeding work at several locations. During the 30-year period from the late 1890s to the early 1930s, agronomic information was obtained on the management of smooth bromegrass at locations throughout the prairie regions of the United States and Canada.

Breeding programs for bromegrass were initiated at several experiment stations beginning in the mid-1930s to develop grasses for reseeding the lands ravaged by the drought. Typically, the breeders worked on multiple species. The breeding programs used available germplasm resources, primarily domestic germplasm sources such as old plantings that had been in existence for sufficient periods of time to have become naturalized (Vogel et al. 1996). The evaluation work documented the existence of "southern" and "northern" strains of smooth bromegrass (Newell and Keim 1943; Knowles and White 1949). The southern types, which were believed to trace from the Hungarian introductions, were the best adapted to the Central Great Plains and the southern part of the Corn Belt, while the northern strains were best adapted to the Northern Plains, the upper Midwest and northeastern states, and to the adjacent provinces of Canada, and traced to some of the accessions originally collected by Hansen in 1897 and to other accessions from northern Europe (Vogel et al. 1996). The first series of cultivars did not involve any formal breeding work other than selection among existing ecotypes or strains. For example, seed from several old plantings in Nebraska proved superior to other germplasm sources and were traced to a common origin. These fields were certified and were the source of the cultivar "Lincoln" which is still a widely used (Vogel et al. 1996; Casler et al. 2000). Bromegrass breeding and management research have been continued to the present at several locations in both the USA and Canada.

4.7 The Success Story

The early evaluation and management work enabled seed production of crested wheatgrass to be rapidly expanded. Information on establishment practices including planting dates enabled land to be rapidly and efficiently re-vegetated during the 1940s and 1950s. Because it was often the only adapted grass for which seed was available in quantity for its area of adaptation, it was often seeded in monocultures. Crested wheatgrass salvaged vast areas of deteriorated rangelands and abandoned cropland in the central and northern plains of North America that were severely damaged during the "dust bowl" period of the 1930s (Rogler and Lorenz 1983; Lorenz 1986). Many of these plantings remain and are still productive. Crested wheatgrass also has become the most important domestic grass in the arid intermountain West of the United States. In the intermountain West, excessive and unregulated livestock grazing by sheep and cattle in the late 1800s and early 1900s led to severe rangeland degradation and to a change in the vegetation community (Harrison et al. 2003). By 1932, the vegetation in the some areas of the Great Basin region had changed from 49 to 81% perennial grass and 10% sagebrush to mainly sagebrush ground cover (Harrison et al. 2003). Re-vegetation was needed to restore these degraded rangelands and to combat invading noxious weeds. In this region, many of the native cool-season plants do not tolerate heavy grazing (Asay and Jensen 1996). Crested wheatgrasses were found to be well adapted in trials that began in the late 1930s (Harrison et al. 2003). The annual poisonous forb, halogeton became a serious threat to livestock on intermountain rangelands during the 1940s. Crested wheatgrass proved to be an effective biological suppressor of this weed (Mathews 1986; Young and Evans 1986). Agricultural statistics are not available on the land area that has been seeded and is currently vegetated by crested wheatgrasses, but it is estimated that 5.1 million and 800,000 ha had been seeded to crested wheatgrass in the USA and Canada, respectively, by the early 1980s (Rogler and Lorenz 1983; Asay and Jensen 1996). It is primarily used north of 40°N lat. and west of 100°W long. It is used in lower latitudes at higher elevations in the intermountain area. Pasture and rangeland planted to crested wheatgrass are primarily used for grazing.

Similar to the crested wheatgrass, the early work on smooth bromegrass enabled it to be rapidly increased and utilized to restore degraded lands following the 1930s drought. For example, by 1945, almost 500,000 ha had been seeded in the state of Nebraska (Frolik 1945). It was and continues to be primarily utilized in North America in regions north of 40°N lat. and east of 100°W long. that have 500 mm or more annual precipitation, or in areas that have similar temperature ranges because of elevation (Vogel et al. 1996). It is estimated that the Midwest and eastern states of the USA and the adjacent provinces of Canada each have over 500,000 ha of smooth bromegrass. The principal use of smooth bromegrass is as a cool-season pasture grass in this region (Vogel et al. 1996). It is the principal component of these pastures, although legumes, particularly alfalfa (*Medicago sativa* L.), are usually included in the plantings. In addition, it remains the most important grass for conservation purposes in this region where it is used on roadsides, grass waterways, and field borders for erosion control.

4.8 The Alien Problem

Crested wheatgrass and smooth bromegrass are not regarded favorably by everyone even though they saved millions of hectares of land from severe erosion and remain economically important grasses in both pasture and rangelands (Lesica and DeLuca 1996; Christian and Wilson 1999). Most of the criticism is based on the fact that they are not native species and that pure stands of the species lack biological diversity. Native species are considered by many ecologists to be more desirable than the introduced grasses. Both crested wheatgrass and smooth bromegrass can be found on invasive species lists of environmental organizations on the internet (http://www.americanlands.org; http://tncweeds.ucdavis.edu; verified 29 September 2003). Some of these groups also list tall fescue (*Festuca arundinaceae* Schreb.) and white and yellow sweet clover (*Melilotus alba* Medikus and *M. officinalis* Lam., respectively) as invasive plants. Smooth bromegrass and crested wheatgrass are viewed as invaders of native rangeland and remnant prairie sites.

Personal observation and published reports (Blankenpoor and May 1999; Bro-ersma et al. 2000) indicate that neither grasses invades grasslands that are well managed and in good condition. Bromegrass will invade disturbed areas and poorly managed native grasslands especially in the tallgrass prairie ecoregion. Crested wheatgrass is very non-invasive and usually does not spread from where it is planted. The soil quality of long-term crested wheatgrass fields or range sites has been reported to be below that in adjacent, unplowed prairie (Lesica and DeLuca 1996; Christian and Wilson 1999). The fact that the sites seeded to crested wheatgrass had been plowed and eroded before they were seeded to crested wheatgrass is often ignored or remains a confounding factor in research reports. Other reports indicate that grazed crested wheatgrass and native prairies produce similar amounts of root biomass (Krzic et al. 2000).

Crested wheatgrass will need to be used in the intermountain West of the USA and Canada for noxious weed control. Invasion of rangelands by the annual bromegrass, *Bromus tectorum* L., or cheatgrass have created severe problems in-cluding rangeland fires that in the state of Utah alone burn almost 60,000 hectares annually (Harrison et al. 2003). Plants that can be used to control cheatgrass and other noxious weeds are needed, although vigorous cultivars such as Hycrest can effectively compete against noxious weeds such as cheatgrass and halogeton. A rapidly spreading noxious weed, spotted knapweed (*Centurea maculosa auct. non* Lam.), has been shown recently to release a phytotoxin from its roots that triggers the death of the root system of grasses native to the intermountain region in the western USA (Bais et al. 2003). European grasses were more resistant to the toxin and will likely be needed as part of the control strategy for this noxious weed.

Pumpelly's brome (*Bromus inermis* Leyss. ssp. *pumpellianus* (Scribn.) Wagnon) is a perennial bromegrass that is native to Alaska and the western and northern intermountain areas of North America. Based on genetic studies which have demonstrated that it has the same nuclear and cytoplasmic genomes as smooth bromegrass with which it is fully fertile, it has been reclassified as a sub-species of smooth bromegrass (Armstrong 1982; Pillay and Hilu 1990; Vogel et al. 1996). Pumpelly's brome probably migrated during periods when Asia and Northern America were connected via a land bridge between Alaska and Siberia, and because of isolation developed small differences in morphology while retain-ing the genome intact. It can be argued that smooth bromegrass should be re-garded as a native of North America because its genome was already present in Pumpelly's bromegrass.

4.9 Conclusions

The crested wheatgrass and smooth bromegrass are now naturalized species of North America. Their use in re-vegetation and conservation saved the topsoil on millions of hectares of land. They will continue to be used in production agricul-ture and in restoring degraded rangelands that have been invaded by noxious weeds.

4.10 References

Armstrong KC (1982) Hybrids between the tetraploid and octaploid cytotypes of *Bromus inermis*. Can J Bot 60:476-482

Asay KH, Jensen KB (1996) Wheatgrasses. In: Moser LE, Buxton D, Casler MD (eds) Cool-season forage grasses. Agron Monogr ASA, CSSA, SSSA, Madison, WI, pp 691-724

Bais HP, Vepachedu R, Gilroy S, Callaway RM, Vivanco JM (2003) Allopathy and exotic plant invasion: from molecules and genes to species interactions. Science 301:1377-1380

Blankenspoor GW, May JK (1996) Alien smooth brome (Bromus inermis Leyss) in a tall-grass prairie remnant: seed bank, seedling establishment, and growth dynamics. Natural Areas J 16:289-294

Broersma, Krzic M, Thompson DJ, Romke AA (2000) Soil and vegetation of ungrazed crested wheatgrass and native rangelands. Can J Soil Sci 80:411-417

Christian JM, Wilson SD (1999) Long-term ecosystem impacts of an introduced grass in the northern Great Plains. Ecology 80:2397-2407

Casler MD, Vogel KP, Balasko JA, Berdahl JD, Miller DA, Hansen JL, Fritz JO (2000) Genetic progress from 50 years of smooth bromegrass breeding. Crop Sci 40:13-22

Cornelius DR (1950) Seed production of native grasses under cultivation in eastern Kansas. Ecol Monogr 20:1-29

Diamond J (1997) Guns, germs, and steel. Norton, New York, NY

Dillman AC (1946) The beginnings of crested wheatgrass in North America. J Am Soc Agron 38:237-250

Frolik EF (1945) 1,000,000 acres of bromegrass. Nebraska Farmer 87:1-40

Harlan JR (1975) Crops and man. American Society of Agronomy and Crop Science Society of America, Madison, WI

Harrison RD, Chatterton NJ, McArthur ED, Ogle D, Asay KH, Waldron BL (2003) Range plant development in Utah: a historical view. Rangelands 25:13-19

Knowles RP, White WJ (1949) The performance of southern strains of bromegrass in western Canada. Sci Agric 29:437-50

Krzic M, Broersma K, Thomson DJ, Bomke AA (2000) Soil properties and species diversity of grazed crested wheatgrass and native rangelands. J Range Manage 53:353-358

Lesica P, DeLuca TH (1996) Long-term harmful effects of crested wheatgrass on Great Plains grassland ecosystems. J Soil Water Conserv 51:408-409

Lorenz RJ (1986) Introduction and early use of crested wheatgrass in the Northern Great Plains. In Johnson KL (ed) Crested wheatgrass: Its values, problems, and myths. Symp Proc Utah State Univ, Logan, UT, 3-7 Oct 1983. Utah State Univ., Logan, UT, pp 9-20

Lyon TL (1899) Hungarian brome grass. Nebraska Agricul Exp Stn Bull 61 (XII):35-63

Masters RA, Mitchell RB, Vogel KP, Waller SS (1993) Influence of improvement practices on big bluestem and indiangrass seed production in tallgrass prairies. J Range Manage 46:183-187

Mathews WL (1986) Early use of crested wheatgrass seedings in halogeton control. In Johnson KL (ed) Crested wheatgrass: its values, problems, and myths. Symp Proc Utah State Univ, Logan, UT, 3-7 Oct 1983. Utah State Univ, Logan, UT, pp 27-28

Morison SE (1965) The Oxford history of the American people. Oxford Press, New York

Moser LE, Vogel KP (1995) Switchgrass, big bluestem, and indiangrass. In: Barnes RF, Miller DA, Nelson CJ (eds) Forages, vol I, 5th edn. An introduction to grassland agriculture. Iowa State Univ Press, Ames, IA, pp 409-420

Newell LC (1973) Smooth bromegrass. In: Heath ME, Metcalfe DS, Barnes RF (eds) Forages. The Science of grassland agriculture. Iowa State University Press, Ames, IA, pp 254-262

Newell LC, Keim FD (1943) Field performance of bromegrass strains from different seed sources. J Am Soc Agron 35:420-434

Pillay M, Hilu KW (1990) Chloroplast DNA variation in diploid and polyploidy species of Bromus (Poaceae) subgenera Festucaria and Ceratchloa. Theor Appl Genet 80:326-332

Rogler GA, Lorenz RJ (1983) Crested wheatgrass - early history in the United States. J Range Manage 36:91-93

Taylor HJ (1941) To plant the prairies and the plains. The life and work of Niels Ebbesen Hansen. BIOS XII:1-72

Tsvelev NN (1984) Grasses of the Soviet Union, part I. Balkema, Rotterdam. (Russian translations series)

Vogel KP, Moore KJ, Moser LW (1996) Bromegrasses. In: Moser LE, Buxton D, Casler MD (eds) Cool-season forage grasses. Agronomy monograph ASA, CSSA, SSSA, Madison, WI, pp 535-567

Waldron LR, Porter WR (1919) Brome-grass, slender wheat-grass and timothy. North Dakota Agric Exp Stn Circ 24:1-8

Young JA, Evans RA (1986) History of crested wheatgrass in the Intermountain Area. In: Johnson KL (ed) Crested wheatgrass: its values, problems, and myths. Symp Proc Utah State Univ, Logan, UT, 3-7 Oct 1983, Utah State Univ, Logan, UT, pp 21-25

Zerebina ZN (1931) Botanical-agronomical studies of awnless brome grass *(Bromus inermis* Leyss.). Bull Appl Bot Genet Plant Breed 25:201-352

Zerebina ZN (1933) Awnless brome grass. Rastenievodstvo USSR 1:507-518

Zerebina ZN (1938) Brome grass. Rukovod Approb Seljskohoz Kuljt 4:112-123

5 Biogeography, Local Adaptation, Vavilov, and Genetic Diversity in Soybean

T.E. Carter, Jr.[1], T. Hymowitz[2], and R.L.Nelson[3]

[1]USDA-ARS, Raleigh, North Carolina, USA, E-mail tommy_carter@ncsu.edu
[2]University of Illinois, Urbana, Illinois, USA
[3]USDA-ARS, Urbana, Illinois, USA

5.1 Abstract

It is the purpose of this paper to illustrate the impact of geography, climate, and humankind in shaping the present-day genetic diversity in soybean [*Glycine max* (L.) Merr.]. Examination of soybean germplasm collections around the globe reveals that an enormous phenotypic range in genetic traits exists in soybean, which is well beyond the phenotypic range observed in the wild progenitor (*Glycine soja* Seib. et Zucc.). Maturity date, seed coat color, plant height, seed size, and seed yield are noted examples of traits which have a wider phenotypic range in *G. max* than in the wild *G. soja*. The diversity found in domesticated soybean is the result of over 3,000 years of cultivation in which Chinese farmers selected more than 20,000 landraces (defined as cultivars that predate scientific breeding). The extensive range in phenotype embodied in landraces today is the result of the slow spread of soybean throughout geographically diverse Asia (China first, then Korea and Japan), the continual occurrence of natural mutations in the crop, and both conscious and unconscious selection for local adaptation. The more recent spread of soybean out of Asia in the past 250 years, coupled with modern breeding efforts of the past 70, has intensified and globalized the process of local adaptation and increased the phenotypic range in soybean beyond that of landraces. The increased range in phenotype for modern cultivars includes increases in seed yield, elevation of seed protein/oil concentration, and development, only within the past 20 years, of commercial cultivars that are sufficiently tall and adapted to be grown profitably near the equator. The phenotypic range and distribution observed in modern cultivars and antecedent landraces have clear biogeographical interpretations which relate primarily to genetic alteration of photoperiod response (a prerequisite to adaptation to diverse latitudes) and tolerance to climate extremes.

Although the phenotypic range in genetic traits has been expanded in modern soybean through global dispersal and genetic recombination, it is perhaps surprising that that these factors have not had a corresponding positive impact on genetic diversity of modern breeding programs outide of China.. Genetic diveresity in breeding programs is important as a concept, because it is a measure of the potential of a country to develop new and substantially improved cultivars. For the purposes of this paper, genetic diversity in breeding programs is defined, as genetic variation among cultivars found within a particular country or country subregion.   Empirical analysis of DNA marker and pedigree diversity in modern cultivars indicates that diversity is greatest in cultivars developed in China, less in Japan and least in North America. Phenotypic analysis of modern Chinese and North American cultivars follows the same pattern of diversity. Pedigree analyses of Latin American breeding programs, although incomplete, show that these programs are derived primarily from a subset of North American breeding stock and are, thus, likely to be less diverse than the North American breeding program. Decreased diversity of cultivars outside of China was also correlated with a reduction in the number of founding stock used to establish the breeding programs from which the cultivars arose.

Although conscious breeding choices, the high economic costs of breeding, and historical factors can be used to explain the reduced diversity in breeding programs outside of China vs. within, it is important to note that these results, obtained from modern breeding programs, are consistent with (1) Vavilov's principle of crop domestication, which states that genetic diversity will be greatest at the center of domestication (China in the case of soybean), and (2) the concept of Darwinian genetic drift which can be used to infer that genetic relatedness or uniformity will increase within breeding populations that are derived from relatively few founding members. A precaution gleaned from the observed trend in diversity is that all soybean breeding programs outside China, regardless of the phenotypic superiority of their genetic breeding materials, should be examined to determine the adequacy of genetic diversity. The impact of the transgenic glyphosate resistance on genetic diversity in soybean is assessed briefly.

5.2 A Brief History of Soybean

'More than 3,000 years ago, Chinese farmers used genetic diversity found in the wild to produce the crop species that we now know as domesticated soybean. Continued use of genetic diversity by farmers since domestication has improved the soybean crop to a remarkable degree, raising the yield of soybean over that of the original wild soybean by at least an order of magnitude, enhancing its resistance to an array of important diseases, and adapting the crop to grow in extreme climates. This transformation of soybean from wild plant to modern crop is one of the more remarkable achievements in agriculture, past or present' (Carter et al., 2004)

Not surprisingly, the technical details of this achievement are lost in antiquity. However, historical records reveal that soybean reached most of China and the Korean peninsula by the first century A.D., and was grown throughout Asia by the 15th century (Hymowitz and Newell 1980). In the 16th century, European visitors to China and Japan were introduced to soyfoods and, in 1712, Engelbert Kaempfer may have been the first Westerner to describe how to prepare food products from the soybean. Later in the 18th century, soybean was introduced throughout Europe and made its first appearance in North America. Samuel Bowen, a former seaman, brought the soybean to Savannah, Georgia, from China via London in 1765 and produced perhaps the first soy sauce in North America. This small enterprise did not persist beyond Bowen's lifetime, the technology was lost, and soybean was reintroduced to the farmer in North America in the early or mid-19th century, probably as a soil-building, green-manure legume crop. The earliest records of soybean in South America and Africa date to the late 19th or early 20th century (Paillieux 1880; D'Utra 1882; Itie 1910; Don 1911; Faura 1933; Burtt-Davy 1905).

Despite the early attempts to establish soybean for soyfood production in the West, as it long had been in the East, the Western world did not fully understand the connection between the cultivation of soy and value of soy as human food. Thus, soybean was grown and valued more familiarly as a forage crop in the late 19th and early 20th centuries. Advances in chemistry led to the discovery, ca. 1915, that the soybean seed is a rich source of oil and protein. This changed soybean from forage to a cash seed crop by 1945 in the West and drastically increased hectarage to that of a major world crop. Recent medical findings which identify the health benefits of soyfoods have fostered considerable interest among western consumers so that we can say that 240 years after the Western debut of the crop, soyfoods have definitely arrived. Today, the USA, Brazil, Argentina, China and India are the major producers of soybean.

5.3 Domestication of Soybean

It is commonly believed that soybean had been domesticated from wild soybean during the Zhou dynasty in the eastern half of China by 1,100 B.C. (Hymowitz and Newell 1980; Gai 1997; Gai and Guo 2001). This estimate is derived from early references to soybean that appeared in Chinese literature (Qiu et al. 1999). At present, there are no archaeological findings to support an older origin of soybean. Poems from 600 B.C. mention the first recorded uses of soybean as food: soup made from soybean leaves and stew made from seed. The use of soybean as tofu and green vegetable were first documented around 1000 and 1550 A.D., respectively (Gai and Guo 2001), although tofu was probably in use by 0 A.D.

Although wild soybean grows commonly in eastern China (from 24–53°N lat.), Japan, Korea, and the eastern extremes of Russia, current theories suggest that soybean was domesticated in either the southern or central regions of eastern China (Wang et al. 1973; Lu 1977; Hymowitz and Newell 1980, 1981; Xu 1986; Chang 1989; Guo 1993; Zhou et al. 1998). Two recent studies illustrate the argument for eastern China as the center of origin (Carter et al. 2004). In the first study, DNA marker data were employed to compare landraces and wild soybean accessions in order to infer domestication patterns using germplasm from the southern, central, and northeastern areas of China. The premise of the study was that *G. max* would be most closely related to the *G. soja* pool from which it was domesticated. Gai et al. (1999) evaluated 200 *G. max* and 200 *G. soja* accessions for chloroplast and mitochondrial restriction fragment length polymorphism (RFLP) markers. These organellar genomes were selected rather than the nuclear genome because their cytoplasmic nature had the presumed advantage of being uncompromised by potential cross pollination between *G. soja* and *G. max* over millennia and long after domestication. Assuming that the geographical distribution of present-day *G. soja* has not changed appreciably from that of ancient times, the marker data of Gai et al. (1999, 2000) indicated that significant differentiation of wild soybean may have occurred long before domestication of the cultivated soybean. *Glycine max* accessions from all regions of China were more closely related to southern accessions of *G. soja* from the Yangzi River valley than to *G. soja* accessions from any other region. These data support the idea that *G. max* was domesticated from southern *G. soja* types and then spread to other regions (Gai et al. 1999, 2000).

In the second study, Zhou et al. (1999) employed Vavilov's concept that the greatest genetic diversity for a species should be at the center of domestication. They evaluated 22,695 *G. max* accessions from China for 15 morphological and biochemical traits and concluded that the center of diversity for soybean resided within a corridor from southwest to northeast China which included Sichuan, Shaanxi, Shanxi, Hebei, and Shandong provinces. This corridor connected two centers of early agriculture in China, the Yellow River Valley and Yangzi River Valley. These two ancient centers have a long history of agricultural exchange (Gai and Guo 2001).

5.4 Rise of Genetic Diversity in Soybean

By the 15th century, soybean had migrated throughout Asia, following sea and land routes commonly traversed by tribes people of China, including the famous Silk Road connecting China with lands to the West, including India. Along these routes, soybean was adapted genetically by farmers to fit hundreds of diverse ethnic food cultures and varied climates alien to the new crop. During this time, genetically distinct landraces, the first cultivars, emerged in Japan, Indonesia, Vietnam, Korea, India, and China. Distinct landraces were documented by at least 1116 A.D., when Chinese authors listed soybean types that had green, brown, and black seed coats, and large and small seed (Gai and Guo 2001). By the beginning of the 20th century, at least 20,000 landraces may have been in existence in China (Chang et al. 1999). Perhaps 3,000 landraces were developed in Japan and 500 in India based on germplasm collection inventories in these countries (Carter et al, 2004).

5.5 Selection for Perceptual Distinctiveness

All studies to date indicate that DNA marker diversity is greater in *G. soja* than in *G. max*, reflecting the fact that *G. soja* is a much older species, and that *G. max* was probably domesticated from a small subset of the diversity present in *G. soja*

(Dae et al. 1995; Maughan et al. 1996; Li and Nelson 2002). Surprisingly, however, morphological genetic diversity is much greater among *G. max* accessions than in *G. soja*. The genetic ranges in seed coat color, seed size, plant height, plant yield, and maturity date, for example, are all much greater in *G. max*. The greater morphological diversity in *G. max* is likely due to a phenomenon which has been called selection for perceptual distinctiveness (Boster 1985). That is, ancient farmers saved desirable morphological mutants for seed size etc. as they harvested the crop over millennia. Such mutants would probably have been lost had they occurred in *G. soja*, and, thus, soybean production may have led naturally to increased agronomic diversity over that observed in *G. soja*. Farmers probably selected and saved mutations not only visually, but, perhaps, unconsciously as well by simply harvesting the crop. Agronomically useful mutants (e.g. disease resistance) would have had a selective advantage in the field and, therefore, would have had the opportunity to proliferate. Once soybean hectarage became significant, the sheer increase in numbers of plants grown, and related increased odds for mutation, may have played an important role as well. Mutations with neutral selective advantage, such as most DNA markers, may have been impacted less by humankind's husbandry of the soybean in ancient times. The reduced variability in DNA markers in *G. max* in comparison to *G. soja* would presumably have resulted from the genetic bottleneck created during the process of domestication (Hartl and Clark 1997). Agronomic diversity in maize (*Zea mays* (L.) is also larger than in its wild relatives (Wilkes 1967; Brown and Goodman 1977; Sanchez and Ordaz 1987).

5.6 Biogeography, Photoperiod, and the Spread of Soybean in Asia

Biogeography is a late Victorian term which refers to the major influence that geography can exert over botany and zoology (Winchester 2003). Plate tectonic theory, and evolutionary concepts developed by Wallace and Darwin were nurtured, in part, by scientific interest in biogeography. Biogeography is still a useful concept and we invoke the word here in assessing the opportunity for geography to exert its influence and shape patterns of diversity in soybean. Although the concept of selection for perceptual distinction is critical in the analysis of morphological genetic diversity in *G. max*, it is humankind's trade and spread of soybean throughout Asia that has provided the opportunity for selection to be so effective. In this context, the authors theorize the following biogeographical interpretation of diversity in soybean, based on the biology of the soybean, especially its sensitivity to photoperiod.

We can assume that soybean was not genetically diverse at domestication (not withstanding Vavilov's principles of crop domestication) in the narrow sense that it probably lacked the genetic flexibility to be farmed successfully at latitudes that diverged more than about 2° (a few hundred kilometers) from the center of domestication. This is a surprisingly narrow zone of adaptation compared to most crops, but holds for all individual soybean landraces and modern cultivars. This cropping restriction is brought about by the extreme sensitivity of soybean to small changes in photoperiod. For example, moving an adapted soybean genotype out of its zone 2° north could delay its maturity sufficiently to risk frost damage, while moving the same genotype south 2° would reduce plant height to a degree unacceptable for farming. Movement of soybean north or south through human migration or trade must, therefore, have been predicated on the discovery and use of rare genetic variants which extended the genetic range in photoperiod response (i.e., time of flowering and maturity) and allowed farmers to fit the life cycle of the crop to a warmer or cooler season, as the soybean gradually moved north or south. Once a soybean type was adapted to a more extreme photoperiod than was possible previously, the stage may have been set for further genetic improvements in productivity and function. It is theorized that these "first adapted" local landraces may have been sufficiently improved over later arriving, less adapted soybean types that first adapted landraces were preferred by local inhabitants. Thus, the first adapted landraces may have become a relatively isolated colony or gene pool. This initial adaptation to local conditions, especially at the edge of a soybean production region, could have preserved the genetic distinctiveness of soybean in a small area and, thereby, might have tended to preserve further genetic improvements in a colony, which would arise through subsequent mutation. This phenomenon of initial local adaptation and then isolation of colonies or gene pools would have been aided by the self-pollinating nature of soybean (>99%) which would limit outcrossing with later arriving soybean types. Outcrossing would also have been limited by the altered photoperiod response of established local landraces, by posing a temporal barrier to hybridization. That is, locally adapted and newly arriving landraces would likely not flower at the same time. In that regard, the striking contrasting photoperiod responses observed among landraces today are such that landraces developed from the northern extreme of Asia mature before landraces from the southern extreme flower, when planted adjacent to each other on the same day. (Gizlice et al. 1996). Such extremes in common in modern cultivars as well.

In the scenario outlined above, colonization, selection, and the noted hybridization barriers would lead to isolated pockets of soybean that diverged over millennia and came to carry unique and agronomcally important alleles. Selecting out of stress tolerance alleles would seem likely during this process because of the extensive spread of soybean to extreme environments and climates as a result of human migration. For example, soybean production through millennia in central and southern China has exposed soybean to the long-term stresses of heat and drought, and to poor, acidic, mineral soils which restrict root growth. Production in northern China, by contrast, has subjected soybean to a much different environment that is cooler, has a shorter growing season, and more productive soils. Selection pressure induced by these conditions, in the long span of time since soybean's initial exposure to them, may have adapted soybean, aided by mankind, to tolerate drought or cold. A putative example of this phenomenon is found in the use of landraces from the northern islands of Japan (northernmost growing region for soybean landraces) as sources of cold tolerance in breeding. Cold tolerance has been the basis for substantial yield improvement of soybean in Canada (Voldeng et al. 1997).

It should be noted that although the above scenario is attractive in explaining the derivation of the present extremes in photoperiod response in *G. max* described above, it is possible that they could have been generated through an alternate scenario involving one or a series of outcrosses with the progenitor *G. soja*, which, as a species, presumably had a wider range of photoperiod response and a more extensive geographic habitat than did *G. max* at the time of domestication. The alternate scenario for adaptation, in effect, requires multiple domestication events for *G. max*. However, the distinctiveness of nuclear DNA marker diversity in *G. soja* relative to *G. max* (i.e. very little overlap) coupled with the great difficulty in obtaining progeny that appear similar to *G. max* in F_2 populations derived from *G. max* and *G. soja* make this second scenario less likely than the first. In addition, breeding experience in North America in the 20th century has shown that natural mutations are quite common and that hundreds have been selected in the field, including maturity variants. Most experienced breeders have preserved several mutations in their careers, indicating that maturity mutants could probably be selected out easily by farmers in the centuries following domestication.

5.7 Global Adaptation of Soybean to Local Conditions in the 20th Century Through Breeding

Landraces derived from extensive dispersal of soybean in Asia and those arising from farmer selection nearer the center of domestication are collectively the reservoir of diversity that we have in soybean today. This diversity has formed the basis for modern breeding over the past 70 years. The breeding programs of Japan, China, and North America have all produced a large number of modern cultivars and have achieved their successes in relative isolation from each other (Carter et al. 1993; Cui et al. 2000a,b; Zhou et al. 2000, 2002).

The global spread of soybean, combined with extensive breeding efforts, has made the 20th century the most intense in the long history of soybean in terms of selection pressure for local adaptation to new climates, to new geographical niches and to new cropping systems involving mechanical harvest. To illustrate the large effect of modern breeding on adaptation, perhaps the best example is the development of novel cultivars which now grow commercially near the equator. Prior to the rise of modern breeding, few landraces grew closer to the equator than perhaps 20° except in Indonesia, the Phillipines, Malaysia and southern India. About 1980, photoperiod response mutants were discovered by breeders in a few landraces and subsequently backcrossed into improved cultivars. This breeding process allowed soybean, after 3,000 years, to finally achieve high productivity at the equator and realize global adaptation (Carter et al. 1999). Soybean now grows on several million hectares near the equator and breeding programs in this zone are in the process of adapting soybean further to local conditions.

5.8 A Note on Hybridization and its Importance in Modern Breeding in Comparison to Farm Development of Landraces

It is important to note that adaptation of soybean to the Brazilian equator is a very rapid, efficient, organized, and scientific version of exactly the same process that led to the spread of soybean throughout the northern and southern reaches of Asia in ancient times. However, modern breeders have one essential advantage which separates them qualitatively from ancient farmers who selected out improved soybean landraces: that advantage is hybridization. Although it can be argued that genetic improvement in modern cultivars is due in part to new mutations which have occurred in soybean since its spread from Asia, most of the improvements in modern soybean trace to alleles derived from the traditional Asian landraces themselves (Carter et al. 2003). Breeding, the intentional hybridization of land races (and their derivatives), in the 20th century has made possible for the first time in the history of soybean an efficient process for accumulating into a single new cultivar the benefits from independent local adaptation events (landraces) which had been separated genetically and physically for millennia through geographical and botanical barriers. New cultivars and large scale production of soybean adapted to the equator proceeded rapidly only because of hybridization.

5.9 Genetic Diversity in Modern Cultivars Around the Globe

Pedigree analysis has shown that although more than 20,000 landraces have been preserved in germplasm collections globally, less than 2% of these have actually been used in modern breeding programs. Eighty percent of the genetic bases in China, Japan, and North America could be accounted for by only 190, 53, and 13 landraces/ancestors, respectively (Carter et al. 2003). Most landraces used in North American breeding are from China, with cultivars from the northern and southern parts of North America tracing primarily to landraces from northern and southern China, respectively. Because only a few of the many Chinese landraces became ancestors of modern breeding in North America or China, it is improbable that the ancestors for the modern North American and Asian breeding programs overlap greatly. A survey of random amplified polymorphic DNA (RAPD) markers for ancestors bears this point out, indicating that the genetic bases of Chinese and North American breeding are quite distinct (Li et al. 2001b).

Simple sequence repeat (SSR) marker analyses indicates that Japanese cultivars are strikingly different from North American and Chinese cultivars. Chinese and North American cultivars also form distinct populations (Carter et al. 2000). Northern and Southern regional distinctions in North America are also apparent based on these same analyses. Modern Japanese cultivars tend to be far less diverse for DNA markers than cultivars from China, indicating the possibility that

Japanese soybean may have experienced a genetic bottleneck when soybean was introduced from the mainland of Asia. The Midwestern and Southern breeding sub-pools of North America are also much less diverse than in China and are on a par with those in Japan.

Diversity of modern Chinese, Japanese, and North American cultivars has also been assessed through analysis of pedigrees, known as coefficient of parentage (Carter et al. 2003). Scores of 1, 0.5, 0.25, or 0 indicate a relation between cultivars of identical twins, brother and sister, half brother and sister, or no relation, respectively. The average coefficient of parentage between Chinese cultivars was near 0, that of Japanese cultivars was slightly greater (less than 0.10), but that of North American cultivars was 0.20 within the southern and midwestern regions, or near that of half brother and sister relation. This level of coefficient of parentage is sufficiently high to warrant concerns regarding long-term sustained breeding progress in North America.

Phenotypic diversity of modern Chinese and North American cultivars has been compared directly. Forty-seven Chinese and 25 North American cultivars were evaluated for 25 phenotypic traits in growth chambers (Cui et al. 2001). Both sets showed phenotypic diversity, though the Chinese cultivars were more diverse than North American cultivars for 24 of the 25 traits. Multivariate analysis of these phenotypic traits indicated that, with a few exceptions, phenotypic diversity in US cultivars was a subset of the diversity found in Chinese cultivars. Cultivars from the southern region of China were the most unique having, among other traits, the highest levels of seed protein content, thinnest leaves, and lowest leaf N and chlorophyll content (Cui et al. 2001).

To summarize the genetic diversity patterns in modern breeding programs, DNA marker and pedigree analyses of diversity in modern cultivars indicate that diversity is greatest in cultivars developed in China, less in Japan and least in North America. Phenotypic analysis of modern Chinese and North American cultivars suggests the same pattern. Pedigree analyses of Latin American soybean breeding programs, although incomplete, show that these programs are derived primarily from a subset of North American breeding stock and are, thus, likely less diverse than the North American breeding program (Vello et al. 1984). . Decreased diversity of cultivars outside of China was also correlated with a reduction in the number of founding stock used to establish the breeding programs from which the cultivars arose.

Although conscious breeding choices, the high economic cost of field breeding, and human historical factors can be used to explain the reduced diversity in breeding programs outside of China vs. within, it is important to note that these results, obtained from modern breeding programs, are consistent with (1) Vavilov's principle of crop domestication which states that genetic diversity will be greatest at the center of domestication, which in this case is China (Vavilov, 1951), and (2) the concept of Darwinian genetic drift which can be used to infer that genetic relatedness will increase within breeding populations that are derived from relatively few founding members. A caution gleaned from this observed trend is that all soybean breeding programs outside China, regardless of the phenotypic superiority of their genetic breeding materials, should be examined to determine the adequacy of genetic diversity.

5.10 Genetic Diversity and Distribution of Genetically Modified Soybeans

Resistance to the herbicide glyphosate is the only genetically engineered trait which has been commercially successful in soybean. This trait first appeared in commercial soybean cultivars about 1994 and since come to occupy about 80% of the approximately 32 million ha annual production in the USA and Canada (Sneller 2003). Currently, transgenic glyphosate-resistant soybean cultivars are very common in Argentina and indirect reports suggest that they are grown in Brazil as well (although it is not legal to produce glyphosate-resistant soybean in Brazil at present). Transgenic soybean is rare in Asia or India. Japan currently does not accept genetically modified soybean in the soyfood market trade. It appears that few if any *new* transgenic traits will be marketed in soybean in the USA in the near future, because of government agency regulatory costs. It is not likely , at present, that the transgenic glyphosate resistance trait will cause any major shifts in genetic diversity in soybean cultivars. Although intense selection on any one trait has the potential to create genetic bottlenecks in breeding, the owners of the glyphosate resistance trait have required in their licenses that all breeders backcross the trait into their own locally adapted breeding stocks before releasing a glyphosate-resistant cultivar. A recent study of transgenic and conventional soybean cultivars reported that the level of diversity within these two groups was similar based on coefficient of parentage analysis (Sneller 2003). This maintenance of diversity in transgenic soybean cultivars is clearly the result of the backcrossing effort.

The potential to produce glyphosate-resistant weeds by transference of the trait from new transgenic cultivars has received considerable attention in recent years in canola and other crops. In the case of soybean, the only weed which is capable of crossing with *G. max* is *G. soja*. *G. soja* grows only in Asia, transgenic soybean is rare in this region, and, thus, no immediate threat can be posed in terms of creation of super weeds or shifting genetic diversity patterns in the *G. soja*. However, one may ask what level of threat might be expected if all soybean production in China were devoted to glyphosate-resistant cultivars. Fortunately, the impact on *G. soja* as a weed problem is not likely to be large because *G. soja* is not an aggressive weed. Although it can be found in cultivated fields in Asia, it is not common. Of the more than 600 accessions of *G. soja* which exist in the USDA soybean germplasm collection, none were collected from cultivated fields, but instead on river courses, at railway stations, and at recently disturbed building sites. The senior author, although spending some months in China, noted only one *G. soja* plant in a cultivated *G. max* field.

5.11 Conclusion

Landraces developed in Asia are irreplaceable agricultural genetic resources because they *are* the sum of genetic diversity that was amassed by farmers during the long transformation of soybean from wild plant to modern crop. These landraces were selected by the human eye for perceptual distinctiveness, traded north and south throughout Asia to allow further farmer selection for local adaptation, and formed a series of isolated gene pools to develop biogeographical patterns which are present today. Many landraces were collected by plant explorers, germplasm curators, and breeders and are preserved today in extensive germplasm collections. These landraces serve as the basis for almost all modern improvements in soybean. Modern breeding expands the process of local adaptation initiated by ancient farmers in Asia to a global basis. The model for modern soybean breeding is hybridization of landraces (and their derivatives), development of large numbers of progeny, and selection of those rare recombinant progeny which have beneficial characters from both parents. Many breeding successes have been achieved in this way, including adaptation of soybean to the Brazilian equatorial region in the past 20 years. The legacy of genetic diversity resulting from the global spread of soybean has made these successes possible. Thus, dispersal and adaptation to local conditions in the 20th century as well as in ancient times, is a part of soybean's interesting genetic story.

5.12 References

Boster JS (1985) Selection for perceptual distinctiveness: evidence from aguaruna cultivars of *Manihot esculenta*. Econ Bot 39:310-325

Brown WL, Goodman MM (1977) Races of corn. In: Sprague GF (ed) Soybeans: corn and corn improvement. Crop Sci Soc of America, Madison, WI, pp 49-88

Burtt-Davy J (1905) Report of the government agrostologist and botanist for the year ending June 30th 1904. Transvaal Department of Agriculture, Annual Report, pp 261-320. For the year 1903-04 see pp 263, 270-71, 274. 1 plate. Reprinted in part in the Rhodesian Agricultural Journal 3:354, 364 (1906)

Carter TE Jr, Gizlice Z, Burton JW (1993) Coefficient of parentage and genetic similarity estimates for 258 North American soybean cultivars released by public agencies during 1954-88. USDA Tech Bull no 1814. US Gov Print Office, Washington, DC

Carter TE Jr, de Souza PI, Purcell LC (1999) Recent advances in breeding for drought and aluminum resistance in soybean. In: Kauffman H (ed) Proc World Soybean Conf VI:106-125. Academic Press, London, 542 pp

Carter TE Jr, Nelson RL, Cregan PB, Boerma HR, Manjarrez-Sandoval P, Zhou X, Kenworthy WJ, Ude GN (2000) Project SAVE (Soybean Asian Variety Evaluation) -- potential new sources of yield genes with no strings from USB, public, and private cooperative research. In: Park B (ed) Proceedings of the 28th Soybean Seed Res Conf 1998. American Seed Trade Association, Washington DC, pp 68-83

Carter TE Jr, Nelson NL, Sneller C, Cui Z (2004) Genetic diversity in Soybean. In: Boerma HR, Specht JE (eds) Soybean monograph, 3rd edn. Am Soc Agron, Madison, WI pp 303-416

Chang RZ (1989) Study on the origin of cultivated soybean (in Chinese). Oil Crops China 1:1-6

Cui Z, Carter TE Jr, Gai J, Qiu J, Nelson RL (1999) Origin, description, and pedigree of Chinese soybean cultivars released from 1923 to 1995. USDA Tech Bull 1871. US Gov Print Office, Washington, DC

Cui Z, Carter TE Jr, Burton JW (2000a) Genetic base of 651 Chinese soybean cultivars released during 1923 to 1995. Crop Sci 40:1470-1481

Cui Z, Carter TE Jr, Burton JW (2000b) Genetic diversity patterns in Chinese soybean cultivars based on coefficient of parentage. Crop Sci 40:1780-1793

Cui Z, Carter TE Jr, Burton JW, Wells R (2001) Phenotypic diversity of modern Chinese and North American soybean cultivars. Crop Sci 41:1954-1967

Dae HP, Shim KM, Lee YS, Ahn WS, Kang JH, Kim NS (1995) Evaluation of genetic diversity among the *Glycine* species using isozymes and RAPD. Korean J Genet 17:157-168

Don W (1911) Appendix II. Annual report-moor plantation, etc. 1910. Southern Nigeria. Annual Report upon the Agricultural Department 44 p. For the year 1920 see pp 31-32

D'Utra G (1882) *Soja* [Soya]. J Agric (Brazil) 4:185-188

Faura RE (1933) La soja: su historia, cultivo, composicion del grano y de la planta, studio de la material grasa, conclusions. [The soybean - its history, culture, composition of the seed and of the plant, study of its oils, conclusions.] Boletin Mensual del Ministerio de Agricultura de la Nacion (Buenos Aires, Argentina) 33:9-22

Gai J (1997) Soybean breeding. *In* J. Gai (ed.) Plant breeding: Crop species. (In Chinese.) China Agric. Press, Beijing.

Gai J, Guo W (2001) History of Maodou production in China. In: Lumpkin TA, Shanmugasundaram S (eds) Proceedings of the 2nd international vegetable soybean conference (Edamame/Maodou). 10-11 Aug 2001, Tacoma WA. Washington State University, Pullman, WA, pp 41-47

Gai J, Gao Z, Zhao T, Shimamoto Y, Abe J, Fukushi H, Kitajima S (1999) Genetic diversity of annual species of soybeans and their evolutionary relationship in China. In: Kauffman HE (ed) Proc World Soybean Res Conf VI, Chicago, IL, 4-7 Aug 1999. Superior Printing, Champaign, IL, p 515

Gai J, Dong-He D-H, Gao Z, Shimamoto Y, Abe J, Fukushi H, Kitajima S (2000) Studies on the evolutionary relationship among ecotypes of *G. max* and *G. soja* in China. Acta Agron Sin 26:513-520

Gizlice Z, Carter TE Jr., Gerig TM, Burton JW (1996) Genetic diversity patterns in North American public soybean cultivars based on coefficient of parentage. Crop Sci 36:753-765

Guo W (1993) The history of soybean cultivation in China (in Chinese). Hehai University Press, Nanjing, Jiangsu, China

Hartl DL, Clark AG (1997) Principles of population genetics, 3rd edn. Sinauer, Sunderland, Mass

Hymowitz T (2003) Taxonomy of soybean. In: Boerma HR, Specht JE (eds), Soybean Monograph, 3rd edn. Am Soc Agron, Madison, WI (~50 pages) (accepted Dec. 2002)

Hymowitz T, Newell CA (1980) Taxonomy, speciation, domestication, dissemination, germplasm resources and variation in the genus *Glycine*. In: Summerfield RJ, Bunting AH (eds) Advances in legume science. Royal Botanic Gardens. Kew, Richmond, Surrey, UK, pp 251-264

Itie G (1910) Le soja: Sa culture, son avenir [Soya: Its cultivation, its future]. Agriculture Pratique des Pays Chauds (Bulletin du Jardin Colonial) 10:137-144

Li Z, Nelson RL (2002) RAPD marker diversity among cultivated and wild soybean accessions from four Chinese provinces. Crop Sci 42:1737-1744

Li Z, Qiu L, Thompson JA, Welsh MM, Nelson RL (2001b) Molecular genetic analysis of US and Chinese soybean ancestral lines. Crop Sci 41:1330-1336

Lu SL (1977) The origin of cultivated soybean (*G. max*) (in Chinese). In: Wang JL (ed) Soybean. Shanxi People's Press, Shanxi, China

Maughan PJ, Saghai-Maroof MA, Buss GR, Huestis GM (1996) Amplified fragment length polymorphism (AFLP) in soybean: species diversity, inheritance, and near-isogenic line analysis. Theor Appl Genet 93:392-401

Paillieux A (1880) Le soya, sa composition chimique, ses variétés, sa culture et ses usages [The soybean, its chemical composition, varieties, culture, and uses]. Bull Soc Acclimat 27:414-471. Sept 27:538-596

Qiu L, Ruzhen Chang R, Sun J, Li X, Cui Z, Li Z (1999) The history and use of primitive varieties in Chinese soybean breeding. In: Kauffman HE (ed) Proc World Soybean Res Conf VI, Chicago, IL, 4-7 Aug 1999. Superior Printing, Champaign, IL, pp 165-172

Sanchez G, Ordaz SL (1987) Systematic and ecogeographic studies on crop genepools II. El Teocintle en Mexico. International Plant Genetic Resources Institute, Rome, p 50

Sneller CH (2003) Impact of transgenic genotypes and subdivision on diversity within elite North American soybean. Crop Sci 43:409-414

Vavilov NI (1951) The origin, variation, immunity, and breeding of cultivated plants. (Trans. By K.S. Chester) Chronica Botanica, Waltham, Mass. Reprinted by Ronald Press, New York.

Vello NA, Fehr WR, Bahrenfus JB (1984) Genetic variability and agronomic performance of soybean populations developed from plant introductions. Crop Sci 24:511-514

Voldeng HD, Cober ER, Hume DJ, Gilard C, Morrison MJ (1997) Fifty-eight years of genetic improvement of short-season soybean cultivars in Canada. Crop Sci 37:428-431

Wang JL, Wu YX, Wu HL, Sun SC (1973) Analysis of the photoperiod ecotypes of soybeans from northern and southern China (in Chinese). J Agric 7:169-180

Wilkes HG (1967) Teosinte: the closest relative of maize. The Bussey Inst, Harvard Univ, Cambridge, Mass

Winchester S (2003) Krakatoa. Harper Collins, New York

Xu B (1986) New evidence about the geographic origin of soybean (in Chinese). Soybean Sci 5:123-130

Zhou X, Peng YH, Wang GX, Chang RZ (1998) The genetic diversity and center of origin of Chinese cultivated soybean (in Chinese). Sci Agric Sin 31:37-43

Zhou X, Peng Y, Wang G, Chang R (1999) Study on the center of genetic diversity and origin of cultivated soybean. In: Kauffman HE (ed) Proc World Soybean Res Conf VI, Chicago, IL, 4-7 Aug 1999. Superior Printing, Champaign, IL, p 510

Zhou X, Carter TE, Cui Z, Miyazaki S, Burton JW (2000) Genetic base of Japanese soybean cultivars released during 1950 to 1988. Crop Sci 40:1794-1802

Zhou X, Carter TE, Cui Z, Miyazaki S, Burton JW (2002) Genetic diversity patterns in Japanese soybean cultivars based on coefficient of parentage. Crop Sci 42:1331-1342

6 Migration of a Grain Legume, *Phaseolus vulgaris*, in Europe

A.P. Rodiño and J.-J. Drevon

Laboratoire Symbiotes des Racines et Sciences du Sol, INRA, 1 Place Viala, 34060 Montpellier-Cedex, France, E-mail drevonjj@ensam.inra.fr

6.1 Abstract

All common bean lines grown in Europe, as in other continents, are the result of a process of domestication and evolution (mutation, selection, migration and genetic drift), from wild forms (*Phaseolus vulgaris* var. *aborigineus* and *Phaseolus vulgaris* var. *mexicanus*) found exclusively in the Americas. Observations on morphological (e.g., bract shape), agronomic (e.g., seed size), biochemical (e.g., isozyme) and molecular variability for wild and cultivated lines, made it possible to distinguish at least two major centers of common bean domestication, namely Andean and Mesoamerican. Subsequently, new cultivars may have evolved within and between the two gene pools in Spain and Portugal, making southern Europe a secondary center of diversity for the common bean. Indeed, the first documented appearance of common bean in Europe, after the discovery of America, was in 1508 as an ornamental plant in France. This happened before it was possible to introduce common bean lines from the Andes, i.e., before Peru was explored by Pizarro in 1528. Both Andean and Mesoamerican germplasm differ by the diversity of phaseolin, a seed protein. Thus, most of the European germplasm is from Andean locations since the type T phaseolin is found in their seeds. It is thought that Mesoamerican lines were less popular because of their lower adaptability to winter cold and to short-duration summers. However, the type C phaseolin, is most frequently found in the Iberian Peninsula, although recent studies show that the populations found in northwestern Spain and Portugal are predominantly of the T-type phaseolin. Concerning their interaction with the root-symbiont rhizobia, genotypic variation has been recently found among various landraces from this region, including Great Northern, Caparrón, White Kidney and Canellini commercial types, for nodulation and the N_2-dependent plant growth after inoculation with *Rhizobium tropici* CIAT 899. It is concluded that knowledge of the origin, evolution and dissemination of common bean in Europe may be valuable information for the improvement of this legume species.

6.2 Origin and Domestication of Common Bean

Over a period of at least 7000 years, the common bean has evolved from a wild-growing into a major leguminous food crop. During this period, which encompasses the initial domestication phase and subsequent evolution under cultivation, mutation, selection, migration and genetic drift have acted on the raw material provided by wild-growing *Phaseolus vulgaris* (Gepts and Debouck 1991). It is only since the late 19th century that scientists have accepted a New World origin for the common bean and this was contrary to the belief in an Asian origin, which had been held for several centuries (Gepts and Debouck 1991). Several remains of the common bean have been uncovered, not only in the Andes but also in Mesoamerica and North America and consist of seeds (Kaplan et al. 1973), pod fragments (Kaplan 1981) and even whole plants (Kaplan and MacNeish 1960) which date from 10,000 to 8000 B.P. in the Andes and 6000 B.P. in Mesoamerica. Dating methods have been recently the subject of much discussion among archaeologists for New World crops. The oldest records for common bean, 4300 years B.P., revised using accelerator mass spectrometry direct dating (Kaplan and Lynch 1999), is from Ancash, Peru, and 2200 years B.P. from Puebla, Mexico, although much earlier dates have been repeatedly presented using radiocarbon indirect dating (Debouck 2000). These archaeological findings are phenotypically similar to current cultivars grown in the same area. Also, there are the historical and linguistic data in the 16th century Spanish texts mentioning the presence of the common bean in America and the vocabulary of several native Indian languages includes a specific word designating the common bean.

Thus, the common bean (*Phaseolus vulgaris* L.) is a species of American origin derived from wild ancestors distributed from northern Mexico to northwestern Argentina (Gepts et al. 1986; Koenig et al. 1990; Toro et al. 1990). This crop was domesticated in two distinct regions of the New World, one in Mesoamerica (principally Mexico) and another along the eastern slope of the Andes in South America (southern Peru, Bolivia and northwestern Argentina; Gepts et al. 1986; Gepts 1990; Gepts and Debouck 1991; Tohme et al. 1995). An additional minor domestication center has been suggested in Colombia, although it is not entirely clear whether this area constitutes a center of domestication or a region of gene flow between wild and domesticated beans (Gepts and Bliss 1986; Beebe et al. 1997; Islam et al. 2002). The cultivated gene pools of common bean can be distinguished by their morphology and agronomical traits (Singh et al. 1991a), phaseolin seed protein electrophoretic type (Koenig et al. 1990), isozymes (Koenig and Gepts 1989; Singh et al. 1991b), molecular markers (Becerra-Velásquez and Gepts 1994; Haley et al. 1994; Freyre et al. 1996; Galvan et al. 2003) and adaptation traits (Singh 1989; Voysest and Dessert 1991). Most cultivars from either the Mesoamerican or the Andean region contain characteristics that are found in wild accessions from the same area, but not in either domesticated or wild accessions from the other gene pool (Koenig and Gepts 1989). Cultivars from Mesoamerica usually are small seeded (< 25 g 100-seed-weight^{-1}) or medium-seeded (25 to 40 g 100-seed-weight^{-1}) and through electrophoresis, produce S and B phaseolin types that differ from those of their South American counterparts, with large seeds (> 40 g 100-seed-weight^{-1}) and T, C, H and A phaseolin types (Gepts et al. 1986; Singh et al. 1991b). Six races (Singh et al. 1991c) have been proposed for common bean from Mesoamerica (Mesoamerica, Durango and Jalisco races) and from the Andean region in South America (Chile, Peru and Nueva Granada races).

In addition to these two major gene pools, recently discovered wild populations constitute a third gene pool ancestral in the evolution of wild common bean (Kami et al. 1995; Tohme et al. 1996) located in Ecuador and northern Peru (Debouck et al. 1993; Tohme et al. 1996). These ancestral populations were not involved in domestication (Debouck et al. 1993) as shown by their phaseolin type, which is absent from the domesticated gene pool.

During the evolution of common bean, some morphological, physiological and genetically marked changes have occurred in this plant: gigantism (seed, pod, stem, leaves), suppression of seed dispersal mechanism, changed growth form (from climbing to dwarf plants), changed life form, loss of seed dormancy, other physiological changes (e.g., loss of photoperiodic sensitivity) and biochemical changes (Smartt 1988; Gepts and Debouck 1991).

The divergence between the Andean and Mesoamerican gene pools has implications for bean breeding that have not yet been fully explored. Despite their partial reproductive isolation (Singh and Gutiérrez 1984; Gepts and Bliss 1985; Koinange and Gepts 1992), the two gene pools still belong to the same biological species (Papa and Gepts 2003). Viable and fertile progeny can be obtained, and therefore, genes can be transferred between the two pools, although the transfer of quantitative traits appears to be problematic. Attempts to recombine desirable traits between both gene pools, such as the large seed size of the Andean gene pool with the yield potential of the Mesoamerican gene pool, have generally failed (Nienhuis and Singh 1986). However, notable exceptions are provided by Chilean landraces which showed signs of introgression (Paredes and Gepts 1995) from the Mesoamerican gene pool. This introgression may have taken place between Andean genotypes of race Chile and Mesoamerican genotypes of race Durango and was not due to breeding efforts.

6.3 Dispersal Routes of Common Bean Cultivars

Some limited bean germplasm exchange took place in pre-Columbian times between Mesoamerica and South America, but much more extensive seed movement occurred after the 1500s. Thus, outside the American centers of primary diversity, one can identify several secondary centers which bean collectors should consider in search of diversity. The different genotypes found in these secondary zones were introduced from the Americas, either soon after the Spanish conquest or more recently. The data of their introduction are key facts, although it cannot be established in every case. As secondary centers, one can tentatively suggest East Africa and Europe, since the *Phaseolus* beans were introduced in those regions by Spaniards and Portuguese in the 16th and 17th centuries. Concerning the origin of the European beans, McClean et al. (1993) suggested that the germplasm dispersed to Europe was probably domesticated in the South American Andes since the Mesoamerican cultivars are not currently very popular in Europe. Gepts and Bliss (1988) suggested that the bean grown on the Iberian Peninsula was introduced from a different area (Chile) with respect to those of the rest of Europe. In any case, the introduction of beans in Europe is unclear and currently under discussion.

Seed exchanges with Europe must have happened from the first visits of Europeans to the Americas. Sailors and traders 500 years ago could have brought the nicely colored, easily transportable bean seeds with them as a curiosity, only for fun, as children do currently at home and school in the Andean region. Within Europe there was likely a quick distribution of seeds as curiosities (Zeven 1997). Thus, it is probable that the initial common bean accessions introduced in Europe (into the Iberian Peninsula) were mainly from Mesoamerica around 1506 (Ortwin-Sauer 1966), as we recall that Columbus arrived in Central America in 1492 and Cortés reached Mexico in 1518. Castiñeiras et al. (1991) proposed the introduction of seeds from Cuba, since Columbus landed there in 1492. Evidence exists that common bean reached France in 1508, probably as an ornamental plant without known value for human consumption at that time (Zeven 1997). In 1528, Pizarro explored Peru, opening the possibility of introducing common bean accessions from the Andes (Berglund-Brücher and Brücher 1976; Debouck and Smartt 1995). No records of common bean earlier than 1543 have been found in NW European herbaria, suggesting that the common bean was distributed in NW Europe after 1540. By 1669 it was cultivated on a large scale (Zeven 1997). There is evidence of seed exchange among farmers and gardeners in many countries of Europe for testing some new material or for avoiding the degeneration of cultivars sown year after year (Zeven 1999). Europeans still collect common beans from neighboring and faraway regions. Therefore, the species has undergone an adaptive evolutionary process in those regions for about 400 years, resulting in today's very important additional variations (Debouck 1988; Hidalgo 1988).

In the 16th century harbors in northwestern Spain (Galicia) maintained active commerce with the New World. The introduction of some crops such as bean and maize and the distribution to other areas could have occurred in/from this area. The traditional cropping systems for the bean crop similar to those used in many areas of America (Santalla et al. 1994) are strong arguments to support this hypothesis. The sensitivity to photoperiod and low temperature during the growing season could have been a limiting factor for cultivated bean in many European latitudes in the early times. In fact it is possible to grow primitive Andean landraces and wild populations innorthern Spain (Pontevedra, 42°N), but only under greenhouse conditions during the fall-winter-spring period (de Ron et al. 1999).

Thus, subsequently, new cultivars may have evolved within and between the two gene pools in southern Europe (mainly Spain and Portugal), making this region a secondary center of diversity for the common bean (Santalla et al. 2002). Because historical and linguistic information provide little evidence regarding the origin and dissemination of common bean in Europe, phaseolin protein pattern, an evolutionary marker (Gepts et al. 1986), was used to complement morphological and agronomic data. This phaseolin protein analysis was useful in identifying gene pools and the origin of accessions in Mesoamerican or Andean domestication centers.

6.4 Phaseolin Is an Important Evolutionary Marker

Phaseolin is the major seed storage protein of *Phaseolus vulgaris* L and it can be used to trace the evolutionary origin of common bean genotypes. The electrophoretic variability of phaseolin of wild-growing common beans from Mesoamerica and the Andes was compared with that of landraces of the same region. The wild common bean accessions of different geographic origin could be distinguished by their phaseolin type. In Mesoamerica, the wild forms showed both the S type (Brown et al. 1981) as well as M types. The Colombian wild common bean exhibited the C, H and B types, whereas in the southern Andes, wild forms showed only the T type (Brown et al. 1981). There was a correspondence in the geographic distribution of phaseolin types between wild and cultivated common bean. The cultivars with S and T phaseolin patterns predominated in Mesoamerica and in the southern Andes, respectively. The B phaseolin type was present only in wild and cultivated common bean from Colombia. On the other hand, the C, H and A phaseolin types were found not only among landraces of the Andes, but also among wild forms (Koenig et al. 1990; Vargas et al. 1990). Therefore, multiple domestication is thought to be the primary cause for parallel geographical phaseolin variation between wild and cultivated common bean forms. Secondarily, occasional outcrosses between wild and cultivated common beans may also contribute to this parallel distribution of phaseolin types. Given the low frequency of B phaseolin cultivars, Colombia might, however, only be a minor or more recent domestication region (Gepts et al. 1986).

A relationship was observed between phaseolin pattern and seed type. Cultivars with T, C, H and A phaseolin types on the average had larger seeds than cultivars with S and B phaseolin types (Gepts et al. 1986). This relation provides evidence for exchange of germplasm between Mesoamerica and the Andes. It makes it possible to follow the worldwide dispersal of common bean cultivars. Mesoamerican cultivars became the major component of the cultivar complement of two regions, lowland South America and southwestern USA. They were also introduced into northeastern USA, the Iberian Peninsula, western Europe and Africa. In these regions, they formed, however, only a minor component compared to the cultivars of Andean origin. The latter may have had a competitive advantage over genotypes of other origins because of a more adequate photoperiodic adaptation due to similar latitudes.

Previous studies have indicated that most of the Iberian cultivars may have been introduced from Chile due to a high frequency of the C phaseolin pattern (Gepts and Bliss 1988; Gil and de Ron 1992). However, recent studies (Escribano et al. 1998; Rodiño et al. 2003) have indicated a high frequency of the T phaseolin pattern among Iberian cultivars, The T phaseolin type was also observed in western Europe (Gepts and Bliss 1988). Subsequently, the Iberian Peninsula landraces could have been introduced in other parts of Europe such as Greece, Cyprus and Italy, as indicated by the high proportions of T and C types in these areas (Lioi 1989; Limongelli et al. 1996). These studies have provided evidence for the existence in the Iberian Peninsula of the two major gene pools, Andean and Mesoamerican. The variation in bean-growing environments, cropping systems and consumer preferences for seed types in this area might have played a significant role in the common bean crop diversity and could give rise to the preservation of a large variation in the domesticated common bean's characteristics.

6.5 Diversity of Rhizobia

A possibility is that the microorganisms associated with the common bean plant for its symbiotic nitrogen fixation may exhibit a similar arrangement of their genetic diversity in Mesoamerica and Andean gene pools. In the centers of domestication, *R. etli* bv. p*haseoli* has been found as the predominant nodule occupant (Martinez-Romero 2003). Isolates belonging to *R. etli,* are predominant in soils of Mesoamerican countries (Eardly et al. 1995; Martinez-Romero and Caballero-Mellado 1996) and in Argentinean soils (Aguilar et al. 1998). A large genetic diversity has been documented for *R. etli* bv. *phaseoli* from the domestication centers. Occasionally bacteria other than *R. etli* have been encountered in bean nodules in Mexico and they correspond to *R. gallicum* (Silva et al. 2003). Whereas Andean cultivars form large number of nodules with *R. tropici* strains (Nodari et al. 1993). Mesoamerican beans with high capacities to fix nitrogen nodulated poorly with *R. tropici* strains and in these beans *R. tropici* blocked *R. etli* nodulation when both strains were tested together (Martinez-Romero et al. 1998). Bean nodule isolates from Ecuador and Peru proved to be very diverse and could be divided into clusters distinct from the Mexican isolates (Bernal and Graham 2001). Ecuadorian and Mexican beans selected different *R. etli* strains both from Ecuadorian and Mexican soils (Bernal and Graham 2001) and efficiency in nodulation and nitrogen fixation was higher when both partners were from the same region. Might it be that domestication and other human selections of beans indirectly affected host range (Martinez-Romero 2003)? What would be the consequence, in terms of N_2 fixation, of inoculating a Mesoamerican cultivar with Mesoamerica *versus* Andean rhizobia strain?

In Europe, *Rhizobium* spp. strains have a narrow genetic diversity that is correlative to beans being an introduced crop (Laguerre et al. 1993). It seems that in some of introduced sites bean is nodulated by other species in addition to diverse *R. etli* bv. *phaseoli* and the co-occurrence of several species is common. Remarkably, in a single soil in Spain, five rhizobial species were found to nodulate *P. vulgaris* among which, *R. leguminosarum* bv. *phaseoli* and *R. tropici* have been found in European and American soils (Herrera-Cervera et al. 1999). Martinez-Romero and Caballero-Mellado (1996) revealed the existence of one of the highest levels of genetic diversity encountered among *Rhizobium* spp; isolates recovered from root nodules of *Phaseolus vulgaris* L. from Spain. The most abundant species appeared to be *R. etli,* brought into Spain from the Americas after introduction of beans about four centuries ago. The results of Herrera-Cervera et al. (1999) indicate that extensive interspecific symbiotic gene exchange has taken place in this site and presumably, *R. etli* strains could have been the original gene donors. The other species, which have also been found at different locations in Europe, probably represent bacteria that pre-existed in European soils when beans were first introduced and have received genetic material from the introduced *R. etli.* Thus, the Spanish soil studied represent a unique case where a donor and the putative DNA recipients co-exist and probably compete for the same ecological niche. In a recent study, phenotypic features of 90 French strains isolates from *Phaseolus* spp. nodules and previously assigned to one of the two genomic species or to one of the three previously named species of rhizobia that nodulate beans (*R. leguminosarum, R. tropici, R. etli*) were compared to the phenotypic features of reference rhizobial strains by numerical taxonomy. As a result of the present and previous studies, Amarger et al. (1997) proposed that two new *Rhizobium* species should be recognized, *R. gallicum* and *R. giardinii.*

The high N_2-fixing potential found in a recent study reveals a genotypic variability for traits associated with N_2 fixation in common bean of the European germplasm collection of Mision Biologica de Galicia – CSIC, that has diversified in the Iberian Peninsula (Rodiño and Drevon, in rev.). This study confirms the large nodulation potential of the studied germplasm. This genetic variability identified is adequate for the improvement of N_2 fixation in cultivars through plant breeding. A significant potential exist to further improve symbiotic and associative N_2 fixation by breeding genotypes with a greater capacity to sustain these interactions with rhizobia.

6.6 Conclusions and Perspectives

The morphological, physiological and genetic characteristics of present common bean cultivars are the result of the evolutionary history of the species before, during and after domestication. A better knowledge of this evolutionary history gives us a deeper understanding of the current characteristics of the cultivated gene pool of the common bean, which in turn, should lead to better management of common bean genetic resources and breeding programs. Evidence available so far indicates two major domestications, in Mexico and in the Andes, which led the two groups of cultivars with contrasting agronomic characteristics. The landraces of common bean present a high diversity for morphological, agronomical and biochemical characters, and this variability can be used in breeding programs. The germplasm dispersed to Europe was probably domesticated in the South American Andes since the Mesoamerican cultivars are not currently very popular in Europe. Details of the introduction of beans in Europe are unclear and currently under discussion. The introduction routes of African common bean cultivars are even more difficult to ascertain. Whereas a majority of the cultivars ultimately originated in the Andes, it is not known by which route they were introduced. They could have been introduced directly from the Andes, indirectly through the Iberian Peninsula or through Western Europe during the colonial period. Thus, it is important to continue the exploration and collection of landraces of common bean, both from Europe and Africa to clarify the dispersal routes in these countries.

The study of rhizobia diversity provides valuable ecological information by defining host preferences and predominance of strains, the dynamics of exchange of genetic material and the basis for the proposal of evolutionary trends. Diversity of rhizobia isolated from common bean has been examined almost worldwide in both centers of origin of bean and in introduced areas with a variety of techniques and criteria. The diversity studies reveal that there is no unique rhizobia strain highly adaptable and efficient for all soils, environmental conditions and bean genotypes. On the other hand, large variations in N_2 fixation among bean cultivars has been shown and those with high capacities to fix nitrogen have been identified. As the nitrogen-fixing capacity of some commercial beans is among the lowest of the widely cultivated legumes, it is important that it be increased through crop management and plant selection. It seems to be desirable to follow the search for common bean landraces expressing high N_2 fixation in order to improve the N-dependent yield of bean in regions of intensive culture, with crop rotation in view. It will be interesting to study the interaction of different rhizobia with those cultivars that have been identified with high capacities to fix nitrogen. The improvement of bean nitrogen fixation is an important goal since biological nitrogen fixation not only lowers production costs but is also environmentally sound.

6.7 References

Aguilar OM, Lopez MV, Riccillo PM, Gonzalez RA, Pagano M, Grasso DH, Pühler A, Favelukes G (1998) Prevalence of the *Rhizobium etli*-like allele in genes coding for 16S rRNA among the indigenous rhizobial populations found associated with wild beans from the southern Andes in Argentina. Appl Environ Microbiol 64:3520-3524

Amarguer N, Macheret V, Laguerre G (1997) *Rhizobium gallicum* sp. Nov. and *Rhizobium giardinii* sp. Nov., from *Phaseolus vulgaris* nodules. Int J Syst Bacteriol 47:996-1006

Beebe S, Toro O, Gonzalez AV, Chacon MI, Debouck DG (1997) Wild-weed-crop complexes of common bean (*Phaseolus vulgaris*) in the Andes of Peru and Colombia, and their implications for conservation and breeding. Genet Res Crop Evol 44:73-91

Becerra-Velásquez VL, Gepts P (1994) RFLP diversity in common bean (*Phaseolus vulgaris* L.). Genome 37:256-263

Berglund-Brücher O, Brücher H (1976) The south American wild bean (*Phaseolus aborigineus* BurK) as ancestor of the common bean. Econ Bot 30:257-272

Bernal G, Graham PH (2001) Diversity in the rhizobia associated with *Phaseolus vulgaris* L. in Ecuador, and comparisons with Mexican bean rhizobia. Can J Microbiol 47:526-534

Brown JWS, Ma Y, Bliss FA, Hall TC (1981) Genetic variation in the subunits of globulin-1 storage protein of French bean. Theor Appl Genet 59:83-88

Castiñeiras L, Esquivel M, Lioi L, Hammer K (1991) Origin, diversity and utilization on the Cuban germplasm of common bean (*Phaseolus vulgaris*). Euphytica 57:1-8

De Ron AM, Rodiño AP, Menéndez-Sevillano MC, Santalla M, Barcala N, Montero I (1999) Variation in wild and primitive Andean bean varieties under European conditions. Annu Rep Bean Improv Coop 42:95-96

Debouck DG (1988) Recoleccion de germoplasma de Phaseolus en Bolivia, abril 23-mayo 14. Informe de viaje. CIAT, Cali, Colombia

Debouck DG (2000) Biodiversity, ecology and genetic resources of Phaseolus beans – Seven answered and unanswered questions. Proceedings of the 7th MAFF International Workshop on Genetic Resources, part I. Wild legumes. AFFRC and NIAR, Japan, pp 95-123

Debouck DG, Smartt J (1995) Bean. In: Smartt J, Simmonds NW (eds) Evolution of crop plants, 2nd edn. Longman Scientific and Technical, Harlow, Essex, pp 287-296

Debouck DG, Toro O, Paredes AM, Johnson WC, Gepts P (1993) Genetic diversity and ecological distribution of *Phaseolus vulgaris* (Fabaceae) in northwestern South America. Econ Bot 47:408-423

Eardly BD, Wang FS, Whittam TS, Selander RK (1995) Species limits in *Rhizobium* populations that nodulate the common bean (*Phaseolus vulgaris*). Appl Environ Microbiol 61:507-512

Escribano MR, Santalla M, Casquero PA, de Ron AM (1998) Patterns of genetic diversity in landraces of common bean (*Phaseolus vulgaris* L.) from Galicia. Plant Breed 117:49-56

Freyre R, Rios R, Guzman L, Debouck DG, Gepts P (1996) Ecogeographic distribution of *Phaseolus* spp. (Fabaceae) in Bolivia. Econ Bot 50:195-215

Galvan MZ, Bornet B, Balatti PA, Branchard M (2003) Inter simple sequence repeat (ISSR) markers as a tool for the assessment of both genetic diversity and gene pool origin in common bean (*Phaseolus vulgaris* L.). Euphytica 132.297-301

Gepts P (1990) Biochemical evidence bearing on the domestication of *Phaseolus* (Fabaceae) beans. Econ Bot 44:28-38

Gepts P, Bliss FA (1985) F1 hybrid weakness in common bean differential geographic origin suggest two gene pools in cultivated bean germplasm. J Hered 76:447-450

Gepts P, Bliss FA (1986) Phaseolin variability among wild and cultivated common beans (*Phaseolus vulgaris*) from Colombia. Econ Bot 40:469-478

Gepts P, Bliss FA (1988) Dissemination pathways of common bean (*Phaseolus vulgaris,* Fabaceae) deduced from phaseolin electrophoretic variability. II. Europe and Africa. Econ Bot 42:86-104

Gepts P, Debouck DG (1991) Origin, domestication and evolution of the common bean (*Phaseolus vulgaris* L.). In: van Schoonhoven A, Voysest O (eds) Common beans: research for crop improvement. CAB Int Wallingford, Reino Unido and CIAT, Cali, Colombia, pp 7-53

Gepts P, Osborn TC, Rashka K, Bliss FA (1986) Phaseolin-protein variability in wild forms and landraces of the common bean (*Phaseolus vulgaris*): evidence for multiples centres of domestication. Econ Bot 40:451-468

Gil J, De Ron AM (1992) Variation in *Phaseolus vulgaris,* L. in the northwest of the Iberian Peninsula. Plant Breed 109:313-319

Haley SD, Miklas PN, Afanador L, Kelly JD (1994) Random amplified polymorphic DNA (RAPD) marker variability between and within gene pools of common bean. J Am Soc Hortic Sci 119:122-125

Herrera-Cervera JA, Caballero-Mellado J, Laguerre G, Tichy HV, Requena N, Amarguer N, Martinez-Romero E, Olivares J, Sanjuan J (1999) At least five rhizobial species nodulate *Phaseolus vulgaris* in a Spanish soil. FEMS Microbiol Ecol 30:87-97

Hidalgo R (1988) The *Phaseolus* world collection. In: Gepts P (ed) Genetic resources of *Phaseolus* beans. Kluwer, Dordrecht, pp 67-90

Islam FMA, Basford KE, Redden RJ, Gonzalez AV, Kroonenberg PM, Beebe S (2002) Genetic variability in cultivated common bean beyond the two major gene pools. Genet Res Crop Evol 49:271-283

Kami J, Becerra-Velasquez V, Debouck DG, Gepts P (1995) Identification of presumed ancestral DNA sequences of phaseolin in *Phaseolus vulgaris*. Proc Natl Acad Sci USA 92:1101-1104

Kaplan L (1981) What is the origin of common bean? Econ Bot 35:240-254

Kaplan L, Lynch TF (1999) *Phaseolus* (Fabaceae) in archaeology: AMS radiocarbon dates and their significance for pre-Colombian agriculture. Econ Bot 53:261-272

Kaplan L, MacNeish RS (1960) Prehistoric bean remains from caves in the Ocampo region of Tamaulipas, México. Bot Mus Leafl Harv Univ 19:33-56

Kaplan L, Lynch TF, Smith CE (1973) Early cultivated beans (*Phaseolus vulgaris*) from an intermontane Peruvian valley. Science 179:76-77

Koenig R, Gepts P (1989) Allozyme diversity in wild *Phaseolus vulgaris*: further evidence for two major centres of genetic diversity. Theor Appl Genet 78:809-817

Koenig R, Singh SP, Gepts P (1990) Novel phaseolin types in wild and cultivated common bean (*Phaseolus vulgaris*, Fabaceae). Econ Bot 44:50-60

Koinange EMK, Gepts P (1992) Hybrid weakness in wild *Phaseolus vulgaris* L. J Hered 83:135-139

Laguerre G, Geniaux E, Mazurier SI, Rodriguez-Casartelli R, Amarguer N (1993) Conformity and diversity among field isolates of *Rhizobium leguminosarum* bv. *viciae*, bv. *trifolii*, and bv. *phaseoli* revealed by DNA hybridization using chromosome and plasmid probes. Can J Microbiol 39:412-419

Limongelli G, Laghetti G, Perrino P, Piergiovanni AR (1996) Variation of seed storage protein in landraces of common bean (*Phaseolus vulgaris* L.) from Basilicata, Southern Italy. Plant Breed 119:513-516

Lioi L (1989) Geographical variation of phaseolin patterns in an Old World collection of *Phaseolus vulgaris*. Seed Sci Technol 17:317-324

Martinez-Romero E (2003) Diversity of *Rhizobium-Phaseolus vulgaris* symbiosis: overview and perspectives. Plant Soil 252:11-23

Martinez-Romero E, Caballero-Mellado J (1996) *Rhizobium* polygenies and bacterial genetic diversity. Crit Rev Plant Sci 15:113-140

Martinez-Romero E, Hernandez-Lucas I, Peña-Cabriales JJ, Castellanos JZ (1998) Symbiotic performance of some modified *Rhizobium etli* strains in assays with *Phaseolus vulgaris* beans that have a high capacity to fix N_2. Plant Soil 204:89-94

McClean PE, Myers JR, Hammond JJ (1993) Coefficient of parentage and cluster analysis of North American dry bean cultivars. Crop Sci 33:190-197

Nienhuis J, Singh SP (1986) Combining ability analyses and relationships among yield, yield components and architectural traits in dry bean. Crop Sci 26:21-27

Nodari RO, Tsai SM, Guzman P, Gilbertson RL Gepts P (1993) Towards an integrated linkage map of common bean III. Mapping genetic factors controlling host-bacteria interactions. Genetics 134:341-350

Ortwin-Sauer C (1966) The early Spanish man. Univ California Press, Berkeley, pp 51-298

Papa R, Gepts P (2003) Asymmetry of gene flow and differential geographical structure of molecular diversity in wild and domesticated common bean (*Phaseolus vulgaris* L.) from Mesoamerica. Theor Appl Genet 106:239-250

Paredes OM, Gepts P (1995) Extensive introgression of Middle American germplasm into Chilean common bean cultivars. Genet Res Crop Evol 42:29-41

Rodiño AP, Santalla M, De Ron AM, Singh SP (2003) A core collection of common bean from the Iberian Peninsula. Euphytica 131:165-175

Santalla M, de Ron AM, Escribano MR (1994) Effect of intercropping bush bean populations with maize on agronomic traits and their implications for selection. Field Crop
Res 36:185-189

Santalla M, Rodiño AP, de Ron AM (2002) Allozyme evidence supporting Southwestern
Europe as a secondary center of genetic for the common bean. Theor Appl Genet
104:934-944

Silva C, Vinuesa P, Eguiarte L, Martinez-Romero E, Souza V (2003) *Rhizobium etli* and
Rhizobium gallicum nodulate common bean (*Phaseolus vulgaris,* Fabaceae). Econ Bot
45:379-396

Singh SP (1989) Patterns of variation in cultivated common bean (*Phaseolus vulgaris* Fabaceae). Econ Bot 43:39-57

Singh SP, Gutiérrez JA (1984) Geographical distribution of the Dl1 and Dl2 genes causing
hybrid dwarfism in *Phaseolus vulgaris* L., their association with seed size, and their
significance to breeding. Euphytica 33:337-345

Singh SP, Gutierrez JA, Molina A, Urrea C, Gepts P (1991a) Genetic diversity in cultivated
common bean II. Marker-based analysis of morphological and agronomic traits. Crop
Sci 31:23-29

Singh SP, Nodari R, Gepts P (1991b) Genetic diversity in cultivated common bean I. Allozymes. Crop Sci 31:19-23

Singh SP, Gepts P, Debouck DG (1991c) Races of common bean (*Phaseolus vulgaris* Fabaceae). Econ Bot 45:379-396

Smartt J (1988) Morphological, physiological and biochemical changes in *Phaseolus* beans
under domestication. In: Gepts P (ed) Genetics resources of *Phaseolus* beans: their
maintenance, domestication, evolution and utilization. Kluwer, Dordrecht, pp 543-560

Tohme J, Jones P, Beebe S, Iwanaga M (1995) The combined use of agroecological and
characterisation data to establish the CIAT *Phaseolus vulgaris* core collection. In:
Hodgkin T, Brown ADH, van Hitum TJL, Morales EAV (eds) Core collections of
plant genetic resources. IPGRI/Wiley, Chichester, pp 95-107

Tohme J, Gonzalez DO, Beebe S, Duque MC (1996) AFLP analysis of gene pools of a wild
bean core collection. Crop Sci 36:1375-1384

Toro O, Tohme J, Debouck DG (1990) Wild bean (*Phaseolus vulgaris* L.): description and
distribution. Centro Internacional de Agricultura Tropical, Cali, Colombia

Vargas, J., J. Tohme, and D. Debouck. 1990. Common bean domestication in thesouthern
Andes. Annual Repport Bean Improvement Cooperative 33: 104-105Voysest O, Dessert M (1991) Bean cultivars classes and commercial seed types. In: van Schoonhoven
A, Voysest O (eds) Common bean research for crop improvement. CAB Int Wallingford, UK and CIAT, Cali, Colombia, pp 119-162

Zeven AC (1997) The introduction of the common bean (*Phaseolus vulgaris* L.) into western Europe and the phenotypic variation of dry beans collected in the Netherlands in
1946. Euphytica 94:319-328

Zeven AC (1999) The traditional inexplicable replacement of seed and seed ware of landraces and cultivars: a review. Euphytica 110:181-191

7 Forest History in Europe

Elisabeth Johann

Institute for Socioeconomics, University of Natural Resources and Applied Life Sciences, Vienna; Austria, E-mail elis.johann@utanet.at

7.1 Abstract

The time when the forest surface became altered by human activities conceals a lot of uncertainties. However, it is a fact that the distribution of tree species of European forests of the present day deviates remarkably from the natural vegetation cover, considering that the climate has remained fairly steady over the past 4,500 years (except for relatively wet and cold events such as experienced about 2,600, 1,400 and 700 to 200 years ago that reduced tree growth across Europe and possibly elsewhere). Historical land use and forest utilization practises still have an immense and long-lasting impact on the present landscape, the condition of the soil, and the quality and structure of forest stands. This paper emphasizes the historical process which led to the present state and which generally can be retraced in all parts of Europe. Alteration in the composition of tree species happened either by non-intentional activities such as litter harvesting, forest grazing, selective cutting or browsing, or was carried out intentionally by the implementation of silvicultural techniques such as selective cutting and the afforestation with seeds or/and seedlings. In describing the history of human impact on forests over all of Europe, the driving forces responsible for the conversion are discussed.

7.2 Introduction

The central European woodland in its present state has developed under continuously changing climatic conditions over the course of time by means of the interaction of various determinants such as geological processes, the succession of plants, and the evolution of animal species. Since the last ice age human beings have taken advantage of the possibilities offered by the surrounding landscape, although their impact on vegetation was initially confined to a small area. While agriculture apparently had been present in Turkey and eastern Greece around 8,000-9,000 years ago, elsewhere in Europe the only human communities present seem to have been hunter-gatherers. Although the natural conditions for the growth varied within a wide range across the different geographical regions, one can conclude that, before human beings started to change the natural landscape surface, broadleaved trees played a far larger, almost predominant, role in the natural composition of tree species in European forests compared to the present day. Pollen analyses indicate the wide spread of pure or mixed broad-leaved forests with a high variation of different species. At this time autochthonous conifers were growing at higher altitudes, in regions with lower temperature (boreal forests), and on sandy and poor soils. Meanwhile, the natural forest cover in central and western Europe were mesophytic deciduous and mixed forests (coniferous and broad-leaved species).

7.3 Definition of Woodland History and Forest History

According to the definition of Mantel (1990, pp. 39-40), woodland history deals with the development of virgin forests from the very first beginning until the time when anthropogenic influence started to change the natural ecosystem. Forest history has been defined as the history of human activities in the forest, which came into being in prehistoric times. However, the history of forestry also has to take into account the natural vegetation prior to human influence in order to be able to evaluate the size and dimension of this influence. Therefore, modern definitions of woodland history – also discussed by Mantel – reflect on the whole complexity of the historical development of forests and woodland, with and without anthropogenic impact in prehistoric and historic times. Thus, this overall definition of forest and woodland history includes a wide range of subjects such as palaeobotany, archaeology, sociology, cultural history, and economics which pay attention to the various impacts of human beings on the forest. Forest history takes into account the results of pollen analysis as well as written historical documents related to economic activities which have been stored in archives for 600 years and more. The Roman Period is to some extent highlighted by records from contemporary writers such as Plinius and Tacitus, which have been underlined by archaeological excavations, dendrochronology, and radiocarbon analysis. Thereby woodland history is able to illustrate forest history's starting position.

7.4 Post-Glacial Migration of Tree Species

Central European forests have experienced a generally similar development from the final glacial period onwards (Rudolph 1930; Firbas 1949). Whereas in North America and eastern Asia the migration of shrubs and tree species progressed along the main mountain chains, in Europe this development was slowed down by the Pyrenees, the Alps, and the Carpathian mountains. Thereby many species died off in the Quaternary, and Middle Europe has a reduced number of tree species compared to North America and East Asia (Küster 1995, 1996a,b; Lang 1994; Straka 1975).

During the last ice age and early post-glacial period central Europe became totally forested except for steep slopes and wet and rocky sites (Gradmann 1950; Küster 1996a).

After the warming, the migration of birch and pine happened rather quickly in central Europe. This is the case of Switzerland, whereas birch forests started to grow about 12,500 years ago (Della Casa 2003). Birch was followed by hazel, spruce, elm, oak, lime tree, ash, and alder about 10,000 years ago, later by fir (Küster 1996a). Beech, hornbeam and yew trees migrated to central Europe not before 4,000 B.C.

With regard to the occurrence of different tree species, a north-south and west-east gradient developed in central Europe over the course of time. Whereas in northern Germany oak-birch forests developed in the western part, pine forests remained quite stable in the east. In southern Germany hazel dominated in the west in contrast to the eastern part with its high percentage of spruce. In the Alps forest development was characterized by the increase in several conifers (pine, cembra pine, larch) and a permanent raising of the timberline, 9,000 years ago, reaching for the first time an altitude of 2,000 m above sea level. Minor climatic changes caused oscillation of between 100 and 300 m from this figure. Till 5,000 B.C. the forests were largely dense virgin forests with only some natural gaps and open areas.

7.5 Anthropogenic Influence on the Distribution of Tree Species in European Forests Before the Time of Permanent Settlement

Though apparently present in Turkey and eastern Greece by around 8,000-9,000 years ago, the impact of agriculture on vegetation seems to have been very localized. Elsewhere in Europe, the only human communities present seem to have been hunter-gatherers, but by 5,000 years ago agriculture had spread to most parts of Europe. At this later time, evergreen Mediterranean vegetation was more widespread through Greece and Italy than around 8,000 years ago, replacing deciduous forest. In southern France, however, pollen evidence indicates closed deciduous forest present in areas that are now evergreen scrub, although with some indications of sporadic clearance. Widespread clearance and replacement by maquis seem to have occurred only around 2,000 years ago. Approximately 5,000 years ago, the pine/evergreen-oak forest cover was decreasing in the Ebro Basin of northeastern Spain. On Crete, pollen evidence also suggests that ca. 5,000 years ago the original deciduous forest cover was lost, to be replaced by maquis. Some parts of the English chalklands may have been largely cleared for agriculture at the same time. However, these localized occurrences did not influence the overall pattern. Willis and Bennett (1994) argue, on the basis of their review of pollen evidence from around Europe, that agricultural impact on the vegetation almost everywhere was negligible before 4,000 years ago. Even in Greece, significant soil erosion due to deforestation does not seem to have occurred until about this time, when the clearance of the original forest cover for agriculture also started in western Europe (Great Britain, France, Netherlands, Denmark; Linnard 1985; Moriniaux 2000). In central Europe, however, the composition of tree species and the spreading out of woodland remained quite stable until the early Medieval period, from which time onward in this region much of the original forest cover was likewise cleared for agriculture.

From 4,000 to 3,500 B.C. in the foothills of the Alps, woodland was cleared and turned to farmland, particularly in the surroundings of settlements. The forest surface was remarkably altered by forest grazing of sheep, goats, and cattle, which were additionally fed with the leaves of lime tree, elm, and ash. The preference for a particular tree changed the natural vegetation. This was the cause of the increase in alder, which came into being with the increased agricultural activities. Two additional factors influenced the natural composition of tree species, particularly in the mountains: alpine pasturage, which was closely connected to the lowering of the timberline by clearing; and fuelwood harvest for the high requirements of the mining industry (copper, gold, silver, iron ore, and salt), which has been carried out since about 1,800 B.C. (Kral 1988/1989, 1991). However, the influence of these factors on any specific area was only temporary because prehistoric settlements moved to another place after a certain period of farming. Abandoned farmland became forested again by secondary successions of birch and other pioneer species (Hvass 1982; Kossack 1982; Küster 1996a). Still, often the secondary forest did not develop to the type of forest which would have been growing on the site without anthropogenic influence. Due to altered site conditions, other species migrated. This was true of beech, whose percentage in the distribution of tree species increased all over Europe, thereby becoming regionally the dominant tree species (Küster 1988; Pott 1992). In other areas such as in the foothills of the Swiss Alps spruce establishment was sped up by human activities, and as a result also spread naturally into higher mountain regions (Markgraf 1970, 1972). This is also the case in the Vosges, the Black Forest, and the Hessian mountain region (Lang 1973a). Fir spread naturally in central Europe thereby pushing back hazel in the Black Forest and elm in the foothills of the Alps (Speier 1994).

Relatively wet (and cold) events of 700 B.C. had a remarkable impact on the distribution of tree species. During this period, the percentage of beech increased, particularly in the mountain areas, fir disseminated in the south of Germany, spruce in the eastern part. Thereby mixed oak forests and pine forests were reduced. Mixed oak forests, however, kept the dominant position at warm sites at lower altitudes in western Germany. The proportion of spruce in European forests started to increase continuously in early A.D. In the second half of the Christian era, the distribution of tree species became quite stable and was rather similar to the potential natural vegetation, that is, the vegetation that would be present today if humans had not altered it by agriculture and forestry (Firbas 1949).

7.6 Anthropogenic Influence on the Natural Distribution of Tree Species in Historic Times

7.6.1 Effects of the First Permanent Settlements

In Europe, the first permanent settlements were established during the Roman period. Because of the continuous utilization of the surrounding forest surface the role of natural succession became less important, apart from over-burning practises. In the neighbourhood of villages, the dissemination of beech came to an end, whereas on the other side of the Limes line beech continued to spread out. This raises the question as to whether the dissemination of beech was more supported by human activities than by climatic or other supra-regional influencing parameters.

The Roman period was marked by the expansion of farmland at the expense of the forest surface and by the increase in wood-consuming economic entities such as the construction of houses and ships, the production of bricks and charcoal, and the mining industry. These activities caused remarkable changes in the forest ecosystem. On the one hand, oak and beech forests were intensively cut; on the other hand, alien species were introduced such as chestnut and walnut tree. Strabo cited the export of resin and tar from the Alpine region, whereas Plinius mentioned the floating of larch to northern Italy. Alpine larch was also favoured for fine buildings as far away as Rome (Della Casa 2003).

At the same time when forests were widely cleared in Italy, Tacitus and Plinius described central European forests as dense, close, and shadowed. Apart from coppicing, clear cuts were scarcely practised, single tree-felling methods being preferred, particularly for the harvesting of construction timber. Already at this time some of the forest stands became over-utilized, mainly when there was a high demand for a certain kind of wood (e.g. fuel wood for the soldiers, shingles and construction timber for military camps). Other regions, which were not accessible, remained quite untouched. Management practises such as shelterwood coppice system or stored coppice system (high stand with coppice in the understory) put beech to disadvantage because of that tree's reduced ability to withstand frequent cutting. Conversely, migration of oak, elm, lime tree, hazel, and hornbeam was furthered (Pott 1993).

Oak, ash, elm, poplar, spruce, pine, fir, and larch trees were in popular use for the construction of houses. Nut and maple trees were used for beams, larch for the construction of bridges and water pipes, beech and oak trees for manufacturing pots and bowls, and the bark of beech, lime, and spruce trees for making baskets (Johann 1996). In some mining districts, beech was intensively harvested for use in mining, and its percentage in the natural composition of tree species was reduced on this side of the Limes line. Two thousand years ago fir and beech became rarer species and the percentage of spruce increased in some regions of central Europe. In the foothills of the Alps former mixed forests of spruce, fir, and beech became pure spruce forests.

With the end of the Roman Empire, settlements were abandoned and the percentage of beech increased again in European forests. This burgeoning finally came to an end during the Medieval period due to the renewed establishment of permanent settlements. At this time beech had migrated to the southeast of England, to southern Scandinavia, and to the Polish coastal fringe along the Baltic Sea, reaching its present extent at around 1000 A.D. (Küster 1996a). Including temporary gaps and cleared areas or forest pastures sparsely stocked with oak, birch, and/or pine, central European woodlands probably still covered three fourths of the total area at the beginning of the Medieval period (Hausrath 1911; Firbas 1949; Mantel 1990).

7.6.2 Conversion Processes

In central Europe, the period from the Middle Ages to the beginning of the 19th century was characterized by intense forest use, sometimes even heavy exploitation. Depending on the available resources and demographic evolution, the development varied with regard to time and region. In the course of centuries a variety of driving forces led to the general decline of broadleaved species in European forests and the promotion of conifers, mainly spruce and pine, and partly larch. Most pronounced was the elimination of beech either intentionally or non-intentionally, giving rise to the growth of secondary spruce forests as well as to the degradation of the forest area with encroachment of wasteland.

On one hand, most pronounced in western and southern Europe, forests were widely cleared because of the high demand for arable land which had to serve the needs of the increasing population. Thus, in some countries, the forest area decreased markedly (e.g., in Scotland to 4% in 1750; Smout and Watson 1997). From the 14th century onward these practices also came to involve many parts of central Europe. In Germany, for instance, the forest area was reduced from 75% total land cover at the time of the migration of nations to 26% in 1400 (Mantel 1990). On the other hand, intense forest utilization caused a progressive change of the former natural forests with respect to density, structure, and distribution of tree species. Coppicing with short rotation periods caused the decline of beech and the increase in oak, hornbeam, hazel, elm, lime tree, and chestnut in mixed forests (Pott 1981, 1993). Finally, the percentage of these species also decreased in coppice forests due to intense and frequent harvesting. Altogether, the percentage of forests was reduced in almost every European country.

The reasons for over-utilization were multiple. The most important driving forces were; (1) economical aspects: high demand of the growing industry (glassworks, salt and other mineral mines, forges and furnaces), ship building and timber trade; (2) ecological aspects: degraded soils caused by concentrated forest grazing and litter harvesting, regionally the increase in deer; (3) political aspects (war and its effects); and (4) forest management practises.

7.6.3 The Implementation of Forest Science in Management Practices

The introduction of conifers during the 200 years of restoration changed European forests decisively. The effects of reforestation on the landscape differ depending on the natural conditions of the site and on the method practised for conversion. The increase in conifers in European forests can primarily be traced back to afforestation programs carried out in the 19th and 20th centuries: Afforestation of clearcuts with only one tree species (spruce or pine) and conversion from mixed forests to pure stands, conversion from coppice forests and high beech forests to pure stands of Norway spruce and Scots pine, or afforestation of wasteland and abandoned farmland.

The method was influenced by scientific findings and by progress in silvicultural techniques. In particular, the implementation of the doctrines of the German classic school of forestry had a huge impact (Johann 2001, 2002).

7.7 Effects of Human Activities on the Present Forest Surface

At the beginning of the 19th century, two principal types of management still existed in European forestry: (1) traditional woodland management, principally taking the form of coppice woodland or high stands with coppice in the understory; and (2) plantation forestry, which had developed from the early 18th century onwards. Forests as a whole have increased from the depression in 1800 up to the present due to monoculture plantations. This increase is today noticeable in almost every European country, even if the share of forest land differs remarkably from about 10% to more than 55%. The variety of tree species, however, was reduced remarkably. In countries where afforestation activities took place only to a minor degree, such as in the Mediterranean, France, and Great Britain, manifold phenomena of erosion have been observed.

In general, the share of spruce has increased in almost every European forest as a consequence of intended or uncontrolled decrease in species diversity. Changes in composition of tree species have been less remarkable in forests with a high natural proportion of Norway spruce, that is, in central European mountain forests at higher altitudes and in boreal forests in Sweden. Forests with a high proportion of natural occurrence of fir and beech, or pure broadleaved forests such as those in Germany, the Czech Republic, Slovenia, and Italy, underwent a much more pronounced change.

In Germany over the period from 1883 to 1913, the reduction of the share of broadleaved trees was mainly caused by the decrease in coppice forests, which lost 40% of their former extent. Among the main coniferous species the proportion of pine increased from 42.6 to 45.5%, and that of spruce from 22.5 to 24.2% (Heyder 1983). In Austria, compared to natural forests, broadleaved trees have been reduced by half, and conifers play a significant dominating role in contemporary forests (Kral 1974; Grabherr et al. 1998).

The present share of *Picea abies* in the total European forest area (EU30 forest area) amounts to 21%, of *Pinus sylvestris* 31%, of *Pinus pinaster* 2.5%, of *Fagus sylvatica* 7.1%, of *Betula pubescens* 4.7%, and of *Quercus ilex* 2.4%. Altogether, the percentage of coniferous species exceeds 60% (Köble and Seufert 2001). In some parts of Europe the conversion of mixed or broadleaved forests to monocultures of Norway spruce (or Scots pine) has been carried out for about 400 years. This has produced a forest landscape of relatively young forest stands, dominated by only one tree species and typically even-aged. This converted forest lacks many important ecological qualities and the remaining old-growth patches are often fragmented (Axelsson and Östlund 2001).

7.8 Conclusions

Due to the distribution of interglacial tree species and post-glacial migration, relevant silvicultural conclusions can be drawn regarding (1) the extent and purposefulness of human-induced alteration of the natural distribution of tree species, and (2) present silvicultural measures (Mantel 1990, pp. 39-40). In fashioning the present cultural landscape, on the one hand, people were straitened by the environment. On the other, for more than 1,000 years human beings increasingly made great efforts to establish their own environment. Several millennia elapsed in the transition of open landscape to forest. The subsequent differentiation of various types proceeded more rapidly involving a lesser number of successive stages than the "basic succession in forest development" would suggest. During the time of development, the modulation of ecosystems has permanently altered. Thereby, the importance of particular tree species increased, whereas others lost importance. However, even in the past, stable conditions never really existed regardless of whether human beings influenced the development of tree species (Küster 1995, 1996a).

7.9 References

Axelsson A-L, Östlund L (2001) Retrospective gap analysis in a Swedish boreal forest landscape using historical data. For Ecol Manage 147:109-122
Della Casa P (2003) Wald. Kap. 1. Ur- und Frühgeschichte. In: Historisches Lexikon der Schweiz. Bern, Elektronische Version 9, 2003
Firbas F (1949) Spät- und naheiszeitliche Waldgeschichte Mitteleuropas nördlich der Alpen, Bd 1. Allgemeine Waldgeschichte. Jena
Grabherr G, Koch G, Kirchmeir H, Reiter K (1998) Hemerobie Österreichischer Waldöko-Systeme. In: Österreichische Akademie der Wissenschaften (Hg) Veröffentlichungen des Österreichischen MaB-Programmes, Bd 17. Universitätsverlag Innsbruck, Innsbruck
Gradmann R (1950) Das Pflanzenleben der Schwäbischen Alb, 4 Aufl. Schwäb Albverein, Stuttgart

Hausrath H (1911) Pflanzengeographische Wandlungen der deutschen Landschaft. Leipzig Berlin

Heyder JC (1983) Waldbau im Wandel. Zur Geschichte des Waldbaus von 1870 bis 1950, dargestellt unter besonderer Berücksichtigung der Bestandesbegründung und der forstlichen Verhältnisse Norddeutschlands, Institut für Waldbau (ed.) TU Tharandt

Hvass S (1982) Ländliche Siedlungen der Kaiser- und Völkerwanderungszeit in Dänemark. Offa 39:189-195

Johann E (1996) Zur Geschichte der Waldbewirtschaftung im Alpenvorland. Österreichische Forstzeitung 12:8-11

Johann E (2001) Zur Geschichte des Natur- und Landschaftsschutzes in Österreich. Historische „Ödflächen" und ihre Wiederbewaldung". In: Weigl N (Hg) Faszination Forstgeschichte. Institut für Sozioökonomik der Forst- und Holzwirtschaft (ed.), Schriftenreihe Bd 42. Wien: 41-60

Johann E (2002) Zukunft hat Vergangenheit. 150 Jahre Österreichischer Forstverein. Österreichischer Forstverein (ed.). Wien

Köble R, Seufert G (2001) Novel maps for forest tree species in Europe. In: Hjorth J, Raes F, Angeletti G (eds) 8th European Symposium on the Physico-chemical behaviour of atmospheric pollutants, 17-20 Sept 2001, Turino, TP 35

Kossack G (1982) Ländliches Siedelwesen in vor- und frühgeschichtlicher Zeit. Offa 39:271-279

Kral F (1974) Grundzüge einer postglazialen Waldgeschichte des Ostalpenraumes. In: Mayer H (ed) Wälder des Ostalpenraumes. Fischer Stuttgart

Kral F (1988/89) Botanische Beiträge zu Fragen der Umwelt im Mittelalter. Beitr Mittelalterarchäol Österr 4/5:243-250

Kral F (1991) Die postglaziale Entwicklung der natürlichen Vegetation Mitteleuropas und ihre Beeinflussung durch den Menschen. Veröff Komm Humanökologie Österr Akad Wiss 3:7-36

Küster H (1986) Vom Werden einer Kulturlandschaft im Alpenvorland. Pollenanalytische Aussagen zur Siedlungsgeschichte am Auerberg in Südbayern. Germania 64 (2): 533-559)

Küster H (1995) Geschichte der Landschaft in Mitteleuropa. Von der Eiszeit bis zur Gegenwart. Beck'sche Verlagsbuchhandlung, München

Küster H (1996a) Auswirkungen von Klimaschwankungen und menschlicher Landschaftsnutzung auf die Arealverschiebung von Pflanzen und die Ausbildung mitteleuropäischer Wälder. Forstwiss Centralbl 115:301-320

Küster H (1996b) Die Stellung der Eibe in der nacheiszeitlichen Waldentwicklung und die Verwendung ihres Holzes in vor- und frühgeschichtlicher Zeit. Ber Bayerischen Landesanst Wald Forstwirtschaft 10:3-8

Lang G (1973a) Neue Untersuchungen über die spät- und nacheiszeitliche Vegetationsgeschichte des Schwarzwaldes IV: das Baldenwegermoor und das einstige Waldbild am Feldberg. Beitr Naturkundl Forsch Südwestdeutschland 32:31-52

Lang G (1994) Quartäre Vegetationsgeschichte Europas. Methoden und Ergebnisse. Fischer, Jena

Linnard W (1985) History of mountain forestry in Wales. History of forest utilization and forestry in mountain regions. Symposium ETH Zürich, 3-7 Sept 1984, IUFRO 6.07 Forest History, Zürich: 53-58

Mantel K (1990) Wald und Forst in der Geschichte. Schaper, Alfeld-Hannover

Markgraf V (1970) Palaeohistory of the spruce in Switzerland. Nature 228:249-251

Markgraf V (1972) Die Ausbreitungsgeschichte der Fichte (*Picea abies* H. Karst) in der Schweiz. Ber Dtsch Bot Ges 85:165-172

Moriniaux V (2000) A history of the French coniferous forest. In: Agnoletti M, Anderson S (eds) Forest history. International Studies on Socio-economic and Forest Ecosystem Change. CABI Publishing in association with IUFRO, Wallingford: 363-369

Plinius der Ältere Naturalis Historia XVI. D. Detlefsen (Hg.) 1866-1882

Pott R (1981) Der Einfluss der Niederholzwirtschaft auf die Physiognomie und die floristisch-soziologische Struktur von Kalkbuchenwäldern. Tuexenia 1:233-242

Pott R (1992) Nacheiszeitliche Entwicklung des Buchenareals und der mitteleuropäischen Buchenwaldgesellschaften. Naturschutzzentrum Nordhein-Westfalen. Seminarberichte 12, Recklinghausen: 6-18

Pott R (1993) Farbatlas Waldlandschaften. Ulmer, Stuttgart

Rudolph K (1930) Grundzüge der nacheiszeitlichen Waldgeschichte Mitteleuropas. Beih Bot Centralbl 47:111-176

Smout C, Watson F (1997) Exploiting Semi-natural Woods, 1600-1800. In: Smout TC (ed) Scottish woodland history. Scottish Cultural Press, Edinburgh: 86-100

Speier M (1994) Vegetationskundliche und paläoökologische Untersuchungen zur Rekonstruktion prähistorischer und historischer Landnutzungen im südlichen Rothaargebirge. Abhandlungen aus dem Westfälischen Museum für Naturkunde 56 (3/4), Münster

Strabo, Geographica IV, 6,9

Straka H (1975) Pollen- und Sporenkunde. Fischer, Stuttgart

Tacitus C, Germania 5,1; Polybios, Historiae, III, 55,9

Willis KJ, Bennett KD (1994) The Neolithic transition - fact or fiction? Palaeoecological evidence from the Balkans. Holocene 4:326-330

8 Forest Biological Resources in the Amazon Basin

Heidi Kreibich[1] and Jürgen Kern

Institute of Agricultural Engineering (ATB), 14469 Potsdam, Germany
[1]Present address: Section Engineering Hydrology, GeoForschungsZentrum Potsdam, 13373 Potsdam, Germany, E-mail kreib@gfz-potsdam.de

8.1 Abstract

The Amazon region holds the world's largest area of tropical rain forest which was settled by the first humans during the Late Pleistocene. A strong anthropogenic impact began in the 1960s when large areas of the tropical forest were cleared. To protect this ecosystem, sustainable utilization of its biological resources is urgently needed.

The most productive areas are the floodplains of the Amazon River which are built up by suspended solids deriving from the Andes. The study was focused on the várzea forest, where a relatively high potential for sustainable agro-forestry can be assumed. Since nitrogen may be the limiting factor, N_2 fixation by nodulated legumes was extensively studied. Legumes provided an annual nitrogen gain between 12.9 and 16.1 kg N ha^{-1}. Considering all available data of input and output, the nitrogen balance was positive. The amount of nitrogen stored in the biomass of the várzea forest is about 50 times the annual nitrogen input. The use of N_2-fixing legumes may open new perspectives for a sustainable use and protection of the várzea forest.

8.2 The Amazon Basin

8.2.1 History of Human Settlement

It is supposed that human settlement of the continent America started by the migration from central Asia during the last glacial period via the Bering Strait. During that era, the sea level was low and North America and Asia were not separated by water. South America was reached at the northeast from where early human groups started to disperse across the continent during the Würm VI-Late Wisconsin epoch, between 23,000 and 12,700 B.P. (Ab´Sáber 2001). Evidence for human activities is given in central and northeastern Brazil for some sites which are dated between 14,000 and 10,000 B.P. (Prous and Fogaça 1999). First, the highlands of the Andes and the Guyana shield were settled before migration started into the river valleys of the Amazon basin (Becher 1989).

Due to extraordinary ecological factors, Amazonia is an extreme habitat for humans. Pre- and post-Columbian settling was controlled primarily by geochemical characteristics responsible for the supply of food that could be utilized by humans. At least 90% of Amazonian soils belong to non-fertile Oxisols and Ultisols which fulfil the fixation of plant roots but not the supply of nutrients. Such nutrient-poor areas like the terra firme upland forests are reflected by a low carrying capacity of 0.2 persons per km^2 (Denevan 1976). In contrast, the population density is generally high with 15 persons per km^2 on average in the alluvial landscapes of Andean rivers (várzea). Outside of the várzea and the foothills of the Andes, food supply enabled only small groups to exist. Primarily the demand for proteins can be considered as a bottleneck for human nutrition. Natural protein resources are used most efficiently by many small groups which specialize in fishing, hunting, and gathering.

When Europeans came to South America, indigenous cultures were well developed in big tribes on the course of the Amazon River. From the first contact with Europeans during the voyage of Francisco Orellanas in 1542, people were displaced from river fringes to less productive upland forests. Due to slavery, the Brazilian river banks were deserted by the end of the 17th century (Sioli 1989). The Spanish and the Portuguese conquerors tried to expand their mother country. To build beautiful cities as in Spain and Portugal, the manpower of many native people was needed. Therefore, indigenous people were enslaved. Besides the degrading conditions, diseases imported from Europe caused the death of thousands of them.

Since the 1960s, the Brazilian government has started new programs of National Integration to enhance the development of infrastructure and to support economic growth. Via road construction and the gift of ownerless land to people, a compensation between high and low crowded regions such as northeastern Brazil is being induced. The "empty" Amazonian lowland was filled by people who cleared large forest areas (Kohlhepp 1989). Today, the clearing of tropical rain forest has already reached a threatening dimension with a loss of about 14% in the Amazon basin (Carvalho et al. 2001; Laurence et al. 2001).

After the need for sustainability was proclaimed worldwide, new strategies for integrated rural development were introduced over the last two decades by the World Bank. Part of regional development today means participation of poor farmers, protection of the indigenous population, and control of deforestation. Overall, any strategy for sustainable development must recognize and respect the limits of carrying capacity (Fearnside 1997). From an indigenous population of about 6 million in pre-Columbian times, the population of the Brazilian Amazon has reached about 20 million people today. Thus, to protect the Amazonian forests, sustainable utilization is needed more than ever.

8.2.2 The Várzea Forest

The Amazon region holds about 40% of the world's tropical rain forest. In relation to the entire Amazon forest, the 100,000 km^2 of floodplain forest along the Amazon River and its large tributaries are small, but the regions flooded by nutrient-rich rivers descending from the Andes are the most fertile ones in Amazonia. Enclosure experiments and field studies indicate that nitrogen can be a limiting factor for biomass production because the river water is rich in major solutes, but not in nitrogen compounds (Furch and Junk 1993).

In comparison to non-flooded rain forests, species diversity is relatively low in várzea forests, which is due to both selection and adaptation to periodic flooding. In várzea forests between 40 and 150 woody species per hectare were found (Worbes 1997). In our study area on Marchantaria Island, we counted 1,441 woody plants of 44 species at 2.1 ha. Legumes had an average absolute density of 68 individuals per hectare, representing the third most important family (Table 1). The high importance of the legumes is characteristic for the várzea; other studies have even revealed the legumes as the most important family with relative densities of 12-17% (Worbes 1986; Klinge et al. 1995).

About 88% of the identified legume species were nodulated. The fertile várzea soils provide favourable conditions for nodulation, since a good nutrient availability especially of phosphorus, potassium, sulphur, and micronutrients stimulates nodulation (Oliveira et al. 1998). Nodulated legumes contribute significantly to the nitrogen balance of Amazonian wetlands (Salati et al. 1982; Martinelli et al. 1992), where nodulation of legumes is much more abundant than in the drier terra firme regions. Due to adaptations to permanent or seasonal flooding, nodules appear to be active during the whole hydrological cycle (James et al. 2001).

Table 1. Family importance value (FIV), relative richness, relative density, and relative dominance of the eight most important families in the várzea forest on Marchantaria Island

Plant families	FIV	Relative richness [%]	Relative density [%]	Relative dominance [%]
Capparaceae	59	2	37	20
Verbenaceae	54	2	23	29
Leguminosae	47	27	10	10
Flacourtiaceae	18	5	7	6
Tiliaceae	16	2	2	12
Bombacaceae	14	2	2	10
Myrtaceae	11	2	7	2
Euphorbiaceae	11	7	3	1
remaining 17 Families	69	50	9	10

8.2.3 Nitrogen Dynamics in the Várzea Forest

The ^{15}N natural abundance method revealed, that nodulated legume species of the várzea fixed between 2 and 70% of their plant nitrogen from the atmosphere. Combining these %Ndfa values with results from stand structure analyses, the average percentage of plant nitrogen derived from the atmosphere for the várzea forest was calculated to be between 4 and 5% (Table 2). Since the net biomass production of the várzea forest is on average 322.7 kg N ha^{-1} year^{-1} (Furch 1999), the nitrogen gain via nodulated legumes is estimated to range between 12.9 and 16.1 kg N ha^{-1} year^{-1}. Considerably lower was the rate in tropical secondary vegetation where Thielen-Klinge (1997) found an input between 0.1 and 4.7 kg N ha^{-1} year^{-1}. This might be due to more abundant nodulation in floodplain forests compared to terra firme forests, although N$_2$ fixation in a disturbed area such as secondary vegetation would be expected to be higher than in natural forests. All these N$_2$ fixation rates are clearly surpassed by mono-specific agricultural stands of legume crops. Lentils, for example, may fix 127-139 kg N ha^{-1} year^{-1}, clover plants fix 27-205 kg N ha^{-1} year^{-1}, and faba beans fix between 200 and 360 kg N ha^{-1} year^{-1} (Moawad et al. 1998; Kilian et al. 2001).

Table 2. Importance value index (IVI) and %Ndfa (minimum and maximum values) for legume and non-legume species. From this data the weighted average %Ndfa of the várzea forest was calculated

Woody plants	IVI	%Ndfa (min) [%]	%Ndfa (max) [%]
Dalbergia riparia	10	54	62
Lonchocarpus sp.	1	47	57
Machaerium aristulatum	2	44	55
Mimosa pigra	2	42	53
Zygia inaequalis	1	41	52
Campsiandra comosa	1	38	50
Albizia multiflora	8	31	48
Dalbergia inundata	1	35	48
Pterocarpus amazonum	5	6	24
Crudia amazonica	1	0	0
Macrolobium acaciifolium	9	0	0
Non-legumes	259	0	0
Weighted average		**4**	**5**

Non-symbiotic N$_2$ fixation by heterotrophic diazotrophs was monitored using the acetylene reduction assay as described by Kreibich and Kern (2003). In contrast to symbiotic N$_2$ fixation, seasonal variation of the microbiological activity was found in the soil of the floodplain forest. During the aquatic period, N$_2$ fixation was negligible. During periods of receding water when soils are exposed to air, nitrogenase activity of free-living N$_2$-fixing bacteria increased. Sufficient organic matter in the surface soil layer supported nitrogenase activity, which was apparently not suppressed by NH$_4^+$ and a relatively low pH. Negligible N$_2$ fixation during the aquatic period is in accordance with previous studies on Marchantaria Island where high NH$_4^+$ concentrations in aquatic sediments did not support N$_2$ fixation (Kern and Darwich 1997). Sum of the mean monthly nitrogen gains during the 5 months of the terrestrial period was 3.6 kg N ha^{-1}, and the amount of nitrogen fixed during the 7 months of the aquatic period was only 0.5 kg N ha^{-1}. Consequently, the high biomass production which occurs in the várzea forest during the terrestrial period profits by the enhanced non-symbiotic N$_2$ fixation.

One important path of nitrogen loss is denitrification, as reported by Kern et al. (1996). Compared to non-symbiotic N_2 fixation, denitrification showed an inverse pattern in the forest soil. High rates were obtained during the aquatic and the transition periods and low rates during the terrestrial period. It may be suggested that denitrification is stimulated primarily by the supply of riverine NO_3^-. The sum of the mean monthly nitrogen losses during the 7 months of the aquatic period was 10.9 kg N ha^{-1}, and the amount denitrified in the 5 months of the terrestrial period was 1.6 kg N ha^{-1}. Thus, nitrogen loss via denitrification is remarkably high and may limit the nutrient supply in the várzea forest.

8.2.4 Nitrogen Balance of the Várzea Forest

In order to obtain an estimate of nitrogen fluxes within the Amazon floodplain forest, data from Kreibich (2002) and earlier studies (Kern 1995; Furch 1997) were used to calculate the input and output of nitrogen. All nitrogen fluxes could not be included, and therefore this balance is provisional.

Symbiotic N_2 fixation by nodulated legumes is between 12.9 and 16.1 kg N ha^{-1} year^{-1}. Non-symbiotic N_2 fixation in the surface soil layer is 4.1 kg N ha^{-1} year^{-1}. Since not all nodulated legume species could be included in our calculation, the real nitrogen gain via N_2 fixation might generally be higher in the várzea forest. Further input derives from the atmosphere by dry and wet deposition, which was estimated to be 2.6 kg N ha^{-1} year^{-1} on Marchantaria Island (Kern and Darwich 1997).

Due to the elevational range in the study area, the Amazon River inundates the forest on average between 4.7 and 7.6 months every year. During this period, processes such as sedimentation, diffusion, re-suspension, fish migration, and the exchange of coarse organic detritus and macrophytes may play an important role for the nitrogen exchange between river and floodplain, as shown by Kern (1995). Sedimentation of particulate nitrogen was estimated using a sedimentation rate of 3.14 mg N m^{-2} day^{-1}. Related to an inundation period of 4.7 to 7.6 months per year, sedimentation would account for 4.4 to 7.2 kg N ha^{-1} year^{-1}. Considering dissolved nitrogen but no other input pathways as mentioned above (fish migration, etc.), the total nitrogen input via the river can be calculated to 10.5 kg N ha^{-1} year^{-1}. Thus, by this balance we assume an input via the river between 4.4 and 10.5 kg N ha^{-1} year^{-1}. Nitrogen output from the várzea to the river due to re-suspension of sediments and outflow of dissolved nitrogen accounts for 6.0 kg N ha^{-1} year^{-1}. Some nitrogen loss is assumed by the feeding of fish, which are migrating into the forest with rising water feeding on fruits, and therefore extracting nitrogen when they return to lakes and rivers (Gottsberger 1978). With receding water, on the one hand, organic detritus from the trees is carried away with the water; on the other hand, dead wood and macrophytes are deposited on the forest floor. However, these exchange processes as well as the nitrogen loss due to fire or extraction could not be included in the balance, due to the lack of suitable estimation models.

Denitrification in the surface soil layer is 12.5 kg N ha^{-1} year^{-1}. Since both, N_2 fixation and denitrification in deeper soil layers were neglected, the gaseous nitrogen fluxes have to be considered as minimum values. Nitrogen leached from the soil at 0-20 cm depth was measured to be 1.2-3.8 kg N ha^{-1} year^{-1} in the study area (Kreibich et al. 2003). Erosion processes were not studied. They are suggested to be low, due to the protective vegetation cover with dense root-systems in litter and the surface soil layers. Nitrogen loss via NH_3 volatilization can be excluded due to the low soil pH.

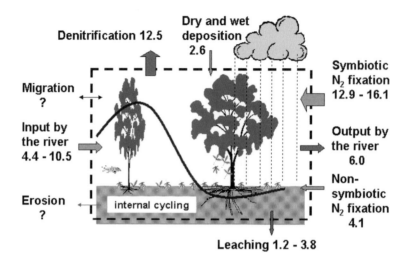

Fig. 1. Average nitrogen fluxes in the várzea forest (in kg N ha^{-1} year^{-1}). The width of the *arrows* reflects the amount of nitrogen input (*gray*) and nitrogen output (dark). Nitrogen fluxes via animal migration and erosion cannot be estimated yet

Although the balance shown in Fig. 1 has a preliminary character, the following conclusions can be drawn for the floodplain forest under study. The gaseous nitrogen turnover plays an important role in the forest area, supplying 60-71% of the total input and 56-63% of the total output. The high abundance of N_2-fixing legumes provides the main nitrogen source, even surpassing the gain of nitrogen via the river water. Denitrification is the main pathway for nitrogen loss from the várzea soil, which is in accordance with earlier results reported by Kern and Darwich (1997). Since total nitrogen input is assumed to be 24-33 kg N ha^{-1} year^{-1}, and total nitrogen output only 20-22 kg N ha^{-1} year^{-1}, there is a surplus of nitrogen in the várzea forest.

The nutrient balance of a tropical forest shows whether and how far a forest can be utilized on a sustainable basis. Due to the considerable nitrogen input via N_2 fixation and the flooding water, an open nitrogen cycle of the floodplain forest is suggested according to the definition of Baillie (1989). The author defined a closed nutrient cycle as such, when the ecosystem is sustained entirely by atmospheric inputs and internal cycling of nutrients. In open nutrient cycles, other nitrogen sources also exist. In the várzea forest, the amount of nitrogen stored in the biomass (1,479 kg N ha^{-1}; Furch 1997) is about 50 times the annual nitrogen input;

in non-flooded Amazonian forests this ratio is about 500 (Brouwer 1996). Due to a relatively good nutrient supply and a positive nitrogen balance of the várzea forest, it may be concluded that a sustainable use of the várzea forest is possible.

8.2.5 Land Use in the Várzea

During the last few decades, the Brazilian government has supported the agricultural development of non-flooded rain forest areas. Considering the sensitivity of these upland forests, today policy makers and investors try to enhance the development of the Amazon floodplain. Thus, strategies for a sustainable management of this ecosystem are becoming more and more important in order to protect biotic and abiotic resources.

Annual flooding makes the soils rather sensitive to physical compaction and to erosion associated with mechanized agriculture, ranching, and indiscriminate timber-harvesting practices. In central Amazonia an elevational range higher than about 25 m above sea level is considered to be suitable for profitable agriculture. Shifting cultivation and small-scale agriculture have been practised for centuries, providing subsistence while conserving soil, water, and forest resources (Coomes and Burt 1997). Indigenous people adapted to the flood pulse, cultivate the floodplain with banana, cassava, jute, and maize, as long as the water level and the specific flood tolerance of plants allow crop production. However, these systems only work as long as population density and consumption remain within the limits of human carrying capacity. Increasing pressure on resources is already apparent and has been documented in many cases (Fearnside 1997; Etchart 1997; Kolk 1998). For instance, in higher elevational ranges of the floodplain there was extensive forest clearing for large-scale jute cultivation, which was the most economic production system of the várzea between 1940 and 1980. In the 1960s, about 20,000 families were engaged in this production line (Junk et al. 2000). Due to the introduction of synthetic fibers and the import of cheaper fibers from India and Bangladesh, jute production decreased in the 1980s.

Since várzea soils are rich in major solutes but not in nitrogen, the development of more sophisticated, productive and sustainable management concepts necessarily relies on nodulated legumes. In agro-forestry, herbal and woody climbers are regularly used as soil cover, and mainly tree legumes are used in alley-cropping and mixed-cropping systems (Peoples et al. 1995). Hardwood timber production represents the major economic use of leguminous trees, although other roles in providing shelter, land conservation, land reclamation, and medicines are important as well (Ladha et al. 1993; Johnston 1998). For example, wood of *Campsiandra* sp. and *Swartzia* sp. is highly durable, and therefore preferably used for heavy construction purposes (Sprent and Parsons 2000). *Vigna* sp., *Aeschynomene* sp., *Sesbania* sp., *Teramnus* sp. etc. are used as good forage, hay, and green manure. By using these species as soil cover, soil fertility may be maintained and erosion may be prevented (Allen and Allen 1981). Since nodulated legumes generally improve the soil nitrogen status and also transfer fixed nitrogen to adjacent or replacement plants, the development of small-scale agro-forestry systems which support the cultivation of legumes may open new perspectives for the várzea.

8.3 Conclusions

Due to the rapid increase in its population during the last few decades, sustainable management is urgently needed in the Amazon region to stop massive deforestation and to protect the fragile ecosystems. The várzea forest is comparatively small, but rich in nutrients pointing to a high potential for sustainable agroforestry. Plant growth may be limited primarily by the availability of nitrogen which is reduced by denitrification. Taking studies between 1992 and 2000 into consideration, a positive nitrogen balance of 1.7 to 13.6 kg N ha^{-1} year^{-1} is obtained and supports a sustainable utilization of the floodplain forest. The greatest nitrogen input occurs via symbiotic N_2 fixation, with about 4-5% of the forest's nitrogen deriving from atmosphere. Thus, the use of N_2-fixing plants in agroforestry systems may open new perspectives for sustainable agriculture in the central Amazon.

8.4 References

Ab´Saber AN (2001) The prehistoric human geography of Brazil. Amazoniana 16:303-311

Allen ON, Allen EK (1981) The *Leguminosae* - a source book of characteristics, uses and nodulation. Macmillan Publishers LTD, London

Baillie IC (1989) Soil characteristics and classification in relation to the mineral nutrition of tropical wooded ecosystems. In: Proctor J (ed) Mineral Nutrients in tropical forest and savanna ecosystems. Blackwell, Oxford, pp 15-26

Becher H (1989) Gegenwärtiger Kenntnisstand der im Amazonastal siedelnden indianischen Gruppen. In: Hartmann G (ed) Amazonien im Umbruch. Reimer, Berlin

Brouwer LC (1996) Nutrient cycling in pristine and logged tropical rain forest - a study in Guyana. Tropenbos Guyana series 1. Tropenbos Foundation, Utrecht

Carvalho G, Barros AC, Mountinho P, Nepstad D (2001) Sensitive development could protect Amazonia instead of destroying it. Nature 409:131

Coomes OT, Burt GJ (1997) Indigenous market-oriented agroforestry: dissecting local diversity in western Amazonia. Agrofor Syst 37:27-44

Denevan W (1976) The native population of the Americas in 1492. University of Wisconsin Press, Madison

Etchart G (1997) Sustainable resource management in the Brazilian Amazon: the case of the community of Tiningu. Coastal Manage 25:205-226

Fearnside PM (1997) Human carrying capacity estimation in Brazilian Amazonia as a basis for sustainable development. Environ Conserv 24:271-282

Furch K (1997) Chemistry of várzea and igapó soils and nutrient inventory of their floodplain forests. In: Junk WJ (ed) The Central Amazon floodplain - ecology of a pulsing system. Springer, Berlin Heidelberg New York, pp 47-68

Furch K (1999) Zur Biogeochemie eines charakteristischen Überschwemmungsgebietes Zentralamazoniens, der Várzea auf der Ilha de Marchantaria nahe Manaus, Brasilien. Habilitationsschrift, University of Hamburg

Furch K, Junk WJ (1993) Seasonal nutrient dynamics in an Amazonian floodplain lake. Arch Hydrobiol 128(3):277-285

Gottsberger G (1978) Seed dispersal by fish in the inundation regions of Humaitá, Amazonia. Biotropica 10:170-183

James EK, Fatima Loureiro de M, Pott A, Pott VJ, Martins CM, Franco AA, Sprent JI (2001) Flooding-tolerant legume symbioses from the Brazilian Pantanal. New Phytol 150:723-738

Johnston M (1998) Tree population studies in low-diversity forests, Guyana. II. Assessments on the distribution and abundance of non-timber forest products. Biodivers Conserv 7:73-86

Junk WJ, Ohly JJ, Piedade MTF, Soares MGM (2000) The central Amazon floodplain: actual use and options for a sustainable management. Backhuys, Leiden, The Netherlands

Kern J (1995) Die Bedeutung der N_2-fixierung und der Denitrifikation für den Stickstoffhaushalt des amazonischen Überschwemmungssees Lago Camaleão. PhD Thesis, University of Hamburg

Kern J, Darwich A (1997) Nitrogen turnover in the várzea. In: Junk WJ (ed) The central Amazon floodplain - ecology of a pulsing system. Springer, Berlin Heidelberg New York, pp 119-135

Kern J, Darwich A, Furch K, Junk WJ (1996) Seasonal denitrification in flooded and exposed sediments from the Amazon floodplain at Lago Camaleão. Microb Ecol 32:47-57

Kilian S, Berswordt-Wallrabe von P, Steele H, Werner D (2001) Cultivar-specific dinitrogen fixation in Vicia faba studied with the nitrogen-15 natural abundance method. Biol Fertil Soils 33(5):358-364

Klinge H, Adis J, Worbes M (1995) The vegetation of a seasonal várzea forest in the lower Solimões river, Brazilian Amazonia. Acta Amazon 25(3/4):201-220

Kohlhepp G (1989) Verkehrs- Siedlungs- und Wirtschaftsentwicklung und Stand der regionalen Entwicklungsplanung im brasilianischen Amazonien. In: Hartmann G (ed) Amazonien im Umbruch. Reimer, Berlin

Kolk A (1998) From conflict to cooperation: international policies to protect the Brazilian Amazon. World Dev 26(8):1481-1493

Kreibich H (2002) N_2 fixation and denitrification in a floodplain forest in Central Amazonia, Brazil. Forschungsbericht Agrartechnik – VDI-MEG no 398

Kreibich H, Kern J (2003) Nitrogen fixation and denitrification in a floodplain forest near Manaus, Brasil. Hydrol Proc 17:1431-1441

Kreibich H, Lehmann J, Scheufele G, Kern J (2003) Nitrogen availability and leaching during the terrestrial phase in a várzea forest of the Central Amazon floodplain. Bio Fertil Soils 39:62-64

Ladha JK, Peoples MB, Garrity DP, Capuno VT, Dart PJ (1993) Estimating dinitrogen fixation of hedgerow vegetation using the nitrogen-15 natural abundance method. Soil Sci Soc Am J 57:732-737

Laurance WF, Cochrane MA, Bergen S, Fearnside PM, Delamonica P, Barber C, D'Angelo S, Fernandes T (2001) The future of the Brazilian Amazon. Science 291:438-439

Martinelli LA, Victoria RL, Trivelin PCO, Devol AH, Richey JE (1992) ^{15}N natural abundance in plants of the Amazon river floodplain and potential atmospheric N_2 fixation. Oecologia 90:591-596

Moawad H, Badr El-Din SMS, Abdel-Aziz RA (1998) Improvement of biological nitrogen fixation in Egyptian winter legumes through better management of *Rhizobium*. Plant Soil 204:95-106

Oliveira WS, Meinhardt LW, Sessitsch A, Tsai SM (1998) Analysis of *Phaseolus-Rhizobium* interactions in a subsistence farming system. Plant Soil 204:107-115

Peoples MB, Herridge DF, Ladha JK (1995) Biological nitrogen fixation: an efficient source of nitrogen for sustainable agricultural production? Plant Soil 174:3-28

Prous A, Fogaça E (1999) Archaeology of the Pleistocene-Holocene boundary in Brazil. Quat Int 53/54:21-41

Salati E, Sylvester-Bradley R, Victoria RL (1982) Regional gains and losses of nitrogen in the Amazon basin. Plant Soil 67:367-376

Sioli H (1989) Indianer und Europäer - Gedanken zum Aufeinandertreffen. In: Hartmann G (ed) Amazonien im Umbruch. Reimer, Berlin

Sprent JI, Parsons R (2000) Nitrogen fixation in legume and non-legume trees. Field Crops Res 65:183-196

Thielen-Klinge A (1997) Rolle der biologischen N_2-Fixierung von Baumleguminosen im östlichen Amazonasgebiet, Brasilien - Anwendung der ^{15}N natural abundance Methode. PhD Thesis, University of Göttingen

Worbes M (1986) Lebensbedingungen und Holzwachstum in zentralamazonischen Überschwemmungswäldern. Scripta Geobotanica, Bd 17. Goltze, Göttingen, pp 1-112

Worbes M (1997) The forest ecosystem of the floodplains. In: Junk WJ (ed) The Central Amazon floodplain - ecology of a pulsing system. Springer, Berlin Heidelberg New York, pp 223-265

9 Migration of Knowledge Leads to Floristic Development in Myanmar

Aye Kyi

Department of Botany, University of Yangon, Union of Myanmar. E-mail:
botyu@dhelm.edu.gov.mm

9.1 Abstract

Those at the earth summit in Rio de Janeiro warned of the human impact and the
effect of population growth on the natural environment of our planet. As the survival of living things greatly depends on the environment, it is vitally important,
even essential, to improve and sustain the ecosystems on a global level. Globalization brings knowledge and technology to every country, and Myanmar is one of
them. Myanmar is situated in Southeast Asia with a total land area of 676,577
km^2, with a long international border of 5,858 km shared with Bangladesh, India,
China, Laos, and Thailand. Myanmar also has a coastline of 2,832 km and three
major river systems, Ayeyarwady, Sittaung ,and Thanlwin, running from north to
south, which facilitate trade, transport and communications. These unique geo-
strategic positions pose both an asset to and a liability for Myanmar. The topography and climatic conditions favor diversity of flora and fauna in Myanmar. It is
endowed with a rich variation in forest ecosystems that stretch from the tropics in
the south, with an evergreen forest habitat of Malayan flora, to the snow-capped
Khakarborazi Mountain in the north, with Himalayan flora. Flow -- or migration --
of environmental knowledge from developed countries has helped Myanmar to
manage and conserve the natural resources of the country. As universities are the
centers of human resources with academically qualified expertise in various disciplines, research activities have been undertaken in them according to the country's
need. Floristic study is one area of research investigation that reflects the country's
needs and its richness. The ongoing research in Yangon University is conveyed in
this presentation by giving examples of orchids from the northern part of the country; the baseline study of flora in the Monywa District; and the socio-economic
status of mangroves in the Ayeyarwady delta. The aim and objective of this presentation are to give information about the ongoing research in Yangon University,
Myanmar, to international scientists who participate in this symposium, and also
to get suggestions and advice for our further research development. In addition, I
would like to invite research cooperation and collaboration with the Department of
Botany at Yangon University in Myanmar.

9.2 Introduction

Understanding global environmental issues, Myanmar also has to face the challenges of the 21st century by using the advanced knowledge of biodiversity conservation, biotechnology, and computer sciences, as well as information and communication technologies. Since Myanmar's aim is to build a Modern Developed Nation through Education, it is important to educate people of all ages, thus to strengthen our human resources for the future benefits of the country. In doing so, universities as the centers of teaching, learning, and research play a central role. As Yangon University is the oldest and most experienced among the universities in Myanmar, she takes the initiative, creating research activities in various disciplines. The Department of Botany is one such discipline conducting these activities. Three different aspects of ongoing research projects undertaken in the department are presented in this paper.

9.3 Location, Physical Features, and Climate

The union of Myanmar is situated in Southeast Asia, lies between 09°32′N and 28°31′N latitude and 92°10′E and 101°11′E longitude. It is bordered by the People's Republic of Bangladesh and the Republic of India in the west, the People's Republic of China in the north and northeast, the People's Democratic Republic of Laos and the Kingdom of Thailand in the east, and the Andaman Sea and Bay of Bengal in the south (Anonymous 2002). The surface area is approximately 676,578 skm[2]. It stretches for 936 km from east to west and 2,051 km from north to south. The length of the coast line from the mouth of the Naaf River to Kawthaung is approximately 2,832 km. The three major river systems running from north to south are Ayeyarwady, Sittaung, and Thanlwin which facilitate trade, transport and communication. In addition, the large tributary Chindwin joins the Ayeyarwady just above Bagan.

The topography of Myanmar can be divided into three regions, namely, the Western Hills Region, the Central Valley Region, and the Eastern Hills Region. The Himalayan Range links with Myanmar as the Western Yoma Range that runs to the south and serves as a wall separating Myanmar from India. The Khakarborazi is 19,296 feet high, situated at the northernmost part of the Western Yoma Range and is the only snow-capped mountain among southern Asian countries. The Central Valley Region consists of the broad valley of the Ayeyarwady which consists of three parts: the first from the origin of the river to Mandalay, the second from Mandalay to Pyay, and the third part from Pyay to the mouth of the river.

In addition, the Chindwin valley and Sittaung valley are also included in this region. Small mountain ranges lie in the center and the Bago Yoma (range) slopes down from north to south along the river. The Eastern Hills Region is the Shan Plateau which is 3,000 to 4,000 feet above sea level. Unlike the plains, the plateau has high mountain ranges and the river Thanlwin flows through Shan Plateau to the northern part of the Taninthayi Coastal Strip. The four small rivers which have their sources at the Shan Plateau flow into the river Ayeyarwady.

The climate is roughly divided into three seasons, namely, summer, rainy season and cold season. Summer in Myanmar is considered as lasting from March to mid-May, the rain falls from mid-May to the end of October, and the cold season begins in November and ends in February. Generally, the country has a tropical monsoon climate. However, the climatic conditions differ widely from place to place due to the different topographical features. For instance, central Myanmar has an annual rainfall of less than 40 inches, while the Rakhine coast gets about 200 inches. The average highest temperature in central Myanmar during the summer months is above 110°F (43.3°C) while in northern Myanmar it is about 97°F (36.1°C) and on the Shan Plateau it is between 85 and 95°F (29.4 and 35°C) . Temperatures in towns vary according to their location and elevation.

9.4 Vegetation Resources: How to Sustain Them?

Myanmar, due to its unique geographical position, is endowed with a rich biodiversity, ranging from high mountain Himalayan flora in the north to the lowland wet tropical forest, with Malayan flora in the south, and dry, mixed deciduous, bamboo, evergreen, and pine forests in the plains and valleys in between. Many of the plant species in these habitats are well known and utilized by local populations and foresters, yet there remain many species to discover, especially in mountainous and rarely explored or unexplored regions of the country.

Botanical exploration of Myanmar began in the 1800s when the country was under the rule of the British colonial system. Several partial plant lists of the country were produced, including "The Forest Flora of British Burma" by Kurz. During the last half of the 19th century, very few plant collections were made in this area. Of all tropical Asia, Myanmar has had the smallest proportion of its flora collected. The only list of plants specifically recorded for her was first compiled in 1912 by J. H. Lace and published as the *List of Trees, Shrubs, Herbs and Principal Climbers, etc.*, and included 2,483 species. The second edition of the book was published in 1922 by A. Roger. In 1961, H. G. Hundley (a silviculturist in Burma) and U Chit Ko Ko (Curator of the Herbarium in Rangoon) updated the earlier work with more complete treatment of the grasses, orchids, and herbs. The last edition was published in 1987 and included about 7,000 species (Kress et al. 2003).

Since no modern synoptic inventory of the plants of Myanmar exists, the Smithsonian Institution in Washington, D.C., and the Forest Research Institute of Myanmar in collaboration, revised and updated the previous checklists by using modern taxonomic resources and information from plant family specialists. Then, a book called *A Checklist of the Trees, Shrubs, Herbs and Climbers of Myanmar* was published in 2003. The current revised list, containing 273 families, 2,371 genera, and over 11,800 species, will serve as a foundation for a major new *Flora of Myanmar* (Kress et al. 2003).

The vegetation types of the country vary depending on their location. They are roughly classified as follows (Anonymous 1994):

1. Mangrove forest
2. Sand-dune forest
3. Muddy estuarine forest
4. Evergreen forest
5. Semi-deciduous forest

6. Dry forest
7. Indaing forest
8. Subalpine forest
9. Mountain forest
10.Temperate forest

Environmental degradation in Myanmar is still minimal and the major issues facing us at present are :
- Land erosion
- Loss of biodiversity
- Waste disposal
- Water pollution
- Depletion of coastal resources

The main problem in addressing these issues lies, as in other developing countries, in the underdevelopment of the country and the poverty of the people . The people of Myanmar fully understand and appreciate the value of nature and have tried to avoid extremes that might harm nature and the environment. They prefer to live in harmony with the environment, and the protection of nature is ingrained in the hearts of the people. Nevertheless, the growing population, and the increased urbanization and industrialization have exerted increasing pressure on the natural resources of the country. So, the Forest Department set up the following national parks, sanctuaries, and natural parks to protect forests and conserve the environment (Sein Mg Wint 1997).

1. Pidaung Wildlife Sanctuary
2. Khakarborazi National Park
3. Indawgyi Wildlife Sanctuary
4. Kahilu Wildlife Sanctuary
5. Natmataung National Park
6. Kaylatha Wildlife Sanctuary
7. Kyaikhtiyo Wildlife Sanctuary
8. Rakhine Yoma Elephant Range
9. Taunggyi Wildlife Sanctuary
10. Inlay Lake Wildlife Sanctuary
11. Wasa Wildlife Sanctuary
12. Loimawe Wildlife Sanctuary
13. Shwe Udaung Wildlife Sanctuary
14. Pyin Oo Lwin Wildlife Sanctuary
15. Popa Mountain National Park
16. Lawkanandar Wildlife Sanctuary
17. Mularit Wildlife Sanctuary
18. Wethtikan Wildlife Sanctuary

19. Shwesettaw Wildlife Sanctuary
20. Chatthin Wildlife Sanctuary
21. Alaungdaw Kathapa National Park
22. Minwuntaung Wildlife Sanctuary
23. Htamanthi Wildlife Sanctuary
24. Hlawga Wildlife Park
25. Yangon Zoological Garden
26. Moeyungyi Wildlife Sanctuary
27. Seinyay Forest Camp
28. Myainghaywun Elephant Camp
29. Thameehla Kyun Wildlife Sanctuary
30. Meinmahla Kyun Wildlife Sanctuary
31. Moscos Islands Wildlife Sanctuary
32. Lampi Marine Park
33. Hponkan Razi Wildlife Sanctuary
34. Hukaung Valley Wildlife Sanctuary

The renewable plant resources, collectively known as flora in different ecosystems, are very important for human beings and all living creatures. They provide food, medicines, shelter, and clothing. In addition, they also help regulate weather conditions in the environment. Therefore, the diversity of flora such as trees, shrubs, herbs, climbers, orchids, bamboos, rattans (canes), grasses and, mangroves should be emphasized, and data on the many species systematically collected in order to develop and sustain these natural resources of the environment before they disappear from the ecosystem.

9.5 Orchids from Northern Myanmar

Orchids are famous all over the world for their beauty, unique floral structures, easiness to grow in pots for carrying around, and for lasting for months to keep as an indoor plants and for floral arrangements. Most of them are found in tropical and subtropical regions, 75% of them found as epiphytes, growing on trees. Since 50% of the rain forests that covered parts of the world have been destroyed by human activities, the orchid population is at risk of extinction due to habitat destruction. Thus, orchidologists from the Myanmar Floriculturist Association started collecting orchids in 1997 during the Biological Expeditions to northern Myanmar. They also jointly surveyed the orchids from different States and Divisions with researchers from the Forest Research Department until 2002.

Previous collections were made and research on native orchids of Myanmar carried out by various scientists and explorers as follows (Saw Lwin and Myo Han 2002):

1829: Dr. Wallis exported Myanmar orchids to Europe. He introduced *Vanda coerulea* to England.

1860: A Roman Catholic monk, Reverend Charles Parish came to Myanmar and resided for 20 years. *Paphiopedilum parishii, Dendrobium parishii* and *Vanda parishii* were named after him.

1866 to 1868: Col. William Benson collected orchids from Taninthayi and Rakhine Yoma and exported them to Europe. *Vanda bensonii* was named after him.

1873: Botanist Dr. William Griffith arrived in Myanmar

1895: Capt. Bartle Grant compiled all papers on Myanmar orchids and published a book titled *Orchids of Burma*.

1914 to 1956: F. Kingdom Ward, botanist and naturalist, conducted many floristic surveys in Myanmar and discovered *Paphiopedilum wardii* in 1920, at that time named after him. This orchid was rediscovered by U Kyaw Nyunt and his staff in Naung Mung of Kachin State during their orchid expedition. Now it is called Black Orchid and is found to be endemic to Myanmar.

1953 and 1956: Silviculturist ranger, U Tha Hla and U Chit Ko Ko, foresters from the Myanmar Forest Department Herbarium, participated with Kingdom Ward in trips to the Triangle area.

1956 to 1997: No foreign orchidologists were allowed to enter the forests of Myanmar.

In 1973 a pact, the Conservation on International Trade in Endangered Species (CITES) of Wild Fauna and Flora, was signed with the basic goal of controlling international trade of endangered and threatened species of wild fauna and flora. The Union of Myanmar also signed CITES in 1997. According to CITES Appendix I, trade in wild specimens is prohibited., though trade is allowed, subject to licensing, in artificially propagated plants. Lists of endangered and threatened species are given in the Appendix I. Among them , two genera and seven species of Myanmar native orchids are included. They are as follows :

1. *Cattleya trianae*
2. *Laelia jonheana*
3. *Laelia lobata*
4. *Dendrobium cruentum*
5. *Paphiopedilum* genus
6. *Phragmipedium* genus
7. *Peristeria elata*
8. *Renanthera imschootiana*
9. *Vanda coerulea*

As conservation of biodiversity is essential for the sustainable use of resources, the conservation of native orchids has been conducted using in situ and ex situ methods. In situ conservation is conducted in national parks, wildlife sanctuaries, elephant ranges, mountain parks, and protected areas. Ex situ conservation is performed in the Pyin Oo Lwin National Kandawgyi Garden, government and orchid nurseries, commercial and armature collections, and orchid culture laboratories.

9.6 The Baseline Study of Flora in Monywa District

The baseline study of flora was conducted around the S&K Mine project (named from the twin Sabetaung and Kyisintaung deposit) located in Monywa district of Sagaing Division. The mine site is situated on the west bank of the Chindwin River and south of Yama stream. It is 110 km west of Mandalay. This project was undertaken according to the Memorandum of Understanding (MOU) signed between Yangon University and Myanmar Ivanhoe Copper Company Limited (MICCL), who owned the S&K project. The aim of the study has been to assess the potential impact of the acid mist from the heap leach on the vegetation of the hills, and that of the effluence from the mine on the vegetation along Yama stream.

The study areas were from the following habitats:
1. Taungkhamauk T_1, T_2, T_3
2. Kyadwintaung T_4, T_5, T_6
3. Yama upstream   SW-110, SW-220
4. Yama downstream   SW- 111
5. Marsh grass   SK-008, SK-0011
6. Revegetation   A, H, J, O

The first and second collections were made from (19-5-01) to (29-5-01) and (26-11-01) to (3-12-01), respectively. Six permanent monitoring transects, T_1 to T_6 have been established on the two hills. Collections were made by using line transect, quadrat method (Santra et al. 1993). The number of species collected from each quadrat was listed and recorded. Further identifications and classifications were made in the Department of Botany in Yangon University by using available, standard flora texts (Hooker 1897; Lawrence 1968; Aye Kyi et al. 2002). Specimens were also pressed for the preparation of herbarium sheets. Photographs of collected species were taken by using a digital or other available camera. Monitoring of the revegetation area was conducted by using a line intercept method and the heights of plants were also recorded.

In this study, 83 species belonging to 39 families were collected during the first collection and 118 species belonging to 48 families were collected the second time. The composition of flora found in different habitats and their habit status are seen in Table 1. The plant community present in different habitats varies in number and habit status according to the time of collection. Variation in total floral composition according to habitat was compared. It was found that the total number of plants was higher in the four habitats during the first collection, but only a small difference was observed in the Yama upstream (SW-110). This showed that many plants growing in the study area were seasonal and flourish at the beginning of rainy season. Tables 2 and 3 show the evaluation of species present in Taungkhamauk and Kyadwintaung during the first and second collections, respectively. It was found that the number of species present in the same habitat changed with the time of collection but some plants were common on both hills.

There are seven villages in the surrounding mine area. Plants collected from those areas were included in the floral composition of the second collection. Naturally, people living in those areas depend on the surrounding vegetation for food, shelter, firewood, and medicine. The plains surrounding the hills are mostly cultivated for wheat, corn, sugarcane, and various other crops. Most of the plains are covered with *Borassus flabellifer*, which is cultivated to produce a type of sugar known as toddy. Plants like *Hesperethusa caenulata,* are used as sun cream or makeup, and *Grewia hirsuta* is used as shampoo by Myanmar people. Several bamboo species are used as vegetables or construction material.

Table 1. Composition of flora from study area

Habitat	No Spe-cies	Total no. of plants	Habit Status						
			Tree	Small tree	Shrub	Herb	Clim-ber	Grass	Seed-ling
Taungkha-mauk	21	393	20	1	14	305	1	22	16
Kyadwint-aung	23	385	44	18	38	256	14	-	-
Yama up-stream (SW-110)	22	73	4	3	24	37	5	-	-
Yama up-stream (SW-220)	23	148	1	9	12	19	42	35	30
Yama down stream (SW-111)	4	47	4	2	-	-	1	-	40
Total	93	1046	73	33	88	617	63	57	86
Taungkha mauk	25	320	18	1	26	148	26	84	16
Kyadwin-taung	21	214	48	12	41	89	19	-	-
Yama up-stream (SW-110)	18	78	4	3	41	9	6	15	-

Yama up-stream (SW-220)	17	123	1	9	11	29	7	60	6
Yama down-stream (SW-111)	4	7	4	2	-	-	1	-	-
Total	85	742	75	27	119	275	59	159	22

Table 2. Evaluation of plant biodiversity in Taungkhamauk

Sr. no.	Family	Botanical name	Code no.	No. of plants 1st collection	No. of plants 2nd collection
1	Acanthaceae	*Barleria prionitis* L.*	A	2	7
2	Capparidaceae	*Boscia variabilis* Coll.&Hemsl.*	B	6	5
		Capparis flavicans Wall.	C	1	1
3	Combretaceae	*Terminalia oliveri* Brandis.*	D	7	7
4	Convolvulaceae	*Ipomoea maxima* (L.).*	E	0	25
5	Euphorbiaceae	*Fluggea leucopyrus* Willd.*	F	3	2
		Phylanthus simplex Retz.	G	1	0
		Phylanthus urinaria L.	H	370	16
6	Gramineae	*Andropogon ternarius* Michx.	I	0	24
		Cymbopogon clandestinus Stapf.	J	22	60
7	Malvaceae	*Abutilon asiatium* Don.	K	0	2
		Abutilon indicum Sweet.*	L	6	9
		Hibiscus spp.	M	7	30
		Sida cordifolia L.*	N	228	105
		Urena speciosa Wall.	O	60	1
8	Meliaceae	*Azadirachta indica* A.Juss.	P	1	1
9	Minispermaceae	*Cissampelos pareira* L.*	Q	1	1
10	Myrsinaceae	Ardisia colorata Roxb.*	R	2	2
11	Papilionaceae	*Atylosia scarabaeoides* Benth.	S	0	1
		Crotalaria senperforens Vent.*	T	0	5
		Dalbergia paniculata Roxb.*	U	7	8
		Desmodium heterocarpon (L.)DC	V	1	2
		Millettia multiflora Coll. & Hemsl.	W	1	1
12	Tiliaceae	*Corchorus acutangulus* Lam.	X	2	2
		Grewia hirsuta (Korth.) Vahl.*	Y	2	1
13	Verbenaceae	*Tectona hamiltoniana* Wall.*	Z	3	1

*Common plant found on both hills

Table 3. Evaluation of plant biodiversity in Kyadwintaung

Sr.	Family	Botanical name	Code no.	No. of plants 1st collection	No. of plants 2nd collection
1	Acanthaceae	*Barleria prionitis* L.*	A	22	22
2	Anacardiaceae	*Lannea coromandelica* (Houtt.) Mert.	B	4	3
3	Asclepiadaceae	*Leptadenia reticulata* Wight.	C	1	0
4	Asparagaceae	*Asparagus racemosus* Willd.	D	1	0
5	Bombacaceae	*Salmalia malabarica* DC.	E	1	0
6	Caesalpiniaceae	*Bauhinia disphylla* Ham. Buch.	F	2	2
7	Capparidaceae	*Boscia variabilis* Coll. & Hemsl.*	G	5	6
		Capparis horrida L.	G_1	0	2
8	Combretaceae	*Terminalia oliveri* Brandis.*	H	31	33
9	Cucurbitaceae	*Coccinia indica* W & A.	I	5	0
10	Commelinaceae	*Commelina obliqua* Ham.	J	0	1
11	Convolvulaceae	*Ipomoea maxima* (L.).*	K	0	13
12	Euphorbiaceae	*Fluggea leucopyrus* Willd.*	L	1	0
13	Gramineae	*Cephalostachyum flavescens* Kz.	M	15	0
14	Labiatae	*Leucas zeylanica* Br.	N	0	2
15	Malvaceae	*Abutilon indicum* Sweet.*	O	5	0
		Sida cordifolia L.*	O_1	240	75
16	Minispermaceae	*Cissampelos pareira* L.*	P	5	3
		Pericampylus glaucus Lam.	P_1	1	0
17	Myrsinaceae	*Ardisia colorata* Roxb.*	Q	5	5
18	Papilionaceae	*Crotalaria senperforens* Vent.	R	0	4
		Dalbergia paniculata Roxb.	R_1	1	1
		Desmodium triquetrum L.	R_2	0	1
19	Rhamnaceae	*Zizyphus* spp.	S	0	1
20	Rutaceae	*Hesperethusa caenulata* Roxb.	T	9	5
		Paramigyna longispina Hk.f.	T_1	9	8
21	Salvadoraceae	*Azima sarmentosa* Benth.	U	2	0
22	Selaginellaceae	*Selaginella* spp.	V	10	14
23	Tiliaceae	*Grewia hirsuta* (Korth.) Vahl.	W	2	2
24	Verbenaceae	*Tectona hamiltoniana* Wall.	X	7	6

*Common plant found on both hills

Since the vegetation of the two hills is located adjacent to the heap leach pads, there is a potential for the vegetation to be impacted by drift of sulphuric acid that is sprayed. The vegetation of the hills has consisted mainly of the low, sparse shrubs that are fairly uniformly distributed. The number of species recorded in this study is higher than the previous report (Muir 1998). However, there are no unusual or rare vegetation types present in the studied habitat. Many of the species recorded are exotic, herbaceous, and abundantly weedy. No endangered species was recorded, but *Terminalia oliveri* and *Tectona hamiltoniana* species have been significantly observed in the hills. In addition, several species collected from this area have significant local medicinal and social value. Although there is currently no evidence to suggest the vegetation is unique for the region, the hills do have some conservation value in that all other vegetation in the immediate vicinity has been removed. Therefore, it is important that the impact of mining on these hills be minimized (MICCL 1999-2000).

Local residents have established a cottage industry of secondary copper extraction from tailings, released by the flotation plant. The waste from copper extraction has a potential impact on the immediate and wider environment of the area. So it is vital to make the public environmentally aware, in particular of the negative impact of copper extraction on the local people. Only then can the biodiversity be conserved and the environment sustained for future generations.

9.7 Socioeconomic Status of Mangroves in the Ayeyarwady Delta

Mangroves in Myanmar can be seen in three regions, namely Rakhine, Ayeyarwady, and Taninthayi along the coast of the Bay of Bengal and Andaman Sea. The Ayeyarwady delta is the largest among the three regions. The delta is a flat deltaic swamp, with riverbanks slightly raised due to higher silt deposit from the initial river overflow. It receives large volumes of fresh water from the Ayeyarwady River flowing from the north and of saline tidewater from the sea on the south. The pH of the soil is 5 to 7.2 and the dominant soil type is silty clay or clay. The area has luxuriant stands of *Heritiera fomes* and *Excoecaria agallocha* (Myint Aung et al. 2003). In addition, it is the most populous among States and Divisions and known as the biggest granary of Myanmar for its highest rice productions.

The delta comprises extensive mengal areas that are of considerable commercial value. Mangrove forests grow along the seacoast, and sandbank forests can also be seen. The mangrove forests serve as natural habitats for commercial fish and prawns and also provide timber, firewood, charcoal, and other forest products. Local communities depend on the products nurtured by the mangroves. In addition, the mangrove ecosystem enhances the production of soil from storm waves, deluge, landslide, and saline infiltration. The vegetation defines the landscape and takes part directly or indirectly in the ecological developments in the ecosystem (Lugo and Snedaker 1974; Hamilton and Snedaker 1984; Tomlinson 1986).

The remote sensing data of forest cover in Ayeyarwady delta in the years 1974, 1990, and 2001 have been compared by Myint Aung et al. (2003). He classified the mangrove communities in his study area into two categories: riverbank communities and inland communities. He mentioned that the distribution of mangrove community depends on soil character, location, and accretion or erosion of the river banks, and he gave relevant examples of communities. The mangrove formations of the Ayeyarwady delta are determined by one or more important factors, such as riverbanks dynamics. Mangrove species like *Hibiscus tiliaceus* and *Phoenix paludosa* can emerge after clear-cutting, grow rapidly, and prevent light from reaching the ground. Thus it is difficult to regenerate other species after their invasion. The study area is identified as secondary forest due to the prevalence of dwarf trees (and small stem diameter) with coppices. The most suitable species for the site is *Avicennia officinalis*, which has the greatest height, largest stand diameter, and most biomass content.

In this investigation, degradation of mangrove vegetation in the delta was quite obvious. This may be due to the population growth, overexploitation of endangered and vulnerable species, and extensive transformation of natural vegetation into agricultural land. Since mangrove resource has a large economic potential, the destruction and denudation of the mangrove ecosystem would cause a great economic and social loss to the country. Therefore, a basic requirement for socioeconomic development is to provide training and environmental education, as well as to encourage the local people in rehabilitation of mangroves. Continuous monitoring of local and national authorities on these activities is also essential for the development of the mangrove ecosystem.

9.8 Conclusions and Suggestions

The overview of research activities mentioned above leads to an awareness that conservation and sustainable utilization of species are vitally important in all floristic studies. As Myanmar still possesses many unexplored, deep primary forest areas, further expeditions and explorations should be undertaken in collaboration with scientists from developed countries in order to discover new species and new plant communities. Over the last 30 years, hundreds of post-graduate research works in floristic studies have been undertaken in the Botany Department of Yangon University. However, only few have been published due to lack of funding. For in situ conservation, community participation is important. Training and education on the effective utilization and management of plant resources should be given to the local people. Meanwhile, ex situ conservation can be conducted in the research laboratories and botanical gardens by scientists and doctoral candidates. For such purposes the following criteria are essential:

- assessment of the current conservation value, assignment of new priorities, identification of important species that require conservation action
- enhancement of the documentation systems and development of database management systems
- development and establishment of greenhouses and botanical gardens to protect rare and endangered species
- development of seed banks and establishment of seed exchange programs
- cooperation and collaboration with other institutions and universities to receive advanced training techniques and education.

Acknowledgements: My deep gratitude is extended to Professor D. Werner for accepting me in his department and giving me an opportunity to present a paper to the International Conference. I also wish to thank the Alexander von Humboldt Foundation for offering the research study visit sponsorship, without which I would not have had the chance to visit Germany.

9.9 References

Aye Kyi, Yee Yee Win, Win Myint, Yin Yin Mya (2002) The base line study of flora around S&K project in Monywa District. The 2nd Conference on Environmental education and research, Council Chamber in Yangon University

Anonymous (1994) A quarterly journal of Myanmar forestry 2

Anonymous (2002) Myanmar: facts and figures 2002. Ministry of Information, Union of Myanmar Printing and Publishing Enterprise

Hamilton RS, Snedaker SC (1984) Handbook of mangrove area management. Commission on Ecology, IUCN, Gland, Switzerland

Hooker JD (1897) Flora of British India, vol I to VII. Reeve, The Oat House, Brook, Nr. Ashford, Kent

Kress WJ, de Filipps RA, Farr E, Daw Yin Yin Kyi (2003) A checklist of the trees, shrubs, herbs and climbers of Myanmar. Department of Systematic Biology - Botany, National Museum of Natural History, Smithsonian Institution, Washington, DC

Lawrence GHM (1968) Taxonomy of vascular plants, 9th edn. Macmillan, New York

Lugo AE, Snedaker SC (1974) The ecology of mangrove. Annu Rev Ecol Syst 5:39-63

MICCL (1999/2000) Environment, health and safety report of Myanmar Ivanhoe Copper Company

Muir BG (1998) Monywa copper project. Supplementary information, part I. Myanmar Ivanhoe Copper Company, Myanmar

Myint Aung, Fujiwara K, Mochida Y (2003) Physiosociological study of mangrove vegetation in Byone-Hmwe Island, Ayeyarwady Delta, Myanmar – Relationship between floristic composition and habitat. Mangrove Science 3, Japan Society for Mangrove

Santra SC, Chatterjee TP, Das AP (1993) College of botany, practical vol 1. New Central Book Agency Ltd. 8/1 Chintamoni Das Lane, Calcutta. 700,009

Saw Lwin, Myo Han (2002) Diversity of Myanmar native orchid species and conservation. The 2nd conference on environmental education and research, Council Chamber in Yangon University, 26th June 2002

Sein Mg Wint (1997) Overview of protective areas system in Myanmar. FREDA, Technical Paper 4. Yangon, Myanmar

Tomlinson PB (1986) The Botany of mangroves. Cambridge University Press, Cambridge

10 Tracking Parasitoids at the Farmland Field Scale Using Microsatellite Markers

Hugh D. Loxdale and Catherine Macdonald

Plant and Invertebrate Ecology Division, Rothamsted Research, Harpenden, Hertfordshire, AL5 2JQ, UK, E-mail hugh.loxdale@bbsrc.ac.uk

10.1 Abstract

Hypervariable microsatellite markers are currently being designed and applied by us to a range of hymenopterous parasitoids from several families attacking insect pests of oilseed rape (*Brassica napus* L.). The aim is to be able to better understand the population genetic structure and dynamics of these parasitoids within the farmland agro-ecosystem for integrated pest management (IPM) programmes involving parasitoid mass release and/or manipulation of extant populations of these insects. As an example of our approach, we here discuss research on the primary hymenopterous parasitoid *Diaeretiella rapae* (M'Intosh) (Braconidae), attacking populations of the cabbage aphid, *Brevicoryne brassicae* (L.), infesting rape in fields in southern England. We will relate how analysis of spatial and temporal allele and genotype data is providing novel information on movement of the parasitoids at a range of geographic scales from single plant (*i.e.* between racemes) to field scale (~ 7 ha). These approaches thus allow us to obtain insights into behaviour, including searching behaviour by individual female parasitoids, largely impossible by other means. In the case of *D. rapae*, the genetic data obtained so far at one locus (*Dr*283) reveal a high level of polymorphism, and are suggestive of random mating and neutrality of marker alleles, whilst aerial movement of the insect appears to be generally local (< 0.5 km), although some genotypes are shared between main sampling sites some 40 km. apart, as found using three additional polymorphic microsatellites (MacDonald et al. 2004).

10.2 Introduction

As we enter the 21st century, damage to agricultural, horticultural and forestry crops by insect pests continues to be a global problem, despite large amounts of insecticides being used against them. A major consequence of this usage is that insecticide resistance has evolved in a considerable number of species (*i.e.* ~500;

Georghiou 1990; Caprio 1996), sometimes cross-resistance, as in aphids, notably the peach-potato aphid, *Myzus persicae* (Sulzer) (Foster et al. 2000). Because of the potential development of resistance in susceptible species or selection for enhanced levels of resistance in already resistant species, it is preferable to reduce the level of pesticides applied, thereby both reducing the chemical selective pressure as well as environmental pollution, with all this means in terms of human and animal health, including that of pollinators such as honeybees.

In order to minimise the impact of pesticides in the environment, much recent work in the UK and abroad has centred on the use of biological control agents (predators, parasites and pathogens) to combat pest insects below economic thresholds, more especially in the agro-ecosystem and usually as part of integrated pest management (IPM) schemes (Dent 1995; Long 2002). Of the various insect agents employed against one or more herbivorous insect pests, hymenopterous parasitoids (*i.e.* parasites that kill their hosts) have been used worldwide, often to great effect. However, the success of such use often depends on knowledge of the ecology and genetics of the parasitoids and insect hosts concerned. (Hopper et al. 1993). This includes determining the genetic associations of parasitoid and host; whether or not the parasitoid is specific to a given host, *i.e.* is monophagous and is hence a 'specialist' or rather, attacks a number of hosts. If the latter is true, it may be termed oligo- or polyphagous depending on the number of these, and is hence more or less 'generalist' in its host preference requirements. Even so-called generalist species tend to preferentially attack hosts within the same family or genus or which feed on the same plant host (*e.g.* crucifers) and in this sense, are themselves specialist to some degree. Yet so, in the case of the object of this study, the primary parasitoid *Diaeretiella rapae* (M'Intosh) (Hymenoptera: Braconidae: Aphidiidae), it may be truly described as a generalist since it attacks aphids in different genera (see below).

In addition to determining host preference, when a natural parasitoid species population sample is collected from the field, it is also important to know what proportion of the local/regional/national population this represents in terms of genetic variance (effectively, a measure of 'genetic richness' of a particular area, broadly analogous to estimating species richness; *e.g.* Scheiner et al. 2000). This is more especially important because inbred populations lacking variance are likely to have less flexibility in terms of host switching and host range (for example, see Baker et al. 2003).

Another important aspect of choosing and successfully utilising parasitoids is knowing how far they travel when released, and the likely persistence of given genetic strains in the field once released (*i.e.* genetic turnover or revolutions). Perhaps of all things that ecologists do, quantifying population dynamics is amongst the most difficult. This is especially so for small insects whose populations are carried on the wind above their flight speed in still air, sometimes to great heights (Chapman et al. 2004) and which are rapidly diluted as a consequence. This of course makes the recapture of individuals marked with, for example, radioactivity or fluorescent dyes, extremely unlikely (Loxdale 2001). Even with insects such as sand flies, *Lutzomyia longipalpis* (Lutz & Neiva) (Diptera: Psychodidae) which tend to move over small spatial scales, direct tagging with fluorescent dyes has revealed that the proportion of marked individuals recaptured declines as a function of distance from the release site, and of course, time after release (Morrison et al. 1993). With small insects that potentially migrate long distances like aphids, *Phorodon humuli* (Schrank) (damson-hop aphid) and whitefly, *Bemisia tabaci* (Gennadius), use of direct tagging is only effective over relatively local geographic scales (Taimr and Kříž 1978; Byrne et al. 1996), let alone when these insects are carried on jet streams over continental landmasses at heights between 300 - 1000 m (Loxdale et al. 1993). With relatively large migratory insects such as the North American Monarch butterfly, *Danaus p. plexippus* L., the chances of recapture, as with small insects, is related to scale and time after release, here a geographic scale representing hundreds or thousands of kilometres (Urquhart and Urquhart 1977).

An alternative approach to directly tagging insects to elucidate population structure and dynamics, including the spatial scale over which they move, is to use an indirect method, namely the use of molecular markers – allozymes and DNA markers, especially microsatellites and mitochondrial markers. In addition, the population genetic turnover, overwintering (where appropriate) and to some extent, flight direction of the species in question can thus be ascertained, although flight height is rarely if ever shown by such methods. Usually, these indirect methods concern changes in population allele and genotype frequency, which ultimately relate to levels of gene flow concomitant upon the movement of the number of individuals sampled; occasionally however, individual movements can be detected – for example, when an insecticide-resistant genotype is detected for the first time in an otherwise and hitherto susceptible population (Foster et al. 1998; see Loxdale and Lushai 1998 for a review of molecular markers and Loxdale 2001 and Loxdale and Lushai 1999, 2001 for reviews concerning their application in studies of insect population movements).

With this kind of information at hand, in IPM scenarios, the targeting and efficacy of the chosen agent can be enhanced, either through mass rearing and release programmes or by manipulating the behaviour of the insects concerned and involving the target pest on the crop and an intercrop 'reservoir' host, the latter inhabiting nearby field edges or other uncultivated habitats (Hopper et al. 1993; Powell and Wright 1988; Powell 1994; Atanassova et al. 1998).

In this chapter, polymorphic microsatellite markers (Jarne and Lagoda 1996; Goldstein and Schlötterer 1999) were used to examine the population structure and dynamics of hymenopteran (wasp) parasitoids at the field scale within the farmland agro-ecosystem. Presently, we are engaged in a number of studies utilizing these markers derived from a range of parasitoids of different species and/or genera attacking a range of insect pests infesting oilseed rape (*Brassica napus* L.) in the UK, including aphids (Macdonald and Loxdale 2002). The advantage of using microsatellites is that they are co-dominant, single-locus Mendelian markers, apparently selectively neutral and randomly scattered throughout the genome, which allow deviations from Hardy-Weinberg (H-W) equilibrium to be calculated and hence, measures of inbreeding and gene flow estimated (*F*-stats; see below). Here, we concentrate on preliminary data gathered for *Diaeretiella rapae* attacking the cabbage aphid, *Brevicoryne brassicae* (L.), a major pest of crucifers in

Britain and elsewhere (Blackman and Eastop 2000). *D. rapae* is the only member of a geographically-widespread genus and is the natural enemy of over 60 different aphid species worldwide (Pike et al. 1999). Whilst four polymorphic microsatellites have been used during the course of the study, detailed analysis is presented for only one of these (*Dr*283) for which a full data set is currently available.

10.3 Methods and Materials

10.3.1 Insects and Collecting Sites

Samples of *D. rapae* were obtained by hand collecting highly *B. brassicae*-infested racemes of oilseed rape plants at different spatial scales in several large fields in southern England during the summers of 1995 and 1999. Included were fields on the Rothamsted estate, Harpenden, and at Hemel Hempstead, both in Hertfordshire, and at Cockayne Hatley and Woburn Experimental Farm (Broad Mead field), Woburn, both in Bedfordshire. Broad Mead field is approximately 0.25×0.25 km in dimensions and 6.6 ha in area. In this chapter, only the samples collected at Broad Mead field, Woburn, some 40 km north-west of Rothamsted (Fig. 1a,b), are described in detail. Individual racemes were sampled from a single plant in the NW corner of the field, from five plants from five subpopulations several metres apart in the NW corner of the field (*i.e.* NW1 – 5; 25 plants total), as well as from four to five individual plants several metres apart in each of the other three remaining corners of the field (*i.e.* NE, SW, SE; 14 plants total; grand total = 39 plants).

Sampling:

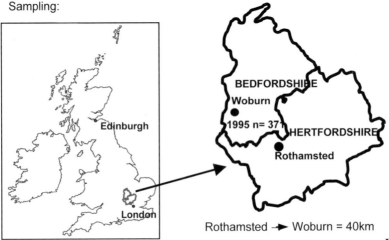

Rothamsted → Woburn = 40km

Fig. 1a

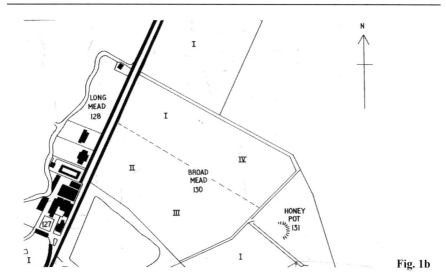

Fig. 1b

Fig. 1. a Map of the southern England (*inset*) with main collecting sites shown: Rotham-sted, Hertfordshire, and Woburn, Bedfordshire; **b** large-scale map of Broad Mead field, Woburn. Rape plants infested with *Brevicoryne brassicae* and parasitized by *Diaeretiella rapae* were collected from the four corners of this field. The number of parasitoid samples (*n*) collected is shown bracketed in parentheses = 'NW' - NW1 (*n*=26), NW2 (*n*=20), NW3 (*n*=46), NW4 (*n*=33), NW5 (*n*=112); total *n*=237 for all NW samples; 'NE' (*n*=47), 'SW' (*n*=37) and 'SE' (*n*=50). For individual racemes taken from a single plant in the NW corner, mean *n*= ~10. Grand total for all plants and field corners *n*=371

Often in the aphid colonies collected, numerous 'mummies' (the pupal stage of the parasitoid within the cuticle of its aphid host) were already apparent. The plant material was placed in polythene bags, labelled in terms of its position on an individual plant and/or the position of the plant in the field and returned to the laboratory in a cool box. Here, transparent plastic specimen tubes (25 mm. external diam. × 90 mm long) were attached and sealed to the open ends of the polythene bags using elastic bands. These tubes were then inserted through holes of the same diameter cut with a cork borer through a large, stout cardboard box such that the bags remained within the interior of the box, whilst the tubes protruded outwards from it. When the lid was closed and the interior thus darkened, emerging adult wasps (~3 mm in length and black-brown in colour) made their way towards the light (Fig. 2). Once a sufficient sample had been collected (10+ insects), the wasps were transferred to new labelled tubes and stored at −80°C until molecular analysis. Several boxes were prepared in this way for the different fields sampled and kept at room temperature (~20°C) near windows lit by natural daylight until a suitable number had emerged. Sample sizes tested electrophoretically per raceme, plant, plants per field corner and site ranged from 4 to 237 (see also Fig. 1 legend).

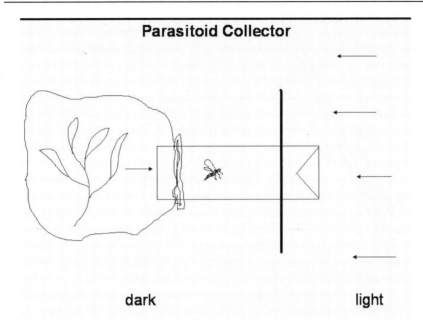

Parasitoid Collector

dark   light

Fig. 2. Simple device comprising a plastic specimen tube, polythene bag and cardboard box for 'extracting' parasitoids from aphid infested plant material

10.3.2 DNA Extraction and Microsatellite Amplification

Individual wasps were homogenized and their DNA extracted and purified by the 'salting out' procedure of Sunnucks and Hales (1996). Thereafter, microsatellites were amplified by the polymerase chain reaction (PCR) in a Hybaid OmniGene thermocycler using four polymorphic primer pairs – $Dr268$ $(TG)_{18}$, $Dr271$ $(TGT)_{15}CGT(TGT)_9$, $Dr280$ $(CT)_{17}$, $Dr283$ $(AAC)_{23}$ as recently detailed by Mac-Donald et al. (2003). The amplification products were separated on sequencing gels and visualized using silver staining, as detailed in Llewellyn (2000) and Llewellyn et al. (2003). For each locus, allele sizes in base pairs were determined by comparison with a sequence of the PGEM 3ZF(+) vector (the control in the Promega silver sequence kit; see Promega 1996). Gels were placed over a light box and 'photographed' using the UVP (Ultraviolet Products) gel documentation system 'Imagestore 7500' (version 7.12).

10.3.3 Population Analysis

Allele and genotype frequencies and deviations of genotype number from H-W expectations and F_{ST} (inbreeding coefficients) were calculated using the program GENEPOP version 3.1b and v3.1c (Raymond and Rousset 1995a; see also http://wbiomed.curtin.edu.au/genepop/). Since the wasps are chromosomally haplo-diploid (Godfray 1994), only data from females was analyzed in the study so that tests such as testing for departures from H-W expectations could be performed. In the case of multilocus genotypes, these were classified as unique (*i.e.* appearing only once in the entire collection) or common (*i.e.* appearing more than once in the entire collection). Departures from H-W equilibrium among complete data sets and in individual samples were tested using the 'exact Hardy-Weinberg test' of Guo and Thompson (1992) using a Markov chain method, the null hypothesis being that populations are in H-W equilibrium (= random union of gametes, in the absence of significant selection, mutation and migration). This is the most powerful test for small sample sizes and large numbers of alleles (Raymond and Rousset 1995b). To test population differentiation based on comparison of allelic and genotypic distributions among samples (pairwise comparisons of genic and genotypic differentiation), an exact nonparametric procedure was employed. Using a Markov chain method, an unbiased estimate of the exact probability test (Fisher exact test) for a contingency table of allelic frequencies was made with the null hypothesis that allelic distribution is identical across populations. This test is accurate, even for very small sample sizes and low-frequency alleles (Raymond and Rousset 1995b). To test the distribution of genotypes across populations, an unbiased estimate of the *p*-value of a log-likelihood-based exact test (G test; Goudet et al. 1996) was used, the null hypothesis in this case being that genotypic distribution is identical across all populations tested. In general, the genotypic test is slightly more powerful than the allelic test, whilst under conditions of unequal sampling, both tests are more powerful than estimators of F_{ST} (Goudet et al. 1996).

F-statistics provides the means to partition departures from random mating into components due to non-random mating within populations and population subdivisions (Wright 1951; Slatkin 1987; Majerus et al. 1996; Neigel 1997). F_{IS} describes the reduction in heterozygosity of an individual relative to the subpopulation in which it exists; F_{ST} reflects the reduction in heterozygosity due to population subdivision; whilst F_{IT} reflects the reduction in heterozygosity in the total population (Majerus et al. 1996). F_{ST} values range from zero to 1, and classically can also be interpreted as the standardized variance (σ^2) in the frequency of an allele (p) among populations (Workman and Niswander 1970):

$$F_{ST} \sim \sigma^2 / p(1-p),$$

and based on the island-model assumption of F_{ST} (Wright 1951). Hence, highly differentiated populations have a large variance in the frequency of a particular allele, and will have a higher F_{ST} values than homogeneous populations. F_{ST} can also be calculated, as here, as pairwise and whole population values by a weighted analysis of variance (Weir and Cockerham 1984). The estimator of F_{ST}, $\hat{\theta}_{WC}$ is defined as

$$\hat{\theta}_{WC} = \frac{\sum_{u-1}^{k} a_u}{\sum_{u=1}^{k} (a_u + b_u + c_u)}$$

where k is the number of alleles at the locus, and a_u, b_u and c_u are the among samples, among individuals within samples and within individual estimates of components of variance, respectively, of a nested analysis of variance on allele frequencies. Accordingly, this estimator of F_{ST} gives the least biased estimator for different sampling strategies (Goudet et al. 1996).

R_{ST} is an analogous measure based on a stepwise mutation model, which may be more applicable to microsatellites (Slatkin 1995), although this measure has a much higher coefficient of variability than F_{ST}. Under the conditions of most investigations, including small sample sizes and relatively few microsatellite loci examined, F_{ST}-based estimates are better than R_{ST} (see Gaggiotti et al. 1999).

Lastly, some of the multilocus data were analyzed by a maximum-likelihood method and Bayesian statistics as detailed by Dawson and Belkhir (2001), an approach that 'partitions the set of sampled individuals, induced by the assignment of individuals to source populations' and which does not make any assumptions about allelic composition or the number of subpopulations represented in the sample, except for first arbitrarily establishing the maximum number of such subpopulations (termed the 'prior'). These results are being published elsewhere (MacDonald et al. 2004).

10.4 Results

Of the approximately 370 *D. rapae* samples collected at Broad Mead field and tested for polymorphism at microsatellite locus *Dr283*, a total of 23 alleles were found and are listed below in terms of their allele size in DNA base pairs, *viz.*

224, 233, 236, 239, 242, 245, 248, 251, 254, 257, 260, 263, 266, 269, 272, 275, 278, 284, 287, 290, 293, 296, 299

From this number of alleles and using the formula: -

$$G = N(N + 1)/2$$

where G is the number of genotypes and N the number of alleles in the population sampled (Loxdale and Brookes 1987), 276 genotypes are expected. However, only 66 genotypes were actually observed at the study site, equivalent to ~24%. On plotting the number of alleles and genotypes graphically as a function of sample size (which includes samples collected from racemes on a single plant, 5×5 plants in the corner of a field (NW1-5), 4–5 plants each in the remaining corners, and the total sample (plants per four corners, NW, NE, SW, SE=39 plants total), a curvilinear relationship is seen in both cases (Fig. 3a,b). From these graphs, a sample size of about 50 individuals chosen at random represents approximately half the variance present in the total field sample. Furthermore, both graphs appear to be reaching an asymptote at the total sample size; hence collection within the field, or probably locally of a much larger sample, is unlikely to recruit many more variants, either alleles or genotypes. In this sense, the graphs are reminiscent of species-richness curves, where the number of species is plotted against sample area (which in effect is equivalent to sample size). The effect of sample size can be further demonstrated when genotype number is plotted *vs.* sample size per the type of sample, *i.e.* whether from racemes, whole individual plants (NW corner), whole plants in the remaining field corners (NE, SW or NE), or the total collection (Fig. 3c). This latter graph clearly shows that even individual plants in one corner (NW) can have a significant proportion of the total variance found within the field.

At the smallest spatial scale sampled from racemes of a single rape plant, the patterns observed in terms of genotype frequency, and here presented graphically in terms of pie charts, is one of extreme heterogeneity; however, it must be emphasized that each sub-sample only represents around ten individuals tested per raceme (Fig. 4a). These data apparently show that individual female wasps, which are thought to only mate once in solitary species such as *D. rapae* ('monoandry'; Godfray 1994) and hence give rise to female offspring of two genotypes, 'work' over the same colony of aphids. They inject an egg (of a total of 300-400) into several individual hosts in the same vicinity, some genotypes being specific to a few racemes, often adjacent ones (Fig. 4a). This being so, the data reflect the searching behaviour of individual wasps, and as such show the level of 'fine-grained' resolution of the hypervariable markers used in terms of elucidating molecular ecological parameters.

In the case of the genotype data for the field corners, Fig. 4b represents the genotype heterogeneity observed, again as pie charts, in five plants a few metres apart in the same corner of Broad Mead, in this case, the NE corner. It is representative of the other three corners. Here, as with the racemes from a single plant, much parasitoid genotype heterogeneity is observed and clearly, different females are working over different plants, so that again, this scale represents the searching behaviour of individual female wasps. Barring sampling effects due to the fairly small population sizes sampled, these data suggest that individual females concentrate on particular colonies of aphids and tend to forage fairly locally rather than infect aphids over a wide area. Having said that, as the figure shows, whilst some *D. rapae* genotypes are unique to some individual plants, others are common to a number of plants in the immediate vicinity (*e.g. Dr*283 homozygote '13:13' common to *D. rapae* samples from three of five rape plants tested; Fig. 4b, pie charts labelled a, b and e), occasionally to all of them, as found at the other field corners (*e.g.* SE corner, 13:13 homozygote). The genotypic pattern obtained for sub-samples (5 plants) per single field corner (NW1-5, *i.e.* 25 plants total) is essentially similar in showing significant heterogeneity at a very localized scale (data not shown).

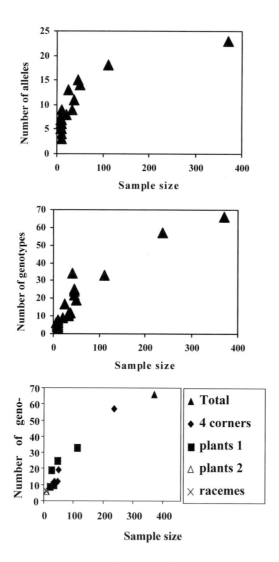

Fig. 3. Graphs plotting microsatellite *Dr*283 allele/genotype number against *D. rapae* sample size; **a** allele; **b** genotypes; **c** genotypes *vs.* sample size shown in relation to plant material. Number of *D. rapae* genotypes from: 12 racemes from an individual plant in the NW corner (×, mean number); individual plants in NW corner (NW1-5; =5×5 plants, =25 in total; ■, labelled 'Plants 1'); plants in the other corners (14 plants in total; NE, SW, SE; Δ, labelled 'Plants 2'; means); plants per all four corners shown separately (♦); and grand total from all plants (*n*=39) sampled (▲)

When the genotypes for all four corners are presented graphically as pie charts (Fig. 4c), again the pattern seen is one of extremely genetic heterogeneity, with a large number of genotypes present, although clearly, some genotypes occur at fairly high frequency and are found at all four corners of the field, *e.g.* the *Dr*283 homozygote '13:13' and the heterozygote '13:14'. The allele frequencies (Fig. 4d) also show much heterogeneity, although whilst a few alleles are dominant, more especially alleles 13, 14 and 15, other alleles are missing from some sites, *e.g.* allele 9 from the south-east corner. Hence, and within the limits of the sample sizes tested, *i.e.* barring sampling effects, the distribution of population alleles in the particular field sampled is not uniform at this spatial scale. In other words, there is some evidence of substructuring, even from the raw allele and genotype data.

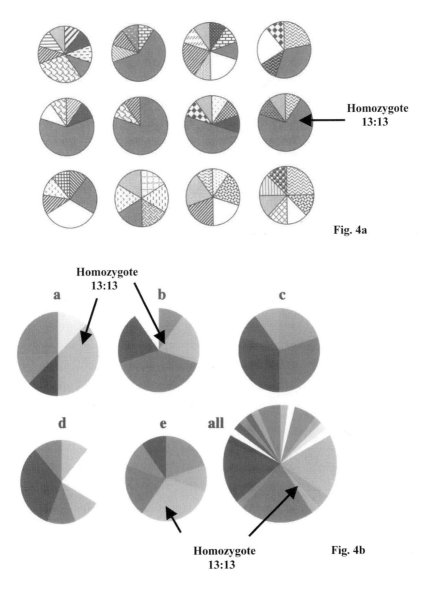

Fig. 4a

Fig. 4b

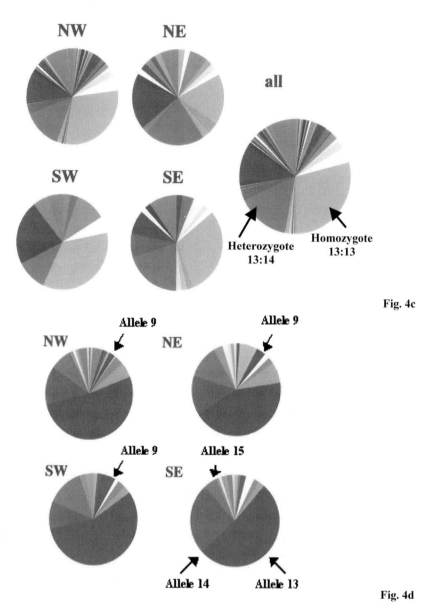

Fig. 4. Pie diagrams showing genotype/allele frequencies as a function of plant material and geographical location in Broad Mead field. **a** Genotype frequencies from 12 individual racemes from a single plant in the NW corner (figure reproduced from Osborne et al. 2002); **b** genotype frequencies from individual plants (*n*=5; labelled *i-v* and *all*, *i.e.* the combined *D. rapae* frequencies from all five plants sampled) in the NE corner of the field; **c** genotype frequencies from plants (*n*=39) in each of the four field corners; **d** ditto allele frequencies

In Fig. 5, the observed genotype frequencies obtained at the *Dr283*locus are plotted *vs.* expected values as if they conformed to H-W expectations (diagonal line). It appears that despite the scatter of points, there is a reasonable correlation between the parameters. The result suggests that mating is thus largely random and in turn, the *Dr*283 microsatellite alleles borne in haploid condition by the male wasps are selectively neutral at this locus.

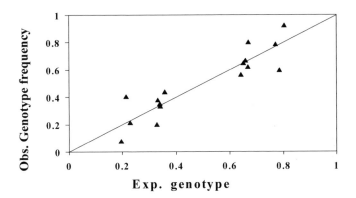

Fig. 5. Observed frequency *vs.* expected frequency of microsatellite *Dr*283 genotypes according to Hardy-Weinberg expectations. *Diagonal line* represents frequencies in equilibrium

Upon analyzing the data using *F*-statistics, F_{ST} values for all sample pairwise correlations were seen to be small, *i.e.* <0.03, showing that the substructuring was minimal (Table 1a). In Table 1b, the data are presented as means for the three main *F*-statistics. All values are again small, <0.09, with F_{ST} being the smallest values, the four corner comparisons being smaller (0.005) than the NW1-5 corner comparisons (0.018).

When tested directly for deviations from H-W expectations, of the eight populations (NW1-5, NE, SW, SE), three differed significantly from H-W expectations *viz.* NE, NW1 and NW3 at the 5% confidence level ($p<0.05$). There was seen to be a heterozygote deficiency (*i.e.*, homozygote excess) in the sample NW5 ($p=0.024$) and overall population samples collectively ($p=0.05$).

When the allele and genotype data were tested statistically for levels of genic and genotypic differentiation, for both parameters, seven of 28 pairwise values differed significantly at the 5% level; of these, five concerned samples from the main corners and two from the NW comparisons. For all eight population samples, the level of significance was found to be $p=0.0012$; hence a significant difference ($p<0.05$) was found to exist in allelic structuring.

Table 1a. F_{ST} estimated as in Weir and Cockerham (1984)

Code for pop names:

1	SE	5	NW1
2	NE	6	NW2
3	NW5	7	SW
4	NW4	8	NW3

Estimates for each locus: Dr283:

pop	1	2	3	4	5	6	7
2	0.0115						
3	0.0100	0.0158					
4	0.0113	0.0023	0.0373				
5	0.0014	0.0082	-0.0069	0.0242			
6	0.0125	-0.0012	-0.0044	0.0254	-0.0003		
7	0.0162	0.0056	-0.0012	0.0325	0.0002	-0.0135	
8	0.0202	-0.0016	0.0192	0.0091	0.0076	0.0035	0.0140

Table 1b. *F*-statistics (means) estimated as in Weir and Cockerham (1984)

Microsatellite locus - Dr283:		
All 8 pops	4 pops (NW, NE, SW, SE)	5 pops NW1-5
$F(\text{is}) = 0.053$	0.059	0.067
$F(\text{st}) = 0.010$	0.005	0.018
$F(\text{it}) = 0.063$	0.063	0.085

10.5 Discussion and Conclusions

Whilst the results presented here are only preliminary and concern one microsatellite locus (*Dr*283), the data and their analyses nevertheless reveal some interesting aspects. To begin with, the locus is seen to be highly polymorphic, with 23 alleles and 66 genotypes identified from a sample size of ~370 individuals collected from a single 6.6-ha field in the UK. Similar results were found for the other three loci tested (data not shown). This observed variation is considerably greater than has often previously been found in studies of Hymenoptera, including parasitic species, all of which generally display lower levels of molecular polymorphisms at both allozyme and nuclear DNA loci compared with fully diploid insects (*e.g.* Menken 1991; Blanchetot 1992). According to calculation, only around a quarter of the genotypes predicted by theory were found. From the plots of allele and genotype number *vs.* sample size (Fig. 3a,b), there is no clear evidence that many more variants exist to be sampled in the immediate environment. The missing alleles and genotypes could be (1) the result of a biased sample/sampling effect, *i.e.* other regions would give more alleles/genotypes; (2) sample-size effect, *e.g.* as with species area effects; or (3) be due to selection against particular genotypes. Considering the large overall sample size tested, the first criterion seems more plausible than the second (see below). The third criterion seems unlikely from what was observed from the H-W data (Fig. 5), although it cannot be ruled out since some microsatellite loci are known to be under selective constraints and chromosomal-interactive influences (Li et al. 2002), whilst it is always possible that neutral alleles are carried by hitchhiking with directly selected alleles, *e.g.* protein-coding alleles such as allozymes (Lushai et al. 2002).

That populations are close to H-W expectations, from which it may be inferred that mating in these wasps may approach randomness, is interesting in light of what is normally said on this topic. It has been proposed (Crozier 1970) that one of the main reasons haplo-diploidy pushes Hymenoptera populations towards homozygosity is that if the hemizygous haploid males bear a slightly selectively deleterious allele at any given locus, they are then more likely to be eliminated from the population than a diploid individual which always possesses two alleles, possibly one selectively advantageous in the case of heterozygotes with a deleterious gene. The present $Dr283$ data do not support this supposition and at the very least show that microsatellite alleles borne by males and females are essentially neutral. On the other hand, their very neutrality may be the primary reason for the large variance seen at this and other polymorphic loci tested to date by us, and suggests a lack of genetic drift and/or other possible mechanisms as proposed for the homozygosity noted in Hymenoptera populations (see Menken 1991 and Baker et al. 2003).

Another interesting aspect is the finding that observed genetic heterogeneity is largely related to sample size and that around 50 individuals represent approximately half of the variance present locally. This means that the total insect sample collected is well within that required to provide an adequate sample of the variance present (Sjörgren and Wyöni 1994), whilst at the same time, showing that a sample as big as 400–500 individuals probably represents most of the variance locally, certainly at the large field scale. From Fig. 3a,b, clearly a huge sample (>1000 individuals) would have to be obtained to collect a significant number of new variants. This is reminiscent of the situation found in species-richness studies (Scheiner et al. 2000), where during sampling most of the abundant species are sampled relatively early on and intensive sampling of a larger area (or regions) is required to collect rare species and generally over a much longer time span in order to collect occasional dispersing (*e.g.* migrant) species.

With regard to what the data reveals about flight behaviour of *D. rapae*, the single raceme and plant data (Figs. 4a) reflects searching behaviour of individual females looking for aphid hosts to parasitize and this is apparent even with plants spaced a few metres apart (Fig. 4b). Perhaps this reflects patch fidelity as a consequence of attracting insect host and plant host odours (Storeck et al. 2000). However, within the field sampled, the discovery of some degree of subpopulation structuring as shown in Fig. 4c,d and the F_{ST} results (Table 1a,b), albeit small; and the genetic/genotypic differentiation analysis, more especially between field corners, shows a clear degree of spatially-related genetic isolation, and thereby a potential restriction of gene flow. In the main year of sampling (1995), both the parasitoid *D. rapae* and its aphid host, *B. brassicae*, were very abundant. Under such circumstances, it is possible that the restricted movement observed for *D. rapae* in Broad Mead field reflects the abundance of immediately locally available aphid hosts on which to feed. In contrast, when hosts are scarce, female wasp parasitoids are likely to fly much further in search of them, as noted by the presence of such parasitoids in 12.2-m-high Rothamsted suction trap samples in the UK (Wilf Powell, pers. comm.).

The restricted aerial movements of *D. rapae* as here inferred supports earlier field scale studies performed by Vaughn and Antolin (1998) in the USA. These authors used random amplified polymorphic DNA (RAPD) markers with heterozygote genotypes revealed following application of single strand conformation polymorphism (SSCP) techniques (see also Sunnucks et al. 2000). Using 11 codominant and 34 dominant RAPD polymorphisms that conformed to Mendelian segregation patterns, they produced a nested analysis of variance. This indicated extensive genetic differentiation among six *D. rapae* populations sampled over two years. They also showed that effective migration rates (*Nm*) between populations (range 1.2 to 1.6 per year) indicated a relatively low dispersal rate. Similarly, analysis of genetic distance estimates between populations and the resulting trees indicated that populations <1.0 km apart were genetically differentiated. Hence they concluded that '*D. rapae* populations are genetically subdivided on a small spatial scale that corresponds to host-use patterns'.

The present results largely agree with this pattern of low aerial mobility and restricted gene flow. As such, they not only demonstrate how highly variable microsatellite markers may be used to gain new insights into movement of very small insects at farmland scales ranging from single plant to large field, but also how the effective range of these parasitoids for use in IPM strategies may be quite small. At the same time, temporal data obtained upon testing a further three microsatellites (data not shown) suggest that whilst some genotypes are indeed shared between sites (Rothamsted and Woburn), due to the apparent high degree of genetic revolution over a relatively short period of time (four years), continual re-stocking of locally adapted genotypes in mass release programmes may be necessary to effect an adequate level of pest control that is below economic thresholds for damage.

Clearly from our preliminary data presented here, more loci need to be tested, and with sampling at greater spatial scales. These aspects are presently being attended to and will be published as a full manuscript in due course (MacDonald et al. 2004). In the meantime, these preliminary data show that the use of microsatellite markers to assess population parameters of hymenopterous parasitoids is a valuable one in terms of better understanding their dynamics.

Acknowledgements. We thank Cliff Brookes for his valuable technical assistance and Dr. Ian Denholm for his helpful comments on the manuscript. This work was financially supported by the Department of Environment, Food and Rural Affairs (DEFRA), UK.

Rothamsted Research receives grant-aided support from the Biotechnology and Biological Sciences and Research Council (BBSRC) of the United Kingdom.

10.6 References

Atanassova P, Brookes CP, Loxdale HD, Powell W (1998) Electrophoretic study of five aphid parasitoid species of the genus *Aphidius* Nees (Hymenoptera: Braconidae), including evidence for reproductively isolated sympatric populations and a cryptic species. B Entomol Res 88:3-13

Baker DA, Loxdale HD, Edwards O (2003) Genetic variation founder effects and phylogeography in *Diaeretiella rapae* (M'Intosh) (Hymenoptera: Braconidae). Mol Ecol 12: 3303 – 3311

Blackman RL, Eastop VF (2000) Aphids on the world's crops: an identification and information guide, 2nd edn. Wiley, Chichester, 466 pp

Blanchetot A (1992) DNA fingerprinting analysis in the solitary bee *Megachile rotundata* - variability and nest mate genetic-relationships. Genome 35:681-688

Byrne DN, Rathman RJ, Orum TV, Palumbo JC (1996) Localized migration and dispersal by the sweet potato whitefly *Bemisia tabaci.* Oecologia 105:320-328

Caprio MA (1996) Resistant pest management, vol 8, no 1, summer 1996. A biannual newsletter of the Pesticide Research Centre (PRC) http://www.msstate.edu/Entomology/v8n1/rpmv8n1.html

Chapman JW, Reynolds DR, Smith AD, Smith ET, Woiwod IP (2004) An aerial netting study of insects migrating at high-altitude over England. B Entomol Res (in press)

Crozier RH (1970) On the potential for genetic variability in haplo-diploidy. Genetica 41:551-556

Dawson KJ, Belkhir KA (2001) Bayesian approach to the identification of panmictic populations and the assignment of individuals. Genet Res 78:59-77

Dent DR (1995) Principles of integrated pest management. In: Dent D (ed) Integrated pest management. Chapman and Hall, London, 356 pp

Foster SP, Denholm I, Harling ZK, Moores GD, Devonshire AL (1998) Intensification of insecticide resistance in UK field populations of the peach-potato aphid *Myzus persicae* (Hemiptera: Aphididae) in 1996. B Entomol Res 88:127-130

Foster SP, Denholm I, Devonshire AL (2000) The ups and downs of insecticide resistance in peach-potato aphids (*Myzus persicae*) in the UK. Crop Prot 19:873-879

Gaggiotti OE, Lange O, Rassman K, Gliddon C (1999) A comparison of two indirect methods for estimating average levels of gene flow using microsatellite data. Mol Ecol 8:1513-1520

Georghiou GP (1990) Overview of insecticide resistance. In: Green MB, Le Baron HM, Moberg WK (eds) Managing resistance to agrochemicals. American chemical Society, Washington, DC, pp 18-41

Godfray HCJ (1994) Parasitoids: behavioural and evolutionary ecology. Princeton University Press, Princeton, New Jersey, 473 pp

Goldstein DB, Schlötterer C (eds) (1999) Microsatellites, evolution and applications. Oxford University Press, Oxford, 352 pp

Goudet J, Raymond M, de Meeüs T, Rousset F (1996) Testing differentiation in diploid populations. Genetics 144: 1933-1940

Guo SW, Thompson E (1992) Performing the exact test of Hardy-Weinberg proportion for multiple alleles. Biometrics 48:361-172

Hopper KR, Roush RT, Powell W (1993) Management of genetics of biological control introductions. Annu Rev Entomol 38:27-51

Jarne P, Lagoda PJL (1996) Microsatellites from molecules to populations and back. Trends Ecol Evol 11:424-429

Li Y-C, Korol AB, Fahima T, Beiles A, Nevo E (2002) Microsatellites: genomic distribution putative functions and mutational mechanisms: a review. Mol Ecol 11:2453-2465

Llewellyn KS (2000) Genetic structure and dispersal of cereal aphid populations. PhD Thesis, Nottingham University, UK

Llewellyn KS, Loxdale HD, Harrington R, Brookes CP, Clark SJ, Sunnucks P (2003) Migration and genetic structure of the grain aphid (*Sitobion avenae*) in Britain related to climatic adaptation and clonal fluctuation revealed using microsatellites. Mol Ecol 12:21-34

Long S (2002) MSU researchers seek biological control options for tackling onion pest, http://www.msue.msu.edu/learnnet/onion_032002.htm

Loxdale HD (2001) Tracking flying insects using molecular markers. Antenna (Bulletin of the Royal Entomological Society) 25:242-250

Loxdale HD, Brookes CP (1987) Use of electrophoretic markers to study the spatial and temporal genetic structure of populations of a holocyclic aphid species *Sitobion fragariae* (Walker) (Hemiptera: Aphididae). In: Holman J, Pelikan J, Dixon AFG, Weismann L (eds) Population structure, genetics and taxonomy of aphids and Thysanoptera. SPB Academic Publishing, The Hague, pp 100-110

Loxdale HD, Lushai G (1998) Molecular markers in entomology (review). B Entomol Res 88:577-600

Loxdale HD, Lushai G (1999) Slaves of the environment: the movement of insects in relation to their ecology and genotype. Philos Trans R Soc B 354:1479-1495

Loxdale HD, Lushai G (2001) Use of genetic diversity in movement studies of flying insects. In: Woiwod IP, Reynolds DR, Thomas CD (eds) Royal Entomological Society 20th international symposium volume: insect movement: mechanisms and consequences. CABI, Wallingford, pp 361-386

Loxdale HD, Hardie J, Halbert S, Foottit R, Kidd NAC, Carter CI (1993) The relative importance of short- and long-range movement of flying aphids. Biol Rev 68:291-311

Lushai G, Markovitch O, Loxdale HD (2002) Host-based genotype variation in insects revisited. B Entomol Res 92:159-164

Macdonald C, Loxdale HD (2002) Investigating the spatial distribution of parasitoids using molecular markers. 'Entomology 2002', Royal Entomological Society National Meeting, Cardiff, 12-13 Sept 2002

MacDonald C, Brookes CP, Edwards KJ, Baker D, Lockton S, Loxdale HD (2003) Microsatellite isolation and characterisation in the beneficial parasitoid wasp *Diaeretiella rapae* (M'Intosh) (Hymenoptera: Braconidae: Aphidiidae). *Molecular Ecology Notes* **3**, 601-603

MacDonald C, Brookes CP, Dawson KJ, Loxdale HD (2004) Population structure and dynamics of the parasitoid wasp, *Diaeretiella rapae* (M'Intosh) (Hymenoptera: Braconidae) studied at the farmland scale using microsatellites (in prep.)

Majerus M, Amos W, Hurst G (1996) Evolution - the four billion year war. Wesley Longman, Harlow, Essex, 340 pp

Menken SBJ (1991) Does haplodiploidy explain reduced levels of genetic variability in Hymenoptera? Proc Exp Appl Entomol 2:172-178

Morrison AC, Ferro C, Morales A, Tesh RB, Wilson ML (1993) Dispersal of the sand fly *Lutzomyia longipalpis* (Diptera: Psychodidae) at an endemic focus of visceral leishmaniasis in Colombia. J Med Entomol 30:427-435

Neigel JE (1997) A comparison of alternative strategies for estimating gene flow from genetic markers. Annu Rev Ecol Syst 28:105-128

Osborne JL, Loxdale HD, Woiwod IP (2002) Monitoring insect dispersal: methods and approaches. In: Bullock JM, Kenward, RE, Hails RS (eds) Dispersal ecology, British ecological symposium. Blackwell, Oxford, pp 24-49

Pike KS, Starý P, Miller T, Allison D, Graf G, Boydston L, Miller R, Gillespie R (1999) Host range and habitats of the aphid parasitoid *Diaeretiella rapae* (Hymenoptera: Aphidiidae) in Washington state. Environ Entomol 28:61-71

Powell W (1994) Nemec and Stary's "Population Diversity Centre" hypothesis for aphid parasitoids re-visited. Norw J Agric Sci Suppl 16:163-169

Powell W, Wright AF (1988) The abilities of the aphid parasitoids *Aphidius ervi* Haliday and *A. rhopalosiphi* De Stefani Perez (Hymenoptera: Braconidae) to transfer between different host species and the implications for the use of alternative hosts in pest control strategies. B Entomol Res 78:683-693

Promega (1996) Silver sequence DNA sequencing system: technical manual Promega corporation (http://wwwpromegacom), 19 pp

Raymond M, Rousset F (1995a) GENEPOP (version 12): population genetics software for exact tests and ecumenicism. J Hered 86:248-249

Raymond M, Rousset F (1995b) An exact test for population differentiation. Evolution 49:1280-1283

Scheiner SM, Cox SB, Willig M, Mittelbach GG, Osenberg C, Kaspari M (2000) Species richness species-area curves and Simpson's paradox. Evol Ecol Res 2:791-802

Sjörgren P, Wyöni P-I (1994) Conservation genetics and detection of rare alleles in finite populations. Conserv Biol 8:267-270

Slatkin M (1987) Gene flow and the geographic structure of natural populations. Science 236:787-792

Slatkin M (1995) A measure of population subdivision based on microsatellite allele frequencies. Genetics 139:457-462

Storeck A, Poppy GM, van Emden HF, Powell W (2000) The role of plant chemical cues in determining host preference in the generalist aphid parasitoid *Aphidius colemani*. Entomol Exp Appl 97:41-46

Sunnucks P, Hales DF (1996) Numerous transposed sequences of mitochondrial cytochrome oxidase I - II in aphids of the genus *Sitobion* (Hemiptera: Aphididae). Mol Biol Evol 13:510-524

Sunnucks P, Wilson ACC, Beheregaray LB, Zenger K, French J, Taylor AC (2000) SSCP is not so difficult: the application and utility of single-stranded conformation polymorphism in evolutionary biology and molecular ecology. Mol Ecol 9:1699-1710

Taimr L and Kříž J (1978) Stratiform drift of the hop aphid (*Phorodon humuli* Schrank). Z Angew Entomol 86:71-79

Urquhart FA, Urquhart NR (1977) Overwintering areas and migratory routes of the Monarch butterfly (*Danaus p. plexippus*; Lepidoptera: Danaidae) in North America with special reference to the western population. Can Ent 109:1583-1589

Vaughn TT, Antolin MF (1998) Population genetics of an opportunistic parasitoid in an agricultural landscape. Heredity 80:152-162

Weir BS, Cockerham CC (1984) Estimating *F*-statistics for the analysis of population structure. Evolution 38:1358-1370

Workman PL, Niswander JD (1970) Population studies of southwestern Indian tribes. II. Local genetic differentiation in the Papago. Am J Hum Genet 22:24-49

Wright S (1951) The genetical structure of populations. Ann Eugenic 15:323-354

11 Interhemispheric Transport of Viable Fungi and Bacteria from Africa to the Caribbean with Soil Dust

Joseph M. Prospero

Cooperative Institute for Marine and Atmospheric Studies, Rosenstiel School of Marine and Atmospheric Science, University of Miami, 4600 Rickenbacker Causeway, Miami, Florida 33149, USA, Email jprospero@rsmas.miami.edu

11.1 Abstract

Studies over the past several decades have shown that aerosols (e.g., particles of pollution, dust, and smoke from fires) can be transported across great distance over the oceans, from one continent to another. There is, however, only anecdotal indirect evidence for the long-range transport (LRT) of viable microorganisms (MOs) on intercontinental scales (Brown and Hovmøller 2002). In particular, there have been no long-term systematic studies of the LRT of MOs. Here, we report on the measurements of aerosols in the trade winds on Barbados, West Indies, over a 2-year period (1996-1997). We made concurrent daily measurements of bacteria and fungi based on cultures developed from filters. Previous studies have shown that large quantities of African dust are present in winds at Barbados during much of the year. Our study shows that cultivable bacteria and fungi are only found in air that contains African dust. Air transported from other regions yields no viable organisms. These results raise questions about the factors that affect the transport and survivability of MOs in the atmosphere (Prospero et al. 2003).

11.2 The Transport of African Dust to the Caribbean

Our study is based on Barbados (13.18°N, 59.43°W) where the aerosol group at the University of Miami has made aerosol measurements continuously since 1965 at a site on the east coast. Daily samples are collected at the top of a 17-m tower located on a 30-m bluff. Sampling is only carried out under on-shore wind conditions so as to minimize the possibility of impacts form local sources.

Mineral aerosol concentrations are determined by extracting the soluble components, ashing the filter, and weighing the residue which is ascribed to dust (Li et al. 1996; Prospero 1999). Figure 1 shows the monthly mean dust concentration measured at Barbados over the period 1965-1998 (Prospero and Lamb 2003). Several patterns are evident in this record. First, there is a strong seasonal periodicity with a summer maximum in June, July, and August; dust concentrations are at a minimum in winter. Second, there is a general increase in dust concentrations beginning in the late 1960s and early 1970s. Third, within the overall record, some years stand out with sharply higher concentration.

There are many lines of evidence that link Barbados dust to African sources, but the most persuasive is satellite imagery. Various satellite products enable us to trace individual dust outbreaks from the time the dust is emitted from sources until the dust enters the Caribbean about one week later. Excellent dust cloud images can be found at a number of websites. The NASA Moderate-resolution Imaging Spectrometer (MODIS) products are available at the MODIS Land Rapid Response website which offers near-real-time "true color" images of land surfaces and adjacent waters at several resolutions.

http://rapidfire.sci.gsfc.nasa.gov/realtime/

The NASA's Visible Earth database is an excellent archive of selected satellite imagery. The section under "Atmosphere/Aerosols/ shows hundreds of dust events.

http://visibleearth.nasa.gov/

The NASA Earth Observatory website also has a excellent collection of images that focus on specific events. They provide supporting text and links to related sources.

http://earthobservatory.nasa.gov/NaturalHazards/

The NASA Total Ozone Mapping Spectrometer (TOMS) website also presents near-real-time aerosol products. The absorbing aerosol product shows the distribution of absorbing aerosols (i.e., dust and smoke) measured in terms of an aerosol index (Herman et al. 1997).

http://jwocky.gsfc.nasa.gov/aerosols/aerosols.html

Taken as a whole, these satellite products (and many others) clearly demonstrate that dust storms and wind-borne dust clouds are the most visible, persistent, and widespread atmospheric aerosol phenomena visible from space and that dust and associated materials are routinely transported over long distances, even between continents.

The seasonal cycle of dust concentrations on Barbados as seen in Fig. 1 reflects the geographic variability of dust source activity in North Africa coupled with the seasonal shift in the large-scale wind system (Herman et al. 1997; Husar et al. 1997). Dust is emitted from Africa most of the year. In winter, dust is carried across the equatorial Atlantic to South America. Indeed, African dust is believed to play an important role in supplying nutrients (especially phosphorous) to the Amazon basin (Swap et al. 1992). As the year progress, the transport path shifts northward until it reaches its northernmost position in the latitudes of the Caribbean in July and August.

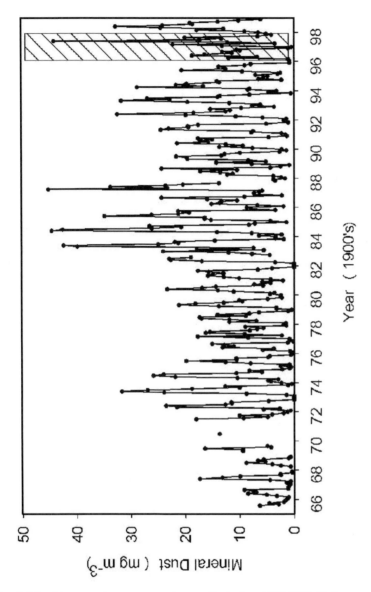

Fig. 1. Monthly mean dust concentrations on Barbados, 1965-1998. Units: μg m^{-3}. The *inset* shows the period during which the MO measurements were made in this study. The time period of our MO overlaps with the major ENSO event occurring in 1997-1998. Other major events during the period shown in the graph were: 1972-1973, 1982-1983, 1986-1987, and 1991-1992. (After Prospero and Lamb 2003)

The transport of dust to Barbados has been widely studied along with its chemical, physical, and radiative properties (Li et al. 1996; Prospero 1996; Prospero and Lamb 2003). The trade winds subsequently carry the dust into the Caribbean, the Gulf of Mexico, and to Florida (Prospero 1999; Prospero et al. 2001). Every summer significant quantities of African dust are measured throughout the eastern United States as far north as New England (Perry et al. 1997). Indeed, during the summer, because of African dust, the highest concentrations of fine-particle (i.e., less than 2.5 μm diameter) dust are measured east of the Mississippi, not in the arid west as one might expect. Also high concentration of dust are measured on Bermuda (Arimoto et al. 1995), carried by southeast winds.

11.3 Windborne Microorganisms at Barbados

For the microbiological studies we collected daily samples on sterile filters which were subsequently cultured for bacteria and fungi. In cooperation with a group of scientists at the University of the West Indies, Barbados, sampling was carried out under the same conditions as the inorganic (i.e., dust) studies so that the inorganic measurements and those of MOs could be directly compared (Prospero et al. 2003). In this study, we examine and compare the daily temporal record of aerosol and MO concentrations over the period 1996-1997.

We used common nutrient media so that our culture results could be compared with those in the broader literature: blood agar medium for bacteria and Sabouraud's medium for fungi. Given the stringent requirements for the culturing of many MOs, only a small fraction of the airborne MOs present in the samples might respond to these media. Nonetheless, our objective was to study the temporal variability of a specific subset of common species so that we could relate the variability to that of other aerosol species.

On the concurrently collected inorganic aerosol filter we measured an array of inorganic aerosol species: mineral dust from soils; non-sea-salt (nss) $SO_4^=$ (i.e., $SO_4^=$ from sources other than the salts in ocean-water spray droplets, primarily pollution sources and the oxidation of dimethyl sulfide emitted by marine phytoplankton); NO_3^-, derived from both natural and pollution sources; sea-salt from ocean spray droplets. Because of our long history of studies on Barbados, we have a good understanding of the factors that control aerosol composition and concentration in the North Atlantic trade winds. Because we collect daily samples, we can more readily relate our measurements to specific meteorological situations and back-trajectories.

The concentration of culture-forming organisms (CFOs) shows a clear seasonal trend with a minimum in the winter and a maximum in the summer and a range from about 0 to 20 culture forming units (CFU) m^{-3} of sampled air. Concentrations were mostly zero during the winter; they increased sporadically during the spring, remained relatively high through the summer, and then showed sporadic behavior once again in the fall. This seasonal cycle matches that of African mineral dust at Barbados (Li et al. 1996; Prospero and Lamb 2003). On our daily samples we only detected CFOs on filters that contained dust. It is notable that during the winter, when aerosols are transported from equatorial Africa and from the North Atlantic, Europe, and North America, our cultures yield no response.

The culture concentrations were evenly split between fungi and bacteria, the latter almost exclusively *Bacillus*. The dominant fungus was *Mycelia sterilia* which comprised 48% of the fungal colonies. *Arthrinium* (22%) and *Periconium* (12%) were also common along with unidentified black/gray (7%) and brown/tan (3%) colonies. This contrasts with studies over the continents where, according to Lacey (1991), *Cladosporium* is the most abundant spore type on an annual basis in temperate and most tropical regions; *Alternaria* spores are the second most abundant overall and in warm, dry regions can exceed the concentrations of *Cladosporium*. In tropical regions *Curvularia* and *Nigrospora* sometimes make large contributions and *Aspergillus* species are particularly characteristic of humid tropical regions.

The concentrations of CFO that we measured in the trade winds on Barbados are orders of magnitude lower than those measured over the continents (Lighthart and Stetzenbach 1994; Lacey and Venette 1995; Muilenberg 1995). Low concentrations might be expected considering that transport times from Africa are about a week or more, a factor that would affect the survival of MOs.

Our results also differ markedly from those of Griffin et al. (2001) on St. John in the Virgin Islands. Using sampling and culturing techniques similar to ours, they found high concentrations of viable MOs, including pathogens, which they associated with the presence of African dust. They measured CFOs at concentrations about 100 times greater than ours. We note in their paper that they sampled at the extreme western end of St. John. Thus the trade winds will have passed over the entire island before reaching their site. This suggests that their samples were highly impacted by local sources. In test carried out by us on Barbados, we obtained sharply higher concentrations of CFOs even at short distances inland from the coast. At inland sites, concentrations were comparable to continental values.

11.4 Dust, Microorganisms, and Climate

In a recent paper, Prospero and Lamb (2003) show that dust concentrations in Barbados trade winds are highly anticorrelated with rainfall in the Soudano-Sahel region of North Africa, a region that has suffered varying degrees of drought. since 1970. It is notable that in Fig. 1, the largest monthly-mean dust peaks were associated with strong El Niño - Southern Oscillation (ENSO) events: 1972-1973, 1982-1983, 1986-1987, 1991-1992, 1997-1998 (Prospero and Lamb 2003). Our MO measurements were made over the years 1996-1997, overlapping with one of the strongest ENSO events in the 20th century, 1997-1998.

It is notable that the MO concentrations measured by us during the summer of 1997 were sharply higher than those in 1996, roughly in proportion to the differences in the respective dust concentrations for those periods. This suggests that the concentration of cultivable MOs may be linked in some way with dust concentrations or to the processes that lead to increased dust concentrations.

Regression estimates based on long-term rainfall data suggest that dust concentrations were sharply lower during much of the 20th century prior to 1970 when rainfall was more normal. This leads us to wonder if MO transport might be anomalously high as well. If so, future changes in climate could result in large changes in emissions of dust and MOs from African and other arid regions which, in turn, could lead to impacts on climate over large areas.

11.5 Conclusions

There was a clear association between CFOs and African dust. No CFOs were obtained on filters that did not contain dust. This suggests that there may be some characteristics of African MOs that may favor their survival. In particular the association with dust might serve as a protective mechanism, for example, through the shielding of MOs against UV radiation. Further studies are needed to assess the factors that contribute to the survivability of these MOs. These same factors could affect the viability of pathogenic MOs as well.

Although our study was based on Barbados, other studies have documented the transport of African dust deep into the eastern United States. Thus the MOs that we see in dust at Barbados will also be transported into North America.

Our data hint at a link between dust variability, climate, and MO transport. It remains to be seen what this link might be and what the implications are for future climate change.

Acknowledgements. We thank H. Maring, D. Savoie, L. Custals, T. Snowdon (Univ of Miami), and C. Shea (Barbados) for technical support; also E. Manning (Dyserth, North Wales, UK) and family for permitting us to continue to operate our laboratory on their property in Barbados. The biological studies in this program were carried out in cooperation with E. Blades, G. Mathison, and R. Naidu (Univ. West Indies, Barbados). This work was supported by grants from US NSF and NASA.

11.6 References

Arimoto R, Duce RA, Ray BJ, Ellis WG Jr, Cullen JD, Merrill JT (1995) Trace elements in the atmosphere over the North Atlantic. J Geophys Res 100:1199-1214
Brown JKM, Hovmøller MS (2002) Aerial dispersal of pathogens on the global and continental scales and its impact on plant disease. Science 297:537-541
Griffin DW, Garrison VH, Herman JR, Shinn EA (2001) African desert dust in the Caribbean atmosphere: microbiology and public health. Aerobiologia 17:203-213
Herman JR, Bhartia PK, Torres O, Hsu C, Seftor C, Celarier E (1997) Global distribution of UV-absorbing aerosols from Nimbus_7/TOMS data. J Geophys Res 102:16911-16922
Husar B, Prospero JM, Stowe LL (1997) Characterization of tropospheric aerosols over the oceans with the NOAA/AVHRR optical thickness operational product. J Geophys Res 102:16889-16909

Lacey J (1991) Aerobiology and health: the role of airborne fungal spores in respiratory disease. In: Hawkworth DL (ed) Frontiers in mycology. CAB International, Wallingford, Oxon

Lacey J, Venette J (1995) Outdoor air sampling techniques. In: Burge HA (ed) Bioaerosols. Lewis, Boca Raton, pp 407-471

Li X, Maring H, Savoie D, Voss K, Prospero JM (1996) Dominance of mineral dust in aerosol light scattering in the North Atlantic trade winds. Nature 380:416-419

Lighthart B, Setzenbach LD (1994) Distribution of microbial bioaerosol. In: Lighthart B, Mohr AJ (eds) Atmospheric microbial aerosols: theory and applications. Kluwer, Boston, pp 68-98

Muilenberg ML (1995) The outdoor aerosol. In: Burge HA (ed) Bioaerosols. Lewis, Boca Raton, pp 163-204

Perry KD, Cahill TA, Eldred RA, Dutcher DD, Gill TE (1997) Long-range transport of North African dust to the eastern United States. J Geophys Res 102:11225-11238

Prospero JM (1996) The atmospheric transport of particles to the ocean. In: Ittekkott V, Honjo S, Depetris PJ (eds) Particle flux in the ocean. SCOPE report 57. Wiley, New York, pp 19-52

Prospero JM (1999) Long-term measurements of the transport of African mineral dust to the Southeastern United States: implications for regional air quality. J Geophys Res 104:15917-15927

Prospero JM, Lamb PJ (2003) African droughts and dust transport to the Caribbean: climate change implications. Science 302: 1024 - 1027, 2003

Prospero JM, Olmez I, Ames M (2001) Al and Fe in PM 2.5 and PM 10 suspended particles in South-Central Florida: The impact of the long range transport of African mineral dust. J Water Air Soil Pollut 125:291-317

Prospero JM, Blades E, Mathison G, Naidu R (2003) Interhemispheric transport of viable fungi and bacteria from Africa to the Caribbean with soil dust. Aerobiologia (submitted) Swap RM, Garstang M, Greco S, Talbot R, Kallberg P (1992) Saharan dust in the Amazon basin. Tellus B 44:133-149

12 Be a Virus, See the World

Stephan Becker

Philipps-University Marburg, Faculty of Medicine, Virology Institute, Robert-Koch-Str. 17, 35032 Marburg, Germany, E-mail becker@staff.uni-marburg.de

12.1 Abstract

As the recent SARS epidemic has shown, viruses are able to migrate with remarkably high speed, endangering countries around the globe within hours. Since viruses are obligate parasites, their migration speed is dependent on the mobility of the respective host. Several examples of pathogenic viruses with different patterns of migration will be discussed.

Lassa virus is endemic in West Africa and causes a highly pathogenic hemorrhagic fever among humans. Lassa virus is transmitted by urine and feces of rodents that are persistently infected. Lassa virus, dependent on its rodent host, seems not to spread significantly outside its endemic regions. The human infection rather represents a dead end for the virus.

West Nile virus entered the American public awareness in 1999 with the first human cases in the United States appearing in New York City. The virus is transmitted by mosquitoes that parasitize migrating birds, and also, e.g., crows. Transmitted to humans, West Nile virus can cause encephalitis that is especially dangerous for the elderly. West Nile virus spread remarkably fast over the North American continent. Nowadays, even several Canadian provinces are facing a severe problem with infected birds and human cases.

The *SARS coronavirus* caused not only epidemics with local transmissions in mainland China, Hong Kong, Taiwan, and Toronto, but was also spread to almost every country worldwide by infected patients. So far, this was the most impressive example for the speed a virus can achieve by the mobility of its host. The natural host of the SARS coronavirus is still unknown and, thus future outbreaks cannot be excluded.

The risk imposed by an emerging virus to the human population is a product of migration velocity, transmission route, and speed of detection. Most dangerous for the human population are highly pathogenic viruses that are transmitted from human to human via the air (SARS coronavirus). However, equally dangerous are viruses that are highly pathogenic, transmitted by blood to blood contact but have a long incubation period and, thus, detection and surveillance are complicated (human immunodeficiency virus).

12.2 Introduction

The title of the talk goes back to a cartoon by Garry Larson. It shows two people sitting in the lobby of a hotel. The man looks deep into the eyes of a young lady saying: "You are from France? Wow! Say... You have lovely eyes!..." In parallel, the biosphere of viruses is shown, where one can see a couple of these bugs and one of them shouts: "Hey everyone, we are going to Paris!". This little cartoon shows nicely how viruses migrate: they depend on their respective hosts and the migration of the host determines the migration of the virus.

The focus of my talk might already have been adduced without being really named. That is the presence of viruses in most living organisms. If we talk about the migration of plants, microbes or people, it is interesting to have in mind that all of these carry their respective viruses with them.

Viruses are obligate parasites that are highly specialized, each adapted to its particular host. The ideal virus might be the one that you do not notice. In this sense endogenous retroviruses are possibly close to ideal.

Mostly, we can investigate the migration of viruses only by the diseases they cause. The appearance of a specific disease indicates the presence of the causative virus. This paper focuses on emerging viruses that cause dramatic and partly fatal diseases. The concept of emerging viruses evolved in consequence of a phenomenon we have been faced with within the past few decades: Hitherto unidentified viruses appear seemingly out of nowhere and cause very severe and life-threatening diseases.

By definition, emerging viruses are those that have emerged or reemerged in the last two decades and cause serious illness. A prime example of an emerging virus is Ebola virus or Marburg virus; the latter had been identified during an outbreak of hemorrhagic fever here in Marburg in 1967. At that time, Marburg virus was transmitted by infected African green monkeys that had been imported from Uganda in order to grow polio virus on their kidney cells for vaccine production.

The investigation of emerging viruses has revealed that in most cases the natural host of these viruses is not a human but another animal. Additionally, the diseases caused by these viruses often display elements of immunopathology indicating that the human immune response is not adequate. Often there is sort of immune overreaction and consequently the immune response becomes the cause, if unchecked, of fatality. Further on, emerging viruses have very often seemed to be displaced from their original habitat by ecological or social changes, thence to come into contact with a new host. Human infection with an emerging virus is therefore in most cases a kind of accident.

The topic of this conference is migration. Since viruses are parasites that usually cannot survive outside the host cell for any considerable time, the migration of viruses is the history of the migration of the respective hosts. I would like to address three examples of emerging viruses with different migration behaviors. First, Lassa virus, which seems not to migrate very far, if it migrates at all. Second, West Nile virus which migrates quite slowly; and third, SARS coronavirus which migrates exorbitantly fast.

12.3 Lassa Virus

Lassa virus (Peters et al. 1996) belongs to the family Arenaviridae.The family of Arenaviridae comprises several members that are divided into *Old World* and *New World* Arenaviruses, Lassa virus belonging in the Old world category. This virus is endemic in West Africa where millions of people are at risk to acquire a Lassa virus infection. And indeed, there are more than 150,000 known infections per year with approximately 10,000 fatal cases.

Lassa virus is transmitted by rodents, most likely *Mastomys natalensis*, the African soft-furred rat. The transmission route is contact with urine or feces of the rats that are persistently infected by the virus and shed the virus in high amounts.

People in remote areas of Sierra Leone, Guinea, and Ivory Coast often live under quite primitive conditions that lead to close contact between humans and rats that feed on food stored inside the huts. Moreover, *Mastomys* is also part of the diet, and people who hunt rats have the highest risk of infection with Lassa virus (terMeulen et al. 1996). Moreover, Lassa virus can be transmitted from human to human by contact with infected body fluids. This route is the cause of hospital transmissions and the rare Lassa virus outbreaks.

As mentioned before, Lassa virus is endemic in western Africa, in countries like Sierra Leone, Ivory Coast, and Guinea. However, *Mastomys natalensis* is distributed widely over the African continent. Currently, it is not understood why Lassa fever does not spread more widely in Africa. One possibility is that the biology of *Mastomys* is not very well understood and the species that transmits Lassa virus is a special subclass with a more narrow distribution.

Most of the people in the endemic regions seem to survive the infection. However, the convalescence is prolonged and some of the patients display a sustained and pronounced hearing loss whose pathogenicity is not understood. Travelers, however, who have become infected have a very high risk of dying from the infection. Almost the entire imported Lassa virus infections reported during recent years resulted in the death of the patients.

Although Lassa virus is a serious disease, especially for people that are not indigenous, this virus is not of high risk for the global human population because the primary infections are strictly from contact with *Mastomys* species that are indigenous in western Africa, and because human to human transmission requires close contact with infected body fluids.

Taken together, Lassa virus represents a risk for people in endemic regions and in hospitals in western Africa (as a nosocomial infection). Lassa virus is not a suitable bioweapon because of the close contact that is needed to transmit the virus from human to human -- apart from the fact that bioweapons might be most dangerous as psychological weapons.

12.4 West Nile Virus

The next example of an emerging virus with the obvious potential to spread more widely than Lassa virus is *West Nile virus* (Campbell et al. 2002). This pathogen belongs to a group of viruses that have a special way to be transmitted. They are called arboviruses which is the abbreviation for arthropod-borne viruses. Arboviruses are transmitted by ticks, mosquitoes, or sand flies. The virus is taken up by the mosquito during its blood meal and starts to replicate in the mosquito. During the next blood meal, the virus is transmitted to the next animal.

Two different arbovirus life cycles can be distinguished: (1) the virus cycles from infected mammals to arthropods, back to mammals. This is called the sylvan or savanna cycle. (2) The other possibility is that an infected mosquito feeds on a human being, thereby infection of the human takes place. The infected human is then the source for infection of another mosquito that in turn transmits the virus to other humans. This is called the urban cycle and is established if both the arthropod vector that can grow the virus and the human host are present in the near vicinity. The urban cycle presents a considerable risk for the population in less developed countries especially in highly populated cities in western Africa. Thus, an urban cycle of one highly pathogenic arbovirus, namely, the Yellow Fever virus is a nightmare. Although a vaccine against yellow fever is available, it is not routinely used in African countries for economic reasons.

However, not only are less developed countries at risk to be taken over by arboviruses. I would like to talk about another virus that is a nice example of an emerging virus that can migrate and conquer new biospheres even in highly developed countries. This is the West Nile virus. West Nile virus was known to be endemic in the whole African continent, Middle, and South Europe, western parts of Asia, and Australia.

In the summer 1999, however, New York City was shaken by several mysterious encephalitis cases that were obviously of viral origin. The investigations revealed that the causative agent was a virus of the family Flaviviridae, the family of the yellow fever virus. However, West Nile virus turned out to be the causative agent of the disease, and it is much less virulent than the yellow fever virus. In the elderly, West Nile virus causes the mentioned encephalitis with a case fatality ratio of 10% in the severe cases. The life cycle of West Nile virus includes birds as amplifying hosts and several *Culex* species as vectors. Usually, the birds do not develop a severe disease but grow the virus to high titers. There are, however, incidental avian hosts, e.g., crows that are highly susceptible to the virus. In regions where West Nile virus is reported, hundreds of dead crows have been observed. Humans are incidental hosts, as are horses. Humans and horses represent dead end hosts, however. That is, the virus is not further transmitted.

I showed the situation of the West Nile virus distribution in the USA in the summer of 1999 some slides ago. The bet of this summer among virologists was whether West Nile virus would survive the winter in New York or not. It survived. Since 1999, West Nile virus has spread from New York to other parts of the US, and today the West Nile virus is endemic in the entire North American continent, including Canada.

12.5 SARS Coronavirus

The third example is the *SARS coronavirus*. This virus caused several changes in the behavior of people in some of the Asian megacities, including Hong Kong. For one thing, it led to the wearing of protective face masks in public places. SARS stands for severe acute respiratory syndrome.

In November 2002 a significant outbreak of atypical pneumonia was noted in southern China, however, nobody, including the health authorities, expressed any particular alarm. In February 2003 the detection of the highly pathogenic influenza virus H5N1 in Hong Kong concerned the World Health Organization (WHO), and when in mid-March almost simultaneously several cases of an acute respiratory symptom complex appeared in Vietnam and in Canada a very rare worldwide alert was issued. The hot spot of this outbreak was definitively China. A high number of local transmissions were reported in Beijing, the southern province Guangdong, and in Hong Kong. Canada also faced local transmissions, and many other countries were burdened by individual imported SARS cases.

The first chain of transmission of SARS was determined and is very interesting. It started with a doctor from the province Guangdong who traveled to Hong Kong where he stayed in the now notorious Hotel Metropole. Here, he became seriously ill and was admitted to a local hospital. During his stay in the hotel, he transmitted the disease to about ten other people who lived there. These people either became ill rapidly and were admitted to Hong Kong hospitals or traveled abroad. The next wave of infections then had a high impact on the public health sector. Virtually hundreds of health care workers became infected; infections that all could be traced back to the single case in Hong Kong. This example shows clearly how fast a virus can migrate. Within hours the epidemic was spread over the whole globe.

Altogether, about 8,000 cases of SARS had been detected by June 2003 with about 800 fatalities. The case fatality rate was not uniform but increased dramatically with age. People older than 65 years had a 50% risk of death from the disease.

The global outbreak needed the global response that was organized by the World Health Organization. WHO asked about 11 labs worldwide to participate in the outbreak investigation, and it was agreed to have daily teleconferences and to freely share material and information (WHO 2003). And, indeed, the collaboration worked nicely, and in consequence it was possible to check findings in one group immediately with those from patients at other places in the world. I am convinced that this collaboration was the reason for the extremely short time needed to identify the causative agent of SARS. Not more then 2 weeks after the outbreak investigations started, the virus associated with SARS was identified to be a novel coronavirus that was thereafter named SARS coronavirus (SARS-CoV; Drosten et al. 2003; Ksiazek et al. 2003; Rota et al. 2003).

It is still not completely clear where SARS-CoV came from. However, investigations of a Chinese group showed that the novel coronavirus could be detected in a cat species in Guangdong province where the outbreak originated. The sequence of the cat SARS-CoV is not 100% identical to most of the human isolates. In the latter, a 29-nucleotide deletion is found when compared to the cat isolate. Thus, the current hypothesis is that SARS is the product of an interspecies viral transmission with dramatic consequences for the new involuntary human host. Today, it is not understood whether the nucleotide deletion in the genome of the human isolates is related to the high pathogenicity of SARS-CoV (Guan et al. 2003).

It is sometimes questioned whether the measures taken by the WHO and the countries with local transmission chains were justified, considering the relatively low number of global cases and deaths. However, at the time of the outbreak, the public health services were confronted with a highly pathogenic virus that was transmitted via the air and had high impact on the public health sector by infecting mainly health care workers. Thus, the fact that SARS so far resulted in a relatively small number of diseases and deaths is mainly the result of the global efforts to contain the outbreak -- *not* the benign character of the SARS-CoV.

12.6 Conclusion

In conclusion, human populations are threatened by emerging viruses whose danger depends on their migration behavior, transmission route, and speed of detection. Most dangerous are airborne viruses that are transmitted easily from human to human (SARS-CoV). Arboviruses are also of high impact because of the fact that these are difficult to contain. Finally, the emerging viruses that are transmitted by local animals and are transmitted from human to human by blood to blood contact, like Ebola and Lassa viruses, have high impact on indigenous populations but are of only moderate risk to the global population.

12.7 References

Campbell GL, Marfin AA, Lanciotti RS, Gubler DJ (2002) West Nile virus. Lancet Infect Dis 2:519-529

Drosten C, Gunther S, Preiser W, van der Werf S, Brodt HR, Becker S, Rabenau H, Panning M, Kolesnikova L, Fouchier RA, Berger A, Burguiere AM, Cinatl J, Eickmann M, Escriou N, Grywna K, Kramme S, Manuguerra JC, Muller S, Rickerts V, Sturmer M, Vieth S, Klenk HD, Osterhaus AD, Schmitz H, Doerr HW (2003) Identification of a novel coronavirus in patients with severe acute respiratory syndrome. N Engl J Med 348:1967-1976

Guan Y, Zheng BJ, He YQ, Liu XL, Zhuang ZX, Cheung CL, Luo SW, Li PH, Zhang LJ, Guan YJ, Butt KM, Wong KL, Chan KW, Lim W, Shortridge KF, Yuen KY, Peiris JSM, Poon LLM (2003) Isolation and characterization of viruses related to the SARS coronavirus from animals in Southern China. Science 10:1126/Science 10:87139

Ksiazek TG, Erdman D, Goldsmith CS, Zaki SR, Peret T, Emery S, Tong S, Urbani C, Comer JA, Lim W, Rollin PE, Dowell SF, Ling AE, Humphrey CD, Shieh WJ, Guarner J, Paddock CD, Rota P, Fields B, DeRisi J, Yang JY, Cox N, Hughes JM, LeDuc JW, Bellini WJ, Anderson LJ (2003) A novel coronavirus associated with severe acute respiratory syndrome. N Engl J Med 348:1953-1966

Peters CJ, Buchmeier M, Rollin PE, Ksiasek TG (1996) Arenaviruses. In: FEA (ed) Virology. Raven Press, New York, pp 1521-1551

Rota PA, Oberste MS, Monroe SS, Nix WA, Campagnoli R, Icenogle JP, Penaranda S, Bankamp B, Maher K, Chen MH, Tong S, Tamin A, Lowe L, Frace M, DeRisi JL, Chen Q, Wang D, Erdman DD, Peret TC, Burns C, Ksiazek TG, Rollin PE, Sanchez A, Liffick S, Holloway B, Limor J, McCaustland K, Olsen-Rasmussen M, Fouchier R, Gunther S, Osterhaus AD, Drosten C, Pallansch MA, Anderson LJ, Bellini WJ (2003) Characterization of a novel coronavirus associated with severe acute respiratory syndrome. Science 300:1394-1399

Ter Meulen J, Lukashevich I, Sidibe K, Inapogui A, Marx M, Dorlemann A, Yansane M L, Koulemou K, Chang-Claude J, Schmitz H (1996) Hunting of peridomestic rodents and consumption of their meat as possible risk factors for rodent-to-human transmission of Lassa virus in the Republic of Guinea. Am J Trop Med Hyg 55:661-666

WHO (2003) A multicentre collaboration to investigate the cause of severe acute respiratory syndrome. Lancet 361:1730-1733

13 Species Delineation and Biogeography of Symbiotic Bacteria Associated with Cultivated and Wild Legumes

Pablo Vinuesa and Claudia Silva

Centro de Investigación sobre Fijación de Nitrógeno, Universidad Nacional Autónoma de México, Av. Universidad S/N, Col. Chamilpa, AP 565-A, Cuernavaca, Morelos CP62210, México, E-mail vinuesa@cifn.unam.mx

13.1 Abstract

Here, we review key issues in bacterial population genetics and evolutionary biology pertinent to the controversial topics of bacterial species concepts and bacterial biogeography. We present a summary of our results and working hypotheses on the latter topics, based on our population genetic and molecular phylogenetic analyses of diverse populations of rhizobial microsymbionts associated with cultivated and wild legumes. This contribution describes our current understanding and thoughts on the biogeography and nature of rhizobial species associated with (1) common beans (*Phaseolus vulgaris*, L.), one of the major grain legume crops worldwide, and (2) with wild genistoid legumes from the Canary Islands, Morocco, and continental Spain.

13.2 Diversity and Nature of the N$_2$-Fixing Legume-Bacterial Nodule Endosymbioses

Several lineages of soil bacteria grouped in the α-2 and β subclasses of the *Proteobacteria* are capable of establishing endosymbiotic N$_2$-fixing associations with most legume genera, particularly with those in the subfamilies Papilionoideae and Mimosoideae (Sprent 2001; Sawada et al. 2003). The association is characterized by the development of symbiotic organs called nodules on the host plant roots or stems. The organogenetic program for the development of nodules is triggered by specific signal molecules called nodulation factors (NFs), which are produced by the bacterial microsymbionts in response to secondary metabolites such as flavonoids and isoflavonoids present in the root exudates of compatible host plants (Spaink et al. 1998). The particular type of NFs produced by a strain largely determines its host range. Hence, from a genetic perspective, bacterial nodule isolates are unique in having so-called nodulation genes (*nod*, *noe* and *nol*) in their genomes, which are responsible for the biosynthesis and secretion of NFs (Spaink et al. 1998). The symbiotic nodulation and nitrogen fixation (*nif* and *fix*) genes from *Rhizobium* and *Sinorhizobium* strains are carried on large symbiotic plasmids (pSyms), which may be 0.4 to >1 Mb in size (Freiberg et al. 1997; Galibert et al. 2001). In other genera like *Bradyrhizobium* or *Mesorhizobium*, they are generally chromosomally encoded and clustered in so-called symbiotic islands (Kaneko et al. 2000; Göttfert et al. 2001). Transfer of pSyms or symbiotic islands to non-symbiotic recipient strains, even of different genera as the donor strain, can transform the recipients into nitrogen-fixing endosymbionts. This has been demonstrated both under laboratory and field conditions (Martínez et al. 1987; Sullivan and Ronson 1998; Rogel et al. 2001).

The Leguminoseae, with its approximately 700 genera and 20,000 species, is the third largest family of flowering plants, behind only orchids (Orchidaceae) and asters (Asteraceae). However, only a fraction of legume genera and species (about 10%) has been surveyed for nodulation, and endosymbionts have been isolated and characterized from even fewer taxa (Moulin et al. 2001; Sprent 2001).

13.3 Diversity and Phylogeny of Legume N$_2$-Fixing Symbionts

Until recently, legume microsymbionts were collectively called rhizobia, a group that included about 30 species of the genera *Rhizobium*, *Bradyrhizobium*, *Sinorhizobium*, *Azorhizobium*, *Mesorhizobium*, and *Allorhizobium* (Sawada et al. 2003). These "classic" rhizobia were defined as endosymbiotic α-*Proteobacteria* recovered from the root nodules of legumes. It has been known for a long time that rhizobial genera are poorly defined on the basis of 16S rRNA sequences as they are intermingled with other non-symbiotic genera, some of which can be converted into symbiotic bacteria after transfer of pSyms or symbiotic islands. The discovery in 2001 of two novel lineages of root-nodule bacteria shook the classic perception of rhizobia. Methylotrophic diazotrophs isolated from diverse Senegalese *Crotalaria* species were identified as members of the genus *Methylobacterium*, which are also α-*Proteobacteria* (Sy et al. 2001). More spectacular were the reports of root-nodule isolates from the β-*Proteobacteria* grouped in the genera *Burkholderia* and *Ralstonia* (Chen et al. 2001; Moulin et al. 2001), isolated from *Aspalathus carnosa* in South Africa and *Mimosa pudica* in Taiwan, respectively. In 2002 a novel root-nodule bacterial lineage was isolated from the nodules of the aquatic legume *Neptunia natans* in India, which was classified as a novel *Devosia* species (Rivas et al. 2002). As of October 2003, 47 species of legume nodule-bacteria have been validly published, which are grouped in 12 genera. A recent phylogenetic analysis of the 16S rDNA sequences of these and related bacteria have shown that the species of nodule bacteria are grouped in nine well-supported monophyletic clusters or clades (Sawada et al. 2003). All these species and genera are delineated on the basis of standard polyphasic taxonomic practice (Vandamme et al. 1996), including 16S rDNA sequences, DNA-DNA hybridization values, and the analysis of multiple phenotypic features.

13.4 The Nature of Bacterial Species and the Problem of Their Delineation

Bacteria multiply asexually by binary fission. Hence, bacteria are essentially clonal organisms. However, many bacterial species are capable of exchanging genetic material across clonal lineages by three parasexual mechanisms: conjugation, transduction, and transformation. The transferred DNA segments generally range from a few hundred base pairs to several hundred kilobases in length. The differences between bacterial and gametic sex are striking: (1) bacterial sex can be highly promiscuous, that is, DNA can be transferred across highly divergent lineages, and (2) bacterial sex is highly localized, meaning that recombinational events affect only a fraction of the recipient's genome (Levin and Bergstrom 2000). Comparative genomic studies have not only uncovered the tremendous impact of lateral gene transfer (LGT) on the evolution of prokaryotic lineages (Ochman et al. 2000; Jain et al. 2002), but have also shown that there is essentially no gene that is "immune" to it (Gogarten et al. 2002). Ironically, ribosomal genes, once thought to be the most suitable molecular markers to delineate prokaryotic species boundaries and even to infer the universal tree of life (Woese 1987; Woese et al. 1990; Stackebrandt and Goebel 1994), are now known to be highly prone to interspecies recombination (Ueda et al. 1999; Yap et al. 1999; Wang and Zhang 2000; van Berkum et al. 2003). The prevalence of LGT in shaping prokaryotic genomes over evolutionary time has opened a debate on whether or not a universal phylogeny can be inferred, and concerns have been raised even about the mere existence of bacterial species (Doolittle 1999; Woese 2000; Lawrence 2002).

The long-term evolutionary consequences of LGT are less relevant to the issues and topics addressed in this review than the consequences and impact that recombination and migration have on shaping the genetic structure of bacterial populations on an ecological time scale. It was not until the early 1990s that analyses of the genetic structures of free-living soil bacteria, such as *Bacillus subtilis* and different rhizobia along with those of several human pathogens, demonstrated that bacterial species exhibit their own population structures, which can range from strictly clonal to essentially panmictic (Istock et al. 1992; Souza et al. 1992; Maynard Smith et al. 1993). Most bacterial species lie somewhere in between (Feil and Spratt 2001), as is the case for rhizobia (Souza et al. 1992; Silva et al. 1999, 2003; Vinuesa et al. 2004a).

Unfortunately, population genetic and evolutionary criteria have been almost completely neglected in bacterial taxonomy, which nowadays is still strongly dominated by typological thinking and similarity criteria. The main genotypic criteria used to delineate bacterial species are >97% overall 16S rDNA sequence identity and >60-70% DNA-DNA homology (Stackebrandt and Goebel 1994; Rosselló-Mora and Amann 2001; Stackebrandt et al. 2002), along with a polyphasic analysis of multiple phenotypic characters (Vandamme et al. 1996). Thus, the species concept used by bacterial taxonomists is static, without lineage perspective. Noteworthy, bacterial species delineation is largely based on finding phylogenetic clades for a single gene (16S rRNA or *rrs*). However, this gene is often too conserved to provide enough resolution at the species level. Furthermore, due to the prevalence of LGT in the bacterial world, single gene phylogenies should never be taken to represent species phylogenies (Dykhuizen and Green 1991; Nichols 2001; Gogarten et al. 2002), a confusion found throughout the vast majority of publications describing new bacterial taxa. The comparative analysis of the genomes of three different *E. coli* strains has revealed that they share only 39% of their protein coding genes (Welch et al. 2002). This finding indicates that different ecotypes of a species may have strikingly different combinations of accessory (adaptive) genes, and therefore it is probably not adequate to use DNA-DNA hybridization values >70% as the "gold standard" for species delineation in bacteriology, as currently done (Stackebrandt and Goebel 1994; Rosselló-Mora and Amann 2001; Stackebrandt et al. 2002).

In view of the nature of bacterial population genetic structures and evolutionary strategies, we use a combination of population genetics based on multilocus enzyme electrophoresis (MLEE) and phylogenetic analyses of multiple sequence loci (targeting different housekeeping and symbiotic genes) to delineate the evolutionary lineages of genes and species, as discussed below.

13.5 Do Bacteria Have Distinctive Distribution Patterns?

It is well established that almost all animal and plant species, particularly the larger ones, have a limited distribution range, which is determined by historical contingencies and environmental cues (Futuyma 1998). Speciation in these organisms generally requires geographic isolation for the interruption of gene flow (vicariance and allopatric speciation), as exemplified by insular faunas and floras (Carlquist 1974; Baldwin et al. 1998). A contrasting picture has emerged for microbial organisms. The astronomic abundance of individuals in microbial species and their tiny body sizes are the postulated reasons why geographical barriers don't seem to be effective in limiting dispersal of microbial cells (Finlay 1998). This notion is consistent with the lack of evidence that flagellated or ciliated protozoan morphospecies, or planktonic foraminifers have distinctive distribution patterns (Darling et al. 2000; Finlay 2002). Since prokaryotes are several orders of magnitude smaller and more abundant than microbial eukaryotes, they are also postulated to be ubiquitously distributed (Finlay 2002), which is consistent with the lack of species differentiation found in different marine and terrestrial prokaryotes, at least at the taxonomic resolution level provided by ribosomal gene sequences (Fulthorpe et al. 1998; Cho and Tiedje 2000; Massana et al. 2000; Martínez-Romero 2002; Bent et al. 2003; Oda et al. 2003). However, studies performed at finer taxonomic resolution levels, such as those provided by multilocus sequence analyses of protein coding genes, or genomic fingerprinting based on polymerase chain reaction (PCR) amplification of repetitive extragenic palindromic sequences (REP-PCR; with BOX, ERIC, or REP primers) have demonstrated the existence of endemic clonal lineages at spatial scales ranging from different continents to a linear 10 m transect along a coastal marsh, and in different habitats such as soils, hot springs, or human bodies (Fulthorpe et al. 1998; Achtman et al. 1999; Cho and Tiedje 2000; Bala et al. 2003; Bent et al. 2003; Falush et al. 2003; Oda et al. 2003; Papke et al. 2003; Whitaker et al. 2003). The authors of some of these studies claim that their works challenge the presumption that ubiquitous dispersal is a universal trait of free-living microorganisms (Finlay 1998, 2002), and that geographical isolation has no impact on the distribution and evolution of bacterial ecotypes in diverse habitats. Thus, the idea that "everything is everywhere and nature selects" is questionable, at least for some bacterial species found in very disjunct and specialized environments such as hot springs and thermal vents (Fenchel 2003). Clearly, multilocus sequence analysis of populations from different geographical areas is the most powerful analytical tool to shed light on this controversy in a proper phylogeographic context (Achtman et al. 1999; Falush et al. 2003; Vinuesa et al. 2004a; Whitaker et al. 2003).

13.6 Population Genetics, Phylogeny and Biogeography of Sympatric *Bradyrhizobium* SpeciesNodulating Endemic Genistoid Legumes in the Canary Islands

The Canary Islands are situated in the northeastern Atlantic Ocean, very close to northwestern Africa. Endemic Canarian genistoid legumes (ECGLs; Papilionoideae:Genisteae) are of great ecological relevance in the archipelago, as they are dominant members and key indicator species of several native plant communities. Our previous studies have shown that ECGLs are nodulated exclusively by members of the genus *Bradyrhizobium* (Vinuesa et al. 1998, 1999; Jarabo-Lorenzo et al. 2000). We showed that two distinct *Bradyrhizobium* species, *B. canariense* sp. nov. and *B. japonicum* bv. *genistearum* nodulate the same ECGLs (Vinuesa et al. 2004a,b). They were delineated by population genetic and phylogenetic criteria, both forming highly supported clades for ITS (rDNA internally transcribed spacers), *atpD* and *glnII* sequences. Interestingly, these sympatric sister species are sexually isolated despite their overlapping ecological niches, as demonstrated by linkage disequilibrium analysis of MLEE data and split decomposition analyses of concatenated ITS, *atpD* and *glnII* sequences. We take this finding as strong evidence for the existence of species (in an evolutionary sense) in the bacterial world in ecological time.

Noteworthy, the two species share monophyletic symbiotic gene sequences (*nodC* and *nifH*), evidencing an ancient lateral transfer event of the chromosomally encoded conjugative symbiotic island across these lineages (Vinuesa et al. 2004a,b). We also showed that the geographic distribution of *B. canariense* sp. nov. includes at least the nearby continental areas of Morocco and Spain, and probably extends to other regions, as has been shown for *B. japonicum* bv. *gen-*

istearum, which can be found also in North America (Jarabo-Lorenzo et al. 2003; Vinuesa et al. 2004a,b). MLEE and sequence data from insular and continental populations of *B. canariense* revealed that there is no significant genetic differentiation between them, and that a significant amount of gene flow is taking place between these populations. The same conclusion was reached for European and American *B. japonicum* bv. *genistearum* populations based on multilocus sequence analyses. Hence, migration and recombination seem to be the most important cohesive forces maintaining species identity in *B. canariense* and *B. japonicum* bv. *genistearum*. We hypothesize that bradyrhizobia are being transported from NW Africa to the Canaries and other parts of the world along with the massive amounts of dust particles arising from NW African storms (Griffin et al. 2002).

Finally, high resolution REP-PCR genomic fingerprints revealed a high diversity of strains across sampling sites, with different epidemic clones being recovered from each of them. This suggests that periodic selection events or selective sweeps (Cohan 2002) are not important in these populations on an ecological time scale, and that clonal lineages are endemic to particular sites, as found in other studies with soil or sediment *Proteobacteria* (Fulthorpe et al. 1998; Cho and Tiedje 2000; Bent et al. 2003; Oda et al. 2003).

13.7 Diversity and Biogeography of Common Bean Rhizobia from Traditionally Managed Milpa Plots in Mexico

The symbiosis between common bean (*Phaseolus vulgaris*) and rhizobia is one of the most studied bacteria-plant interactions (Spaink et al. 1998). All the available archeological, morphological, and molecular evidence suggests that cultivated common bean evolved from its wild relative, with multiple domestication centers in Mesoamerica and in the Andean region of South America around 4,000 years ago (Singh et al. 1991; Kaplan and Lynch 1999). The discovery of the Americas triggered the rapid spread of bean cultivars to Europe and other parts of the world, probably along with American rhizobia carried on their seed coats (Pérez-Ramírez et al. 1998), which are postulated to be the donors of *R. etli*-like pSyms found in diverse rhizobial species that nodulate beans (Amarger et al. 1997; Martínez-Romero 2002; Segovia et al. 1993). Nowadays, common bean is grown in all continents and has great dietary and economic importance (Graham and Vance 2003). Beans were probably codomesticated with maize in Mesoamerica, since these two crops are grown together in a traditional agroecosystem called "milpa" (Souza et al. 1997; Martínez-Romero 2002). Here, we present the results from the analysis of the genetic diversity and population structure of rhizobia associated with common bean cultivated under the milpa system in Mexico.

The root nodules of common bean plants from six milpa plots were sampled in a single year to address the spatial distribution of the genetic variability measured by MLEE. Five genetically related multilocus genotypes (ETs) were recorded in all six plots, evidencing a high ecological dominance of these genotypes, which occupied half of the nodules. On the other hand, more than half of the ETs were recorded only once, showing some degree of genotype endemicity in each plot (Silva et al. 1999). Cluster analysis revealed two distantly related genetic groups, and linkage disequilibrium analyses showed evidence of genetic exchange within each one. However, linkage disequilibrium was found when analyzing both genetic groups together, suggesting sexual isolation between them (Silva et al. 1999). In order to determine the temporal constancy of the genetic structure, and to further characterize the genetic groups found, one plot was sampled over three consecutive years, and in addition to MLEE, other genetic markers targeting the chromosome and plasmids were assessed (Silva et al. 2003). The genetic structure was found to be stable over time; the same genetically related dominant genotypes were recorded, and again the two distantly related genetic groups showed evidence of genetic exchange within but not across them. Moreover, the molecular markers targeting the plasmid compartment revealed substantial evidence of plasmid transfer within each group, but not across groups. The complete set of molecular markers analyzed along with host range experiments showed that the two distantly related groups corresponded to *R. etli* and *R. gallicum*, two previously described species (Amarger et al. 1997; Segovia et al. 1993; Silva et al. 2003). Taken together these results show that: (1) the milpa beans are nodulated by *R. etli* and *R. gallicum*; (2) their genetic structure is stable on the space and time scales analyzed; (3) genetic exchange is an important evolutionary force shaping their genetic structures; and (4) although these species have the ecological opportunities for genetic exchange, it is not detectable, at least for the populations occupying the bean nodules.

To gain a broader view of the genetic relationships of the Mexican *R. etli* and *R. gallicum* populations, and to set them in a phylogeographic framework, three chromosomally and two plasmid encoded genes were sequenced. The sequence data supported the sexual isolation between the two species and provided evidence for plasmid transfer within populations. Phylogenetic analysis showed a low genetic differentiation among *R. etli* and *R. gallicum* isolates from different continents (unpubl. data), suggesting a high degree of migration and gene flow among geographically distant areas. The inclusion of sequences from related species revealed that isolates from other hosts and geographic regions belong to the *R. gallicum* lineage (unpubl. data). In particular, chromosomal genes (*rrs*, *glnII* and *atpD*) showed that *Medicago ruthenica* isolates from Mongolia are indistinguishable from *R. gallicum* isolates, although their plasmid-encoded symbiotic genes (*nifH* and *nodB*) are related to those from *Sinorhizobium* species nodulating *Medicago* spp. Based on classical taxonomic criteria, these isolates were named as a new species, *R. mongolense* (van Berkum et al. 1998). Our analyses show that the *R. mongolense* isolates belong to the *R. gallicum* lineage with symbiotic genes compatible with *M. ruthenica*, and therefore rather correspond to a *R. gallicum* biovariety and not to a different species. The misclassification of the *M. ruthenica* isolates exemplifies the frequent mistake of creating new species or even genera based on ecological characters, an issue widely documented for some pathogenic bacteria (Lan and Reeves 2001).

13.8 Concluding Remarks

The emerging picture from biogeographic studies with different functional groups of marine and terrestrial prokaryotes is that many species are cosmopolitan, with locally adapted or "endemic" ecotype populations. However, many prokaryotic species are ill-defined due to the poor performance of standard taxonomic practice in delineating coherent evolutionary lineages. Therefore, phylogeographic studies of prokaryotes should be based on the sequence analysis of multiple loci, including adaptive and neutral genes. The population structures of bacterial species are also complex and often difficult to determine, because different subpopulations may have different structures (Feil and Spratt 2001). Furthermore, depending on the relative intensity of recombination rate (r) vs. point mutation rate (m), evolutionary change at neutral (housekeeping) loci may be more likely to occur by the former than by the latter force (Feil et al. 2001). In species with high r/m rates, their long-term evolution is dominated by recombination, which may erase any deep-rooted phylogenetic signal. In such species, recombination is an important cohesive force that maintains species integrity. The r/m rates for highly recombinogenous bacteria are apparently not high enough, however, to prevent the emergence of local adaptive clones. Therefore, in order to make sound assertions on the levels of endemicity of local clones and populations it is crucial to assess the impact of recombination on the temporal stability of such clones, which has been performed in very few cases (Silva et al. 2003). Our studies on different rhizobial genera also highlight the significant impact that migration has on shaping the population structures of these bacteria at local and global scales. However, many more studies are required to unravel the dispersion mechanisms of prokaryotes. Examination by appropriate methods of the microbial populations associated with soil dust particles carried across continents by storms originating in Asia and Africa seems a very promising strategy to unravel the distribution pathways of rhizobia, as documented for other soil-borne bacteria and fungi (Griffin et al. 2002; Prospero, this Vol.).

Clearly, the field of bacterial biogeography is in its very infancy and requires many more judicious studies before general conclusions can be drawn about distribution patterns exhibited by diverse functional groups of prokaryotes at various levels of taxonomic resolution, or about the impact that key evolutionary forces (mutation, recombination, and migration) have on shaping their local and global population structures.

Acknowledgements. We would like to express our gratitude to Luis E. Eguiarte, Esperanza Martínez Romero, Valeria Souza, and Dietrich Werner, our mentors, for their intellectual inspiration and generous support. This work was supported by PAPIIT-DGAPA IN205802 from Mexico and by INCO-DEV ICA-4-CT-2001-10057 from the European Union and DFG in the SFB395 project A6.

13.9 References

Achtman M, Azuma T, Berg DE, Ito Y, Morelli G, Pan ZJ, Suerbaum S, Thompson SA, van der Ende A, van Doorn LJ (1999) Recombination and clonal groupings within *Helicobacter pylori* from different geographical regions. Mol Microbiol 32:459-470

Amarger N, Macheret V, Laguerre G (1997) *Rhizobium gallicum* sp. nov. and *Rhizobium giardinii* sp. nov., from *Phaseolus vulgaris* nodules. Int J Syst Bacteriol 47:996-1006

Bala A, Murphy P, Giller KE (2003) Distribution and diversity of rhizobia nodulating agro-forestry legumes in soils from three continents in the tropics. Mol Ecol 12:917-929

Baldwin BG, Crawford DJ, Francisco-Ortega J, Kim S-C, Sang T, Stuessy TF (1998) Molecular phylogenetic insights on the origin and evolution of oceanic island plants. In: Soltis DE, Soltis PS, Doyle JJ (eds) Molecular systematics of plants II: DNA sequencing. Kluwer, Dordrecht, pp 410-441

Bent SJ, Gucker CL, Oda Y, Forney LJ (2003) Spatial distribution of *Rhodopseudomonas palustris* ecotypes on a local scale. Appl Environ Microbiol 69:5192-5197

Carlquist S (1974) Island Biology. Columbia Univ Press, New York

Chen WM, Laevens S, Lee TM, Coenye T, de Vos P, Mergeay M, Vandamme P (2001) *Ralstonia taiwanensis* sp. nov., isolated from root nodules of *Mimosa* species and sputum of a cystic fibrosis patient. Int J Syst Evol Microbiol 51:1729-1735

Cho JC, Tiedje JM (2000) Biogeography and degree of endemicity of fluorescent *Pseudomonas* strains in soil. Appl Environ Microbiol 66:5448-5456

Cohan FM (2002) What are bacterial species? Annu Rev Microbiol 56:457-487

Darling KF, Wade CM, Stewart IA, Kroon D, Dingle R, Brown AJ (2000) Molecular evidence for genetic mixing of Arctic and Antarctic subpolar populations of planktonic foraminifers. Nature 405:43-47

Doolittle WF (1999) Phylogenetic classification and the universal tree. Science 284:2124-2129

Dykhuizen DE, Green L (1991) Recombination in *Escherichia coli* and the definition of biological species. J Bacteriol 173:7257-7268

Falush D, Wirth T, Linz B, Pritchard JK, Stephens M, Kidd M, Blaser MJ, Graham DY, Vacher S, Perez-Perez GI, Yamaoka Y, Megraud F, Otto K, Reichard U, Katzowitsch E, Wang X, Achtman M, Suerbaum S (2003) Traces of human migrations in *Helicobacter pylori* populations. Science 299:1582-1585

Feil EJ, Spratt BG (2001) Recombination and the population structures of bacterial pathogens. Annu Rev Microbiol 55:561-590

Feil EJ, Holmes EC, Bessen DE, Chan MS, Day NP, Enright MC, Goldstein R, Hood DW, Kalia A, Moore CE, Zhou J, Spratt BG (2001) Recombination within natural populations of pathogenic bacteria: short-term empirical estimates and long-term phylogenetic consequences. Proc Natl Acad Sci USA 98:182-187

Fenchel T (2003) Biogeography for bacteria. Science 301:925-926

Finlay BJ (1998) The global diversity of protozoa and other small species. Int J Parasitol 28:29-48

Finlay BJ (2002) Global dispersal of free-living microbial eukaryote species. Science 296:1061-1063

Freiberg C, Fellay R, Bairoch A, Broughton WJ, Rosenthal A, Perret X (1997) Molecular basis of symbiosis between *Rhizobium* and legumes. Nature 387:394-401

Fulthorpe RR, Rhodes AN, Tiedje JM (1998) High levels of endemicity of 3-chlorobenzoate-degrading soil bacteria. Appl Environ Microbiol 64:1620-1627

Futuyma DJ (1998) Evolutionary biology. Sinauer Associates, Sunderland, Mass

Galibert F, Finan TM, Long SR, Puhler A, Abola P, Ampe F, Barloy-Hubler F, Barnett MJ, Becker A, Boistard P, Bothe G, Boutry M, Bowser L, Buhrmester J, Cadieu E, Capela D, Chain P, Cowie A, Davis RW, Dreano S, Federspiel NA, Fisher RF, Gloux S, Godrie T, Goffeau A, Golding B, Gouzy J, Gurjal M, Hernandez-Lucas I, Hong A, Huizar L, Hyman RW, Jones T, Kahn D, Kahn ML, Kalman S, Keating DH, Kiss E, Komp C, Lelaure V, Masuy D, Palm C, Peck MC, Pohl TM, Portetelle D, Purnelle B, Ramsperger U, Surzycki R, Thebault P, Vandenbol M, Vorholter FJ, Weidner S, Wells DH, Wong K, Yeh KC, Batut J (2001) The composite genome of the legume symbiont *Sinorhizobium meliloti*. Science 293:668-672

Gogarten JP, Doolittle WF, Lawrence JG (2002) Prokaryotic evolution in light of gene transfer. Mol Biol Evol 19:2226-2238

Göttfert M, Rothlisberger S, Kundig C, Beck C, Marty R, Hennecke H (2001) Potential symbiosis-specific genes uncovered by sequencing a 410-kilobase DNA region of the *Bradyrhizobium japonicum* chromosome. J Bacteriol 183:1405-1412

Graham PH, Vance CP (2003) Legumes: importance and constraints to greater use. Plant Physiol 131:872-877

Griffin DW, Kellogg CA, Garrison VH, Shinn EA (2002) The global transport of dust - an intercontinental river of dust, microorganisms and toxic chemicals flows through the Earth's atmosphere. Am Sci 90:228-235

Istock CA, Duncan KE, Ferguson N, Zhou X (1992) Sexuality in a natural population of bacteria - *Bacillus subtilis* challenges the clonal paradigm. Mol Ecol 1:95-103

Jain R, Rivera MC, Moore JE, Lake JA (2002) Horizontal gene transfer in microbial genome evolution. Theor Popul Biol 61:489-495

Jarabo-Lorenzo A, Velazquez E, Perez-Galdona R, Vega-Hernandez MC, Martinez-Molina E, Mateos PE, Vinuesa P, Martinez-Romero E, Leon-Barrios M (2000) Restriction fragment length polymorphism analysis of 16S rDNA and low molecular weight RNA profiling of rhizobial isolates from shrubby legumes endemic to the Canary islands. Syst Appl Microbiol 23:418-425

Jarabo-Lorenzo A, Pérez-Galdona R, Donate-Correa J, Rivas R, Velázquez E, Hernández M, Temprano F, Martínez-Molina E, Ruiz-Argüeso T, León-Barrios M (2003) Genetic diversity of bradyrhizobial populations from diverse geographic origins that nodulate *Lupinus* spp. and *Ornithopus* spp. Syst Appl Microbiol 26:611-623

Kaneko T, Nakamura Y, Sato S, Asamizu E, Kato T, Sasamoto S, Watanabe A, Idesawa K, Ishikawa A, Kawashima K, Kimura T, Kishida Y, Kiyokawa C, Kohara M, Matsumoto M, Matsuno A, Mochizuki Y, Nakayama S, Nakazaki N, Shimpo S, Sugimoto M, Takeuchi C, Yamada M, Tabata S (2000) Complete genome structure of the nitrogen-fixing symbiotic bacterium *Mesorhizobium loti*. DNA Res 7:331-338

Kaplan L, Lynch TF (1999) *Phaseolus* (Fabaceae) in archeology: AMS radiocarbon dates and their significance for pre-Columbian agriculture. Econ Bot 53:261-272

Lan R, Reeves P (2001) When does a clone deserve a nave? A perspective on bacterial species based on population genetics. Trends Microbiol 9:419-424

Lawrence JG (2002) Gene transfer in bacteria: speciation without species? Theor Popul Biol 61:449-460

Levin BR, Bergstrom CT (2000) Bacteria are different: observations, interpretations, speculations, and opinions about the mechanisms of adaptive evolution in prokaryotes. Proc Natl Acad Sci USA 97:6981-6985

Martínez E, Palacios R, Sánchez F (1987) Nitrogen-fixing nodules induced by *Agrobacterium tumefaciens* harboring *Rhizobium phaseoli* plasmids. J Bacteriol 169:2828-2834

Martínez-Romero E (2003) Diversity of *Rhizobium-Phaseolus vulgaris* symbiosis: overview and perspectives. Plant Soil 252:11-23

Massana R, DeLong EF, Pedrós-Alió C (2000) A few cosmopolitan phylotypes dominate planktonic archaeal assemblages in widely different oceanic provinces. Appl Environ Microbiol 66:1777-1787

Maynard Smith J, Smith NH, O'Rourke M, Spratt BG (1993) How clonal are bacteria? Proc Natl Acad Sci USA 90:4384-4388

Moulin L, Munive A, Dreyfus B, Boivin-Masson C (2001) Nodulation of legumes by members of the beta-subclass of Proteobacteria. Nature 411:948-950

Nichols R (2001) Gene trees and species trees are not the same. Trends Ecol Evol 16:358-364

Ochman H, Lawrence JG, Groisman EA (2000) Lateral gene transfer and the nature of bacterial innovation. Nature 405:299-304

Oda Y, Star B, Huisman LA, Gottschal JC, Forney LJ (2003) Biogeography of the purple nonsulfur bacterium *Rhodopseudomonas palustris*. Appl Environ Microbiol 69:5186-5191

Papke RT, Ramsing NB, Bateson MM, Ward DM (2003) Geographical isolation in hot spring cyanobacteria. Environ. Microbiol. 5:650-659

Pérez-Ramírez NO, Rogel MA, Wang E, Castellanos JZ, Martiínez-Romero E (1998) Seeds of *Phaseolus vulgaris* bean carry *Rhizobium etli*. FEMS Microbiol Ecol 26:289-296

Rivas R, Velazquez E, Willems A, Vizcaino N, Subba-Rao NS, Mateos PF, Gillis M, Dazzo FB, Martinez-Molina E (2002) A new species of *Devosia* that forms a unique nitrogen-fixing root-nodule symbiosis with the aquatic legume *Neptunia natans* (L.f.) druce. Appl Environ Microbiol 68:5217-5222

Rogel MA, Hernandez-Lucas I, Kuykendall LD, Balkwill DL, Martinez-Romero E (2001) Nitrogen-fixing nodules with *Ensifer adhaerens* harboring *Rhizobium tropici* symbiotic plasmids. Appl Environ Microbiol 67:3264-3268

Rosselló-Mora R, Amann R (2001) The species concept for prokaryotes. FEMS Microbiol Rev 25:39-67

Sawada H, Kuykendall LD, Young JM (2003) Changing concepts in the systematics of bacterial nitrogen-fixing legume symbionts. J Gen Appl Microbiol 49:155-179

Segovia L, Young JP, Martinez-Romero E (1993) Reclassification of American *Rhizobium leguminosarum* biovar *phaseoli* type I strains as *Rhizobium etli* sp. nov. Int J Syst Bacteriol 43:374-377

Silva C, Eguiarte LE, Souza V (1999) Reticulated and epidemic population genetic structure of *Rhizobium etli* biovar phaseoli in a traditionally managed locality in Mexico. Mol Ecol 8:277-287

Silva C, Vinuesa P, Eguiarte LE, Martinez-Romero E, Souza V (2003) *Rhizobium etli* and *Rhizobium gallicum* nodulate common bean (*Phaseolus vulgaris*) in a traditionally managed milpa plot in Mexico: population genetics and biogeographic implications. Appl Environ Microbiol 69:884-893

Singh SP, Gepts P, Debouck DG (1991) Races of common bean (*Phaseolus vulgaris*, Fabaceae). Econ Bot 45:379-396

Souza V, Nguyen TT, Hudson RR, Piñero D, Lenski RE (1992) Hierarchical analysis of linkage disequilibrium in *Rhizobium* populations: evidence for sex? Proc Natl Acad Sci USA 89:8389-8393

Souza V, Bain J, Silva C, Bouchet V, Valera A, Marquez E, Eguiarte LE (1997) Ethnomicrobiology: do agricultural practices modify the population structure of the nitrogen fixing bacteria *Rhizobium etli* biovar *phaseoli*? J Ethnobiol 17:249-266

Spaink HP, Kondorosi A, Hooykaas PJJ (1998) The Rhizobiaceae. Kluwer, Dordrecht

Sprent JI (2001) Nodulation in legumes. Royal Botanic Gardens, Kew

Stackebrandt E, Goebel BM (1994) Taxonomic note: a place for DNA-DNA reassociation and 16S rRNA sequence analysis in the present species definition in bacteriology. Int J Syst Bacteriol 44:846-849

Stackebrandt E, Frederiksen W, Garrity GM, Grimont PA, Kampfer P, Maiden MC, Nesme X, Rossello-Mora R, Swings J, Truper HG, Vauterin L, Ward AC, Whitman WB (2002) Report of the *ad hoc* committee for the re-evaluation of the species definition in bacteriology. Int J Syst Evol Microbiol 52:1043-1047

Sullivan JT, Ronson CW (1998) Evolution of rhizobia by acquisition of a 500-kb symbiosis island that integrates into a phe-tRNA gene. Proc Natl Acad Sci USA 95:5145-5149

Sy A, Giraud E, Jourand P, Garcia N, Willems A, de Lajudie P, Prin Y, Neyra M, Gillis M, Boivin-Masson C, Dreyfus B (2001) Methylotrophic *Methylobacterium* bacteria nodulate and fix nitrogen in symbiosis with legumes. J Bacteriol 183:214-220

Ueda K, Seki T, Kudo T, Yoshida T, Kataoka M (1999) Two distinct mechanisms cause heterogeneity of 16S rRNA. J Bacteriol 181:78-82

Van Berkum P, Beyene D, Bao G, Campbell TA, Eardly BD (1998) *Rhizobium mongolense* sp. nov. is one of three rhizobial genotypes identified which nodulate and form nitrogen-fixing symbioses with *Medicago ruthenica* [(L.) Ledebour]. Int J Syst Bacteriol 48:13-22

Van Berkum P, Terefework Z, Paulin L, Suomalainen S, Lindstrom K, Eardly BD (2003) Discordant phylogenies within the *rrn* loci of rhizobia. J Bacteriol 185:2988-2998

Vandamme P, Pot B, Gillis M, de Vos P, Kersters K, Swings J (1996) Polyphasic taxonomy, a consensus approach to bacterial systematics. Microbiol Rev 60:407-438

Vinuesa P, Rademaker JLW, de Bruijn FJ, Werner D (1998) Genotypic characterization of *Bradyrhizobium* strains nodulating endemic woody legumes of the Canary Islands by PCR-restriction fragment length polymorphism analysis of genes encoding 16S rRNA (16S rDNA) and 16S-23S rDNA intergenic spacers, repetitive extragenic palindromic PCR genomic fingerprinting and partial 16S rDNA sequencing. Appl Environ Microbiol 64:2096-2104

Vinuesa P, Rademaker JLW, de Bruijn FJ, Werner D (1999) Characterization of *Bradyrhizobium* spp. strains by RFLP analysis of amplified 16S rDNA and rDNA intergenic spacer regions. In: Martínez E, Hernández G (eds) Highlights on nitrogen fixation. Plenum, New York, pp 275-279

Vinuesa P, Silva C, Werner D, Martínez-Romero E (2004a) Population genetics and phylogenetic inference in bacterial molecular systematics: the roles of migration and recombination in *Bradyrhizobium* species cohesion and delineation. Mol Phylogenet Evol (in press)

Vinuesa P, León-Barrios M, Silva C, Willems A, Jarabo-Lorenzo A, Pérez-Galdona R, Werner D, Martínez Romero E (2004b) *Bradyrhizobium canariense* sp. nov., an acid-tolerant endosymbiont isolated from the nodules of endemic genistoid legumes (Papilionoideae:Genisteae) growing in the Canary Islands. Int. J. Syst. Evol. Microbiol. (submitted)

Wang Y, Zhang Z (2000) Comparative sequence analyses reveal frequent occurrence of short segments containing an abnormally high number of non-random base variations in bacterial rRNA genes. Microbiology 146(Pt 11):2845-2854

Welch RA, Burland V, Plunkett G 3rd, Redford P, Roesch P, Rasko D, Buckles EL, Liou SR, Boutin A, Hackett J, Stroud D, Mayhew GF, Rose DJ, Zhou S, Schwartz DC, Perna NT, Mobley HL, Donnenberg MS, Blattner FR (2002) Extensive mosaic structure revealed by the complete genome sequence of uropathogenic *Escherichia coli*. Proc Natl Acad Sci USA 99:17020-17024

Whitaker RJ, Grogan DW, Taylor JW (2003) Geographic barriers isolate endemic populations of hyperthermophilic archaea. Science 301:976-978

Woese CR (1987) Bacterial evolution. Microbiol Rev 51:221-271

Woese CR (2000) Interpreting the universal phylogenetic tree. Proc Natl Acad Sci USA 97:8392-8396

Woese CR, Kandler O, Wheelis ML (1990) Towards a natural system of organisms: proposal for the domains Archaea, Bacteria, and Eucarya. Proc Natl Acad Sci USA 87:4576-4579

Yap WH, Zhang Z, Wang Y (1999) Distinct types of rRNA operons exist in the genome of the actinomycete *Thermomonospora chromogena* and evidence for horizontal transfer of an entire rRNA operon. J Bacteriol 181:5201-5209

14 Micro-spatial Distribution of Bacteria in the Rhizosphere

Yusuke Unno[1], Takuro Shinano[2], Jun Wasaki[1] and Mitsuru Osaki[1]

[1]Graduate School of Agriculture, Hokkaido University, Sapporo, 060-8589, Japan
[2]Creative Research Initiative "Sousei" (CRIS), Hokkaido University, Sapporo, 001-0020, Japan, E-mail takuro@chem.agr.hokudai.ac.jp

14.1 Abstract

Long-term experimental fields (having either no nitrogen, phosphorus, or potassium fertilizer or with complete application of N, P, and K fertilizers since 1914) were used for the investigation of bacterial habitat distribution in the rhizosphere. From the rhizosphere of *Lupinus albus* L., we screened rhizobacteria by their utilization ability of phytate as sole carbon (C) and phosphorus (P) source. Though almost all of the isolates were identified as *Burkholderia* genus, there was a wide variation in their phosphate utilization. By comparing isolates from rhizosphere soil and isolates from the rhizoplane, those isolates from the latter exhibit higher inorganic phosphate uptake ability when compared at the same extracellular phytase level. We propose that phytate-utilizing bacteria may adapt and segregate from their habitat along a spatial gradient of C and P availability in the rhizosphere. That is, rhizobacteria, which live very close to the rhizoplane, utilize phytate mainly as a P source rather than as a C source, because a large amount of C is provided from roots, whereas this may not be true for P. On the other hand, rhizosphere bacteria that live in a remote area from roots utilize phytate as a C source rather than as a P source, because the lack of available C may restrict the growth of bacteria.

14.2 Introduction

The rhizosphere is the "intersphere" between plant and soil, and it is generally considered that rhizosphere soil exists just around the root (Fig. 1). The rhizosphere's definition has thus been proposed to be the soil with high microbial activity around the legume root (Hiltner 1904). Though it has been considered that relatively larger amounts of and more active microorganisms live in the rhizosphere because of the supply of carbon from plant roots, detailed information is restricted to pathogens, rhizobium, and mycorrhiza. This lack of information is due to the difficulty of isolating and identifying microorganisms from the soil. However, recently the introduction of molecular biological techniques, e.g., direct sequencing, restriction fragment length polymorphism (RFLP), amplified fragment length polymorphism (AFLP), polymerase chain reaction-denaturing gradient gel electrophoresis (PCR-DGGE), array technique, etc., to the study of soil microorganisms has yielded more precise determination.

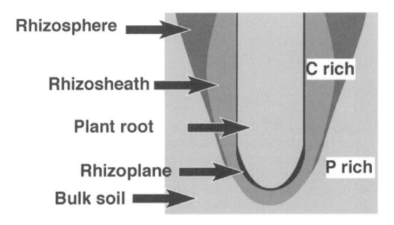

Fig. 1. Ideological model of a rhizosphere. It is considered that as the C supply from plant roots is high, C content is highest near the roots (C-rich). On the other hand, as the plant vigorously takes up P from the soil solution, the level of P close to the root plane is low, thus as distance from the root increases, the soil becomes increasingly relatively P-rich, as shown

14.3 Materials and Methods

We have isolated rhizobacteria of white lupin (*Lupinus albus* L.) grown in a long-term experimental field that had not been fertilized with phosphorus for 90 years. Total rhizosphere bacteria were further divided into rhizoplane-derived and rhizosphere-derived by the water immersion method (Ishizawa et al. 1958). By using inorganic phosphate-limiting selecting culture with only phytate (sodium inositol hexa-phosphate; see Dalal 1977) as the only applied carbon and phosphorus source, we obtained over 300 isolates. DNA isolation was based on Hiraishi (1992), and 16S rDNA sequence was determined by using the following two primers: 5′-GACTACCAGGGTATCTAATC-3′ (positions 805 to 786 of *E. coli* 16S rRNA), and r3L: 5′-TTGCGCTCGTTGCGGGACT-3′ (positions 1111 to 1093 of *E. coli* 16S rRNA). The sequencing reactions were performed using a

DTCS Quick Start kit (Beckman Coulter, CA, USA), and sequenced by CEQ 8000 (Beckman Coulter). Each isolate was cultivated in vitro, then phytase activity and inorganic phosphorus content in the solution were determined based on the methods of Richardson and Hadobas (1997) and Murphy and Riley (1962). One unit (U) of activity was defined as 1 µmol Pi released from Na-IHP per minute. AFLP fingerprinting was performed based on the manufacturer's instructions and Coenye et al. (1999) using *Taq* I and *Apa* I as restriction enzymes.

14.4 Results and Discussion

From the analysis of 16S rDNA sequences, the isolates were determined to be mainly *Burkholderia* species. While based on biochemical analysis, a clear habitat segregation around roots was observed; that is, rhizosphere soil isolates showed low P uptake ability at the same extracellular phytase activity (meaning not all of the phosphate released by phytase was utilized by the isolate itself) when compared with rhizoplane bacteria. The area closest to plant root (rhizoplane) is strongly influenced by the plant's demand for nutrient, although a sufficient supply of C (e.g., polysaccharides, sugars, organic acids, amino acids, etc.) from plant roots existed for the bacteria. There is a close interaction between plant and rhizobacteria regarding phytate as a C and P source under the micro-scale spatial bacterial diversities in the rhizosphere.

The genus *Burkholderia* comprises 29 species, including soil and rhizosphere bacteria as well as plant and human pathogens (Yabuuchi et al. 1993). This genus is very diverse, and includes plant growth promoting rhizobacteria (PGPR) and rhizobia (Moulin et al. 2001), and a significant proportion of maize-, wheat-, and (many more) lupin-associated bacteria have been demonstrated (Balandreau et al. 2001). Sequence analysis of partial 16S rDNA of each *Burkholderia* isolate reveals that the difference between those isolates from rhizoplane and rhizosphere is quite small (sometimes the similarity was 100%); thus, we could not find any species difference based on the C and P nutrition around the root. On the other hand, by using the AFLP analysis, a clear difference was observed among isolates. The AFLP technique can analyze the difference in genome construction, and therefore this result shows a potential for bacterial genome rearrangement in the lupin rhizosphere. In addition, it has been pointed out that the rhizosphere, especially on the rhizoplane, is a "hot spot" for horizontal gene transfer (van Elsas et al. 1998), because this area is normally filled with mucigel, which acts as a nutrient provider at sites that grow microbes abundantly. Thus we assume that there was some mechanism for developing quick genetic variation for adapting to micro-spatial nutrient conditions in the rhizosphere. This might be caused by the rapid migration of functional plasmid or the transfer of more integrated genes, such as observed in symbiotic island transfer (Sullivan and Ronson 1998), or rapid gene rearrangement (LiPuma 1998)

14.5 Conclusions

Although the rhizosphere is a very limited space existing between soil and root, a clear functional distribution of bacteria was found to exist, based on the C and P availability as well as the distance from the root. As the 16S rDNA-sequence-based difference between two isolate groups obtained from rhizoplane and rhizosphere was quite small, we assume that some kind of genetic modification (e.g., plasmid- or island-based genetic migration, genome rearrangement, etc.) occurred within the very limited micro-scale area around the root. Such microbial genetic modification may be occurring more frequently than expected, and may play an important role in nutrient dynamics in the soil.

14.6 References

Balandereau J, Viallard V, Cournoyer B, Coenye T, Laevens S, Vandamme P (2001) *Burkholderia cepacia* genomovar III is a common plant-associated bacterium. Appl Environ Microbiol 67:982-985

Coenye T, Schouls LM, Govan JRW, Kersters K, Vandamme P (1999) Identification of *Burkholderia* species and genomovars from cystic fibrosis patients by AFLP fingerprinting. Int J Syst Bacteriol 49:1657-1666

Dalal RC (1977) Soil organic phosphorus. Adv Agron 29:83-117

Hiltner L (1904) Über neuere Erfahrungen und Probleme auf dem Gebiet der Bodenbakteriologie und unter besonderer Berücksichtingung der Gründüngung und Brache. Arb Dtsch Landw Ges 98:59-78

Hiraishi A (1992) Direct automated sequencing of 16S rDNA amplified by polymerase chain reaction from bacteria cultures without DNA purification. Lett Appl Microb 15:210-213

Ishizawa S, Suzuki T, Sato O, Toyoda H (1958) Studies on microbial population in the rhizosphere of higher plants with special reference to the method of study. Plant Soil Food 3:85-94

LiPuma JJ (1998) *Burkholderia cepaciea.* Management issues and new insights. Curr Opin Pulm Med 4:337-341

Moulin L, Munive A, Dreyfus B, Boivin-Masson C (2001) Nodulation of legumes by members of beta-subclass of proteobacteria. Nature 411:948-950

Murphy J, Riley JP (1962) A modified single solution method for the determination of phosphate in natural waters. Anal Chim Acta 27:31-36

Richardson AE, Hadobas PA (1997) Soil isolates of *Pseudomonas* spp. that utilize inositol phosphates. Can J Microbiol 43:509-516

Sullivan JT, Ronson CW (1998) Evolution of rhizobia by acquisition of a 500-kb symbiosis island that integrates into a phe-tRNA gene. Proc Natl Acad Sci USA 95:5145-5149

Van Elsas JD, Gardener BB, Wolters AC, Smit E (1998) Isolation, characterization, and transfer of cryptic gene-mobilizing plasmids in the wheat rhizosphere. Appl Environ Microbiol 64:880-889

Yabuuchi E, Kosako Y, Oyaizu H, Yano I, Hotta H, Hashimoto Y, Ezaki T, Arakawa M (1993) Proposal of *Burkholderia* gen. nov. and transfer of seven species of the genus *Pseudomonas* homology group II to the new genus, with the type species *Burkholderia cepasia* (Palleronhi and Holmes 1981) comb. nov. Microbiol Immunol 36:1251-1275

15 Colonization of Some Polish Soils by *Azotobacter* spp. at the Beginning and at the End of the 20th Century

Stefan Martyniuk

Department of Agricultural Microbiology, Institute of Soil Science and Plant Cultivation, Czartoryskich 8, 24-100 Pulawy, Poland, E-mail sm@iung.pulawy.pl

15.1 Abstract

The genus *Azotobacter* comprises aerobic, free-living diazotrophic (with the ability to use N_2 as the sole nitrogen source) soil bacteria, occurring also in other habitats, e.g., water, rhizosphere. The genus *Azotobacter* includes six species, with *A. chroococcum* most commonly inhabiting various soils all over the world. The occurrence of other *Azotobacter* species seems to be much more restricted in nature, e.g., *A. paspali* can be found only in the rhizosphere of a grass, *Paspalum notatum*. In temperate climate, soil populations of *Azotobacter* spp. rarely exceed several thousands cells per gram of neutral or alkaline soils, and in acid (pH <6.0) soils these bacteria are generally absent or occur in very low numbers. With respect to Polish soils, in 1923 Ziemiecka published results of her studies on the occurrence of *Azotobacter* spp. in soil samples collected in 1917 and 1918 from 28 locations in the former Polish Kingdom. These studies showed that 50% of the examined soils contained *Azotobacter* spp. In the year 2000 the occurrence of *Azotobacter* spp. in 31 Polish soils was assessed and results of this study were compared with those published by Ziemiecka. Almost 52% of the soils tested in 2000 were colonized by *Azotobacter* spp., indicating that intensification of agricultural practices that took place during the past century did not significantly change (50% in 1917-1918) colonization of Polish soils by the studied bacteria. When present, numbers of these bacteria varied widely from several to almost 10,000 cfu g^{-1} of soil. The results are discussed in relation to the use of mineral fertilizers, particularly nitrogen, and changes in the soil pH.

15.2 Introduction

Aerobic bacteria belonging to the genus *Azotobacter* represent a diverse group of free-living diazotrophic (with the ability to use N_2 as the sole nitrogen source) microorganisms commonly occurring in soil. The genus *Azotobacter* includes six species, with *A. chroococcum* most commonly inhabiting various soils all over the world. The occurrence of other *Azotobacter* species is much more restricted in nature, e.g., *A. paspali* can be found only in the rhizosphere of a grass, *Paspalum notantum* (Tchan and New 1984; Döbereiner 1995). An important ecological feature of bacteria from the genus *Azotobacter* is their sensitivity to soil reaction; in acid (pH <6.0) soils these bacteria are generally absent or occur in very low numbers. Populations of *Azotobacter* spp. in neutral or alkaline soils are also not very high and rarely exceed several thousands cells per gram of soil (Gainey 1923; Zawislak 1973; Mahmoud et al. 1978; Martyniuk and Wagner 1978; Balandreau 1986). Other soil factors or soil management practices, e.g., salinity, phosphorus, and organic matter contents, or soil cultivation, can also affect colonization of soils with *Azotobacter* spp. (Ziemiecka 1923; Strzelczyk 1958; Zawislak 1973; Mahmoud et al. 1978; Martyniuk and Wagner 1978). For example, Zawislak (1973) detected *Azotobacter* spp. more frequently, and in higher numbers, in cultivated soils than in uncultivated (sodded) soils collected from hills in the Olsztyn province, suggesting that agricultural practices might create environmental conditions more favorable for the development and survival of *Azotobacter* in soil.

15.3 Colonization of some Polish Soils at the Beginning of the 20th Century

In 1923 Ziemiecka published results of her pioneer studies on the occurrence of *Azotobacter* spp. in soil samples collected in 1917 and 1918 from 28 locations in the former Polish Kingdom. These studies showed that 50% of the examined soils contained *Azotobacter* spp. (Table 1). Using the most probable number (MPN) technique, Ziemiecka assessed the following numbers of these bacteria in the soils: eight soils contained low populations of *Azotobacter* spp., in five soils moderate numbers were found, and only one soil was colonized by relatively high numbers of the examined bacteria (Table 1).

Table 1. Populations of *Azotobacter* spp. in soils (*n*=28) examined in 1917-1918

	Number of soils
Soils colonized by *Azotobacter*	14
Soils with:	
Low populations (<100 cells g^{-1})	8
Moderate populations (100-1,000 cells g^{-1})	5
High populations (>1,000 cells g^{-1})	1

Most of the soils (28) studied by Ziemiecka (1923) were also characterized for their chemical properties, such as soil reaction and contents of: humus, total N, P_2O_5, and CaO. Among these parameters, soil reaction was found to be the most important environmental factor influencing the occurrence and numbers of *Azotobacter* spp. in the studied soils.

In 2000 we examined populations of various groups of microorganisms, including *Azotobacter* spp., in 31 different soils from Poland. It seemed interesting to compare our data with those reported by Ziemiecka (1923) to see whether intensification of agricultural practices that took place during the course of the past century caused any changes in colonization of Polish soil by *Azotobacter* spp.

15.4 Colonisation of Some Polish Soils at the End of the 20th Century

Numbers of colony-forming units (cfu) of *Azotobacter* spp. in the examined soils were assessed by the dilution-pour plates method (Fenglerowa 1965) on N-free agar medium containing: $K_2 HPO_4$ 0.5 g, $MgSO_4$ 0.2 g, NaCl 0.2 g, $CaCO_3$ 5 g, sucrose 10 g, agar 12 g, H_2O dist. 1,000 ml, and traces of Mn, Fe, and Mo. On this medium *Azotobacter* spp. form large, moist colonies after 48-72 h of incubation at 28°C. All colonies turned dark brown after 5-7 days of incubation, indicating that they belonged to *A. chroococcum* (Tchan and New 1984; Dobereiner 1995). However, we did not aim at isolation or exact identification of these bacteria, and for this reason no species name of *Azotobacter* is used throughout this paper.

Table 2. Colonization of some Polish soils ($n=31$) with *Azotobacter* spp. in the year 2000

	Number of soils
Soils colonized with *Azotobacter*	16
Soils with:	
Low populations (8-97 cfu g^{-1})[a]	4
Moderate populations (130-810 cfu g^{-1})	11
High populations (99,000 cfu g^{-1})	1

[a]Colony-forming units in 1 g of soil dry matter

We detected various populations of *Azotobacter* spp. in 16 of 31 different Polish soils examined in the year 2000 (Table 2). Thus, the percentage of soils with *Azotobacter* (51.6%) found in our studies was only slightly higher than that (50%) reported by Ziemiecka (1923). When present, numbers of these bacteria varied widely, from several to almost 10,000 cells (cfu) in one of the examined soils. However, the extreme numbers were rare and in the majority of the soils with *Azotobacter* spp. populations of these bacteria varied within the range of about 130 to 810 cfu g^{-1} (Table 2). It is interesting to note that even though Ziemiecka used different methodology (liquid media) to detect and assess the MPN of *Azotobacter* spp., her estimates of soil populations of these bacteria were very similar to those presented in our work. For example, in only one soil was the MPN of *Azotobacter* spp. assessed as being "higher than 1,000 cells", while in other soils examined by Ziemiecka populations of these bacteria (when present) were most often placed within the range of 100-1,000 cells g^{-1}. Numbers of *Azotobacter* spp. reported by Zawislak (1973) also rarely exceeded 1,000 cfu per 1 g of soil.

Our results were not intended to represent a detailed and exact comparison with those published by Ziemiecka (1923). This was impossible for several reasons. For example, beside the above-mentioned differences in the methods used to count populations of *Azotobacter* spp., we could not analyze the same soils as Ziemiecka because their exact locations were not given by the author. Having these restrictions in mind, comparison of our results with those described by Ziemiecka (1923) seemed to us interesting and justified, particularly with respect to the percentage of soils colonized by *Azotobacter* spp. during the first two decades of 20th century and over 80 years later. This comparison indicates that intensification of agricultural practices during the course of the past century did not significantly change colonization of Polish soils by the studied bacteria. Detailed discussion on changes in Polish agriculture taking place during the course of the 20th century, and on the effects of these changes on soil quality, is beyond the scope and volume of this chapter. However, the use of mineral fertilizers and their effect on soil chemical properties, particularly soil pH, deserve a brief consideration in relation to *Azotobacter*. It has been well documented in the long-term field experiments that mineral N fertilizers may cause substantial acidification of soil, particularly when used in high doses without liming (Mercik 1994; Filipek and Kaczor 1997). Since *Azotobacter* is very sensitive to soil acidity, one might expect intensification of the use of mineral N fertilizers to cause acidification of agricultural lands and thus a reduction of soil populations of *Azotobacter* spp. or even elimination of these bacteria from some soils. A condensed description of changes in Polish agriculture, including the use of mineral fertilizers in the 19th and 20th centuries, is given by Krasowicz (2002). For example, application of mineral N increased in Poland (on average) from less than 5 kg N ha^{-1} in the 1940s to about 70 kg N in the 1980s. However, data presented by Lipinski (2000) prove that acidification of Polish soils did not occur in the period between 1955 and 1999. On the contrary, slight improvement of this property of Polish soils could be seen in that period, probably as a result of lime application by farmers to counterbalance soil acidification by mineral fertilizers. For example, the area of acidic soils (pH <5.5) decreased from 58% in the decade 1955-1964 to 55% in the period of 1994-1999, and simultaneously the area of soils having the pH >6.5 increased from 17 to 19% in the respective periods. Thus, these data seem to correspond with the slightly higher percentage of soils colonized by *Azotobacter* spp. found in our studies as compared to that presented by Ziemiecka (1923).

More research is needed under different environmental conditions, e.g., in different countries, to see if this group of soil microorganisms can be used as an indicator of changes in soil biological resources related to sustainable agriculture.

15.5 References

Balandreau J (1986) Ecological factors and adaptive processes in N$_2$-fixing bacterial populations of the plant environment. Plant Soil 90:73

Döbereiner J (1995) Isolation and identification of aerobic nitrogen-fixing bacteria from soil and plants. In: Alef K, Nannipieri P (eds) Methods in applied soil microbiology and biochemistry. Academic Press, London, pp 134-141

Fenglerowa W (1965) Simple method for counting *Azotobacter* in soil samples. Acta Microb Pol 14:615

Filipek T, Kaczor A (1997) Dynamika antropogenicznej presji zakwaszenia rolniczej przestrzeni produkcyjnej w Polsce. In: Kukula S et al. (eds) Ochrona i wykorzystanie rolniczej przestrzeni produkcyjnej Polski. IUNG, Pulawy, pp 35-43

Gainley PL (1923) Influence of the absolute reaction of a soil upon *Azotobacter* flora and nitrogen fixing ability. J Agric Res 24:907

Krasowicz S (2002) Produkcja roślinna na ziemiach polskich w XIX i XX wieku - rys historyczny. Pam Pul 130/I:11

Lipiński W (2000) Odczyn i zasobnosc gleb w swietle badan stacji chemiczno-rolniczych. Nawozy i Nawozenie 3:89

Mahmoud SAZ, El-Sawy M, Ishac YZ, El-Safty MM (1978) The effect of salinity and alkalinity on the distribution and capacity of N_2-fixation by *Azotobacter* in Egyptian soils. In: Granhall U (ed) Environmental role of nitrogen-fixing blue-green algae and asymbiotic bacteria. The Swedish Natural Science Research Council/NFR, Uppsala, Sweden, pp 99-109

Martyniuk S, Wagner GH (1978) Quantitative and qualitative examination of soil microflora associated with different management systems. Soil Sci 125:343

Mercik S (1994) Most important soil properties and yielding in long-term static fertilising experiments in Skierniewice. Rocz Glebozn 44 [Suppl]:71

Strzelczyk E (1958) Wplyw roznych roslin uprawnych na liczebnosc azotobaktera i *Clostridium* w ich ryzosferze. Acta Microb Polon 7:112

Tchan Y-T, New PB (1984) Genus I. *Azotobacter* Beijerinck 1901. In: Holt JG, Krieg NR (eds) Bergey's manual of systematic bacteriology, vol 1. Williams and Wilkins, Baltimore, pp 220-229

Zawislak K (1973) Wystepowanie azotobaktera w glebach na stokach w wojewodztwie olsztynskim. Rocz Glebozn 24:344

Ziemiecka J (1923) Wystepowanie azotobaktera w glebach polskich. Rocz Nauk Roln 10:1-78

16 Estimation of Biotic and Abiotic Factors That Affect Migration of Rhizobia

En Tao Wang[1] and Wen Xin Chen[2]

[1]Depto. de Microbiología, Escuela Nacional de Ciencias biológicas, Instituto Politécnico Nacional, México D.F., México, E-mail ewang@bios.encb.ipn.mx
[2]Dept. of microbiology, College of Biological Science, China Agricultural University, Beijing 100094, China

16.1 Abstract

Rhizobia are soil bacteria that form nitrogen-fixing nodules on leguminous plants. Three possible ways have been estimated for the long-distance transport of these bacteria: by introduction of legume seeds or plants that are accompanied naturally by rhizobia, by application of rhizobial inoculants, and by rivers that carry rhizobia downstream. In all these three forms, or any other possible way, rhizobia are moved passively and they have to face biotic and abiotic factors in a new environment. So, a successful migration of rhizobia mainly depends on whether they can colonize in the new environment after transportation. Since rhizobia are symbiotic bacteria associated with legumes, their colonization in new environments depends on both their host plants and environmental factors. The colonization of rhizobia is easier when the new environment is similar to the old and the host plants have restricted nodulation ability, as with the introduction of *Amorpha fruticosa* from USA to China. In contrast, the introduced rhizobia may be eliminated if the new environment is quite different and the hosts can nodulate with a wide range of rhizobia, as in the case of soybean. In any event, a successful migration of rhizobia can affect the local microbial community by the establishment of a new bacterial population and by the lateral gene transfer of symbiotic genes or other genetic elements. Here, some investigations relative to rhizobial migration will be reviewed.

16.2 Introduction

Rhizobia are one of the most important microbial resources. They are Gram-negative bacteria that induce the formation of nitrogen-fixing root and/or stem nodules in leguminous plants. This symbiotic system contributes a major part of the combined global nitrogen input. It also helps legumes to inhabit adverse environments. Since many leguminous plants are valuable crops or woody plants for forestation, many of them have been introduced to other geographic regions from their original habitats. Examples are common bean, soybean, peanut, and *Leu-*

caena leucocephala. Considering the symbiotic relationships between the legume species and the rhizobia, the migration of legumes should affect the migration of related rhizobia, including their long-distance transportation and their colonization in a new environment.

16.3 Transportation of Rhizobia

It has been proposed that rhizobia can be relocated over long distances by their natural accompaniment of transported legume seeds or plants (Pérez-Ramirez et al. 1998), by application of rhizobial inoculants, and by rivers into which rhizobia can be released from soils (Wang and Martinez-Romero 2000).

By seeds. Many legume plants have been introduced from their original geographic location to other regions. In many cases, people do not realize that the rhizobia associated with these plants are also introduced. The seed-bearing rhizobial cells have been documented for *S. meliloti* in alfalfa, *R. etli* in common bean, *Azorhizobium caulinodans* in *Sesbania rostrata*, and *R. leguminosarum* bv. *trifolii* in red clover (see Pérez-Ramirez et al. 1998). In the case of *R. etli* in common bean seeds, the rhizobial cells have been found to survive for several years at room temperature (Pérez-Ramirez et al. 1998). So, it is very possible that some rhizobia have been dispersed by commercial transportation of seeds. The resistance of rhizobial strains to drought may be a determinative factor in the successful relocation, and investigation has indicated that the seed-borne *R. etli* strains that survived transportation were restricted to limited lineages (Pérez-Ramirez et al. 1998).

By utilization of commercial rhizobial inoculants. Because of the great ecologic and economic value of symbiotic nitrogen fixation, commercial rhizobial inoculants have been developed and applied in the fields for many leguminous crops. There is no doubt that the use of these inoculants and their introduction in many countries causes artificial dispersion of rhizobia.

By running water. As soil bacteria, it is easy for rhizobia to be rushed into rivers by rainwater or by irrigation. Thus, running water may be an important medium for long-distance transportation of some rhizobia, especially for those related to aquatic legumes. Examples are *Azorhizobium caulinodans* and *Rhizobium huautlense* associated with *Sesbania* species, *Allorhizobium undicola* and *Devosia natans* associated with *Naptunia natans* (Rivas et al. 2003), and the stem nodule-forming photosynthetic *Bradyrhizobium* associated with *Aeschynomene*. We have detected the existence of *R. huautlense* in water (Wang and Martinez-Romero 2000). Recently, nodulation on *Aeschynomene indica* has been confirmed in a budding and cluster-forming freshwater bacterium *Blastobacter denitrificans* (van Berkum and Eardly 2002) that is isolated from water and has a close phylogenetic relationship with *Bradyrhizobium*. These data indicate that rhizobia can survive in fresh water and can be transported by running water.

Besides the three ways mentioned above, there may be other mechanisms of long-distance transportation of rhizobia. However, in all of the possible ways, the migration of rhizobia is a passive procedure, and they have no influence over their location, regardless of whether or not the new environment is suitable for them. A successful migration of rhizobia depends on their ability to colonize in the new environment.

16.4 Colonization

For colonization in a new environment, rhizobia have to face two types of stress: survival in the new environment, and acquisition and maintenance of predominance in the nodulation of their host plants. To survive in a new environment, the rhizobia need to form a stable population in the local microbial community amid the competition of indigenous bacteria. This is the basis for maintaining rhizobial predominance in nodules. However, colonization in soils and colonization in roots of the host plants may be not related. It is possible that the rhizobial migrants can colonize successfully in soils, but are outcompeted in nodulation by indigenous rhizobia, although we do not have evidence for this case.

To colonize in the soils, rhizobial migrants are in competition with the indigenous microbial community for space and nutrients. Overlapping nutritional requirements of the introduced rhizobia with that of indigenous microbes is one of the important factors affecting this struggle (Schwieger and Tebbe 2000). Another related factor could be the adaptation of rhizobia to environmental factors, such as soil temperature, pH, salt content, humidity, oxygen potential, etc. (Andrade et al. 2002). In Xinjiang, China, a *Bradyrhizobium japonicum* inoculant for soybean has been used, but *B. japonicum* has not been isolated from soybean nodules in that region. The indigenous rhizobia associated with soybean in Xinjiang are *Sinorhizobium xinjiangense* and *Mesorhizobium tianshanense*. This example might be evidence that the environmental factors affect the colonization of rhizobia. The main environmental characteristics in Xinjiang are a very dry climate and highly salted and alkaline soil. The acid-producing rhizobia *S. xinjiangense* and *M. tianshanense* may be more adapted to that region than the alkaline producing *B. japonicum*. So, the introduced *Bradyrhizobium* strains may have been eliminated.

Since rhizobia are symbiotic bacteria, it is very important for them to form nitrogen-fixing nodules with their host pants. For this reason, the nodulation characters of the legume plants are important factors affecting the migration and colonization of rhizobia. It has been reported that some legumes require only the symbiotic gene background of the rhizobia for nodulation, while other plants need both the symbiotic and the genomic gene background (Laguerre et al. 2003). The former type of plants that may nodulate with many different rhizobial species or genera include soybean, common bean, and *Leucaena leucocephala*. The other type of plant that nodulates with restricted rhizobia includes alfalfa and *Amorpha fruticosa*. Therefore, rhizobia associated with alfalfa or *A. fruticosa* colonize more easily in a new environment amid their host plants because fewer indigenous rhizobia can compete with them for nodulation. *A. fruticosa* is a shrub originating in the USA that was introduced into China a half century ago. Now widely distributed in northern China, this plant nodulates only with *Mesorhizobium*, mainly *M. amorphae,* in China and in the USA (Wang et al. 1999, 2002). The Mesorhizobial populations associated with *A. fruticosa* in China and in the USA are closely related based upon the characteristics of their symbiotic plasmids and their chromosomal background (Wang et al. 2002). Thus we presumed that the Chinese isolates were introduced from the USA together with the plant and that they had colonized successfully in China. This successful migration might be due to the similar climate and soil conditions in the original region of *A. fruticosa* (northern parts of the USA) and in the northern parts of China. Also, the restricted nodulation range of *A. fruticosa* may have reduced the competition of indigenous rhizobia. In contrast, the soybean rhizobial migrants are not so lucky; soybean has been introduced into many different geographic regions that are very different from its original center and can nodulate with any rhizobia that contain adequate symbiotic genes, no matter how different their chromosomes. So, when soybean rhizobia are transported from its original center into other zones, these rhizobia have to face competition with indigenous rhizobia for both nutrients and nodulation. For these reasons, soybean nodulates worldwide with *Bradyrhizobium japonicum*, but not in Xinjiang, China. Similar situations have been noted regarding the common bean and its rhizobia.

16.5 Biological Effects of Rhizobial Migration

Some reports have indicated that the survival of rhizobia in soils is not dependent on association with host legumes. Rhizobia have been detected in fossil oil-contaminated fields after 10 years of host plant absence. In such a case, rhizobia survive as saprophytic bacteria. On the other hand, rhizobia can exist as endophytic bacteria in many non-leguminous plants. These reports imply that the migration of rhizobia does not involve only the indigenous symbiotic bacterial community, but also saprophytic and endophytic bacterial populations.

One of the effects of the rhizobial migration is that establishment of introduced rhizobia affects the native bacterial populations by reason of nutrient competition (Schwieger and Tebbe 2000). Another effect is that the introduced rhizobia can disperse their symbiotic genes in the native bacterial community. Comparative sequence analysis has revealed that the different common-bean-nodulating rhizobia have similar nodulation genes. Lateral gene transfer has been found in *Mesorhizobium* species nodulating *Lotus* (Sullivan et al. 1995) and in other rhizobia. Lateral gene transfer by conjugation, transformation and transduction are possible avenues of symbiotic gene dispersion. In these types of cases, the symbiotic genes of the introduced rhizobia can remain in the native bacterial community even though the introduced rhizobia fail in their colonization. Perhaps calling this situation molecular migration, lateral gene transfer is an important phenomenon in the evolution of rhizobia.

16.6 Conclusions

Rhizobia are transported passively, and some of their characters, such as lack of resistance to adverse conditions, may determine that not all of them are suitable for long-distance transport. In any case, transport is only the first step in rhizobial migration. Colonization must occur in order for migration of rhizobia to be successful. Since these organisms are symbiotic bacteria associated with legumes, their colonization in new locations depends on both host plants and environmental factors.

16.7 References

Andrade DS, Murphy PJ, Giller KE (2002) The diversity of *Phaseolus*-nodulating rhizobial populations is altered by liming of acid soils planted with *Phaseolus vulgaris* L. in Brazil. Appl Environ Microbiol 68:4025-4034

Laguerre G, Louvrier P, Allard MR, Amarger N (2003) Compatibility of rhizobial geotypes within natural populations of *Rhizobium leguminosarum* biovar viciae for nodulation of host legumes. Appl Environ Microbiol 69:2278-2283

Pérez-Ramirez NO, Rogel-Hernández MA, Wang ET, Martínez-Romero E (1998) Seeds of *Phaseolus vulgaris* bean carry *Rhizobium etli*. FEMS Microbiol Ecol 26:289-296

Rivas R, Willems A, Suba-Rao NS, Mateos PF, Dazzo FB, Kroppenste RM, Martínez-Molina E, Gillis M, Velazquez E (2003) Description of *Devosia neptuniea* sp. nov. that nodulates and fixes nitrogen in symbiosis with *Neptunia natans*, an aquatic legume from India. Syst Appl Microb 26:47-53

Schwieger F, Tebbe CC (2000) Effect of field inoculation with *Sinorhizobium meliloti* L33 on the composition of bacterial communities in rhizospheres of a target plant (*Medicago sativa*) and a non-target plant (*Chenopodium album*) - linking of 16S rRNA gene-based single-strand conformation polymorphism community profiles to the diversity of cultivated bacteria. Appl Environ Microbiol 66:3556-3565

Sullivan JT, Patrick HN, Lowther WL, Scott DB, Roson CW (1995) Nodulating strains of *Rhizobium loti* arise through chromosomal symbiotic gene transfer in the environment. Proc Natl Acad Sci USA 92:8985-8989

Van Berkum P, Eardly BD (2002) The aquatic budding bacterium *Blastobacter denitrificans* is a nitrogen-fixing symbiont of *Aeschynomene indica*. Appl Environ Microbiol 68:1132-1136

Wang ET, Martínez-Romero E (2000) *Sesbania herbacea-Rhizobium huautlense* nodulation in flooded soils and comparative characterization of *S. herbacea*-nodulating rhizobia in different environments. Microbiol Ecol 40:25-32

Wang ET, van Berkum P, Beyene D, Sui XH, Chen WX, Martínez-Romero E (1999) Diversity of rhizobia associated with *Amorpha fruticosa* isolated from Chinese soils and description of *Mesorhizobium amorphae* sp. nov. Int J Syst Bacteriol 49:51-65

Wang ET, Rogel MA, Sui XH, Chen WX, Martíez-Romero E, van Berkum P (2002) *Mesorhizobium amorphae*, a rhizobial species that nodulates *Amorphae fruticosa*, is native to American soils. Arch Microbiol 178:301-305

17 Indigenous Strains of Rhizobia and Their Performance in Specific Regions of India

Muthusamy Thangaraju[1] and Dietrich Werner[2]

[1]Department of Agricultural Microbiology, Tamil Nadu Agricultural University, Coimbatore 641 003, Tamil Nadu, India, E-mail: mthangaraju@vsnl.net
[2]Department of Biology, Phillips University of Marburg, 35032 Marburg, Germany

17.1 Abstract

The production and application of biofertilizers for leguminous crops, oilseeds, rice, millets, and other important crops (besides forest nursery plants) are very common in India. Currently, more than 95 firms belonging to public and private sectors are involved in the production of biofertilizers, with the annual capacity of 18,000 metric tons (out of which substantial amount is accounted as *Rhizobium*) against the total potential demand of 3.4×10^5 confirmed metric tons. *Rhizobium* is one of the important nitrogen-fixing bacteria, helping the legumes to maintain soil fertility by means of their symbiotic association and nitrogen fixation. Besides the efforts taken by the government and research institutions and agricultural universities, rhizobial technology has still a long way to go to obtain maximum benefit. This may be due to the varied reasons which limit the usage of the inoculant. However, the most important difficulty lies in the performance of the inoculated strain of rhizobia. This problem may be solved by using appropriate strains of rhizobia, i.e., crop/variety-specific, location-specific, and soil-specific, to be properly isolated and screened for effectiveness in nitrogen fixation. Still, variations among the local strains is possible and is evidenced through field experiments conducted at different locations in India. Some strains performed better than the best local strains, though they originated under different agro-ecological conditions. However, in most cases, the best performing strain in one location did not differ significantly at other locations tested. confirmed.

The effectiveness of the rhizobial strain also depends on host genotype, soil factors, presence of antagonistic bacteria in the rhizosphere, coinoculation with phosphate solubilizing bacteria/arbuscular mycorrhizal fungi, plant growth promoting rhizobacteria, etc., which are also discussed. Hence, the effective symbiosis between rhizobia and legume host rests mainly with the strain of microsymbiont used. The diversity of rhizobia is wide, like that of leguminous plants. The diversity of this microsymbiont can be best utilized for crop production in two ways. One is by selection of suitable, efficient situation-specific, location-specific, soil-specific, and crop- (plant-genotype) specific strains of rhizobia; the other is by genetic improvement of the selected strains.

17.2 Introduction

In India, more than 75%of the population depends on agriculture. The majority of the farm holdings are small or marginal. Farmers holding one to 2 ha constitute 18%, while marginal farmers having less than 1 ha account for about 55% (Kannaiyan 2003). It is very difficult for them to purchase and apply recommended doses of inorganic fertilizers at the current prices. They are forced to seek alternative, cheaper sources of nutrients to maximize crop productivity. The production and application of biofertilizers for leguminous crops, oilseeds, rice, millets, and other important crops (besides forest nursery plants) are very common in India.

Rhizobium is one of the important nitrogen-fixing bacteria, helping the legume to maintain soil fertility because of their symbiotic association and nitrogen fixation. A legume plant having effective root nodules not only can meet its own nitrogen requirement, but can also enrich the soil nitrogen content and thus improve its fertility and sustainability. The N_2 fixed by different legumes in association with specific species of *Rhizobium* under ideal and controlled conditions is estimated to amount to: 100-300 kg ha^{-1} in alfalfa, 102-150 kg ha^{-1} in clover, 37-196 kg ha^{-1} in cluster bean, 112-152 kg ha^{-1} in peanut, 68-200 kg ha^{-1} in pigeon pea, 49-130 kg ha^{-1} in soybean, 36-63 kg ha^{-1} in chickpea, 46 kg ha^{-1} in pea, and 53-85 kg ha^{-1} in cowpea (Lee and Wani 1989; Subba Rao 2002).

17.3 *Rhizobium* Biofertilizer Status in India

The first commercial production of rhizobial inoculants in India came in 1956 at the Indian Agricultural Research Institute, New Delhi, and at the Agricultural College and Research Institute (now Tamil Nadu Agricultural University), Coimbatore (Kannaiyan et al. 2001). This was followed by research into and the production of different types of biofertilizers in smaller quantities in State Agricultural Universities. Some state governments of the country also started production units to meet their local demand. In 1978, the Indian Council of Agricultural Research (ICAR) started an All India Coordinated Research Project on Biological Nitrogen Fixation (AICRP-BNF). Subsequently, in 1983, the government of India launched a national project on the development and use of biofertilizers. Later, in 1988-1989, the department of biotechnology government of India initiated a massive project on technological development and demonstration of biofertilizers. With the full support of the government, biofertilizer technology is catching up satisfactorily and people are starting to use the inoculants willingly. Currently, more than 95 firms belonging to public and private sectors are involved in production of biofertilizers with an annual capacity of 18,000 metric tonnes (out of which a substantial amount is accounted to *Rhizobium*) against the total potential demand of 3.4×10^5 metric tonnes (FAI 2001).

Despite the great effort made by the government, research institutions, and agricultural universities, rhizobial technology still has a long way to go to achieve maximum benefit. This may be due to the varied reasons which limit the use of the inoculant or the performance of the inoculated strain of *Rhizobium*. Problems in availability of proper strains of rhizobia for large-scale production, e.g., unavailability of inoculant at the right time, right place, ignorance among farmers as to using the technology, lack of knowledge of quality inoculant production, lack of infrastructural facilities, lack of storage facilities, limit the use of rhizobial technology. Still, the most important one is the performance of the inoculated strain of rhizobia. Do all inoculations turn out positively? No. Then what are the factors arresting or limiting the performance of an introduced strain? To name only a few:

1. Effectiveness of the strain used
2. Ecological competitiveness of the strain at different locations
3. Presence of one or more effective native strains of the particular cross-inoculation group of *Rhizobium*
4. Method of inoculation of the organism – whether a sufficient load of bacteria has been brought to the rhizosphere
5. Spermosphere and rhizosphere effect
6. Interaction of the organism with other soil heterotrophic microflora
7. Effect of antagonistic soil microbes on the introduced organism
8. Soil types – red, clay, calcareous, etc.
9. Soil conditions – acidity, alkalinity, etc.
10. Soil moisture and temperature
11. Agronomic management of soil and agricultural operations
12. Crop rotation
13. Effect of fertilizers, herbicides, and pesticides on the organism.

17.4 Performance of *Rhizobium* Strains

Most of the above problems may be solved by using appropriate strains of rhizobia, i.e., crop-specific, location-specific and soil-specific, which could be properly isolated and screened for their effectiveness in nitrogen fixation. But variations among the local strains is still possible and is evidenced through field experiments conducted at different locations in India.

Responses to *Rhizobium* inoculation have been demonstrated with major grain legumes such as red gram (*Cajanus cajan*), gram (*Cicer aritinum*), and green gram (*Vigna radiata*) in different parts of India (Subba Rao et al. 1993). In most of the cases, up to 80% of the nitrogen requirement of the plants could be obtained through dinitrogen fixation. The cultures of *Rhizobium* inoculant consisting of the strains from black gram and peanut performed better than the single strain in green gram in recording maximum grain yield, whereas this inoculant was not consistent with black gram under field conditions (Oblisami et al. 1976; Balasubramanian et al. 1980). However, both success and failure at field experiments have been documented (Subba Rao and Balasundaram 1971; Subba Rao 1976, 1979; Balasundaram and Subba Rao 1977; Subba Rao and Tilak 1977). Colonization of the rhizosphere by *Rhizobium* sp. is considered to be the most vital factor for efficient N_2 fixation.

The demonstration trials conducted at the farmers' field by Tamil Nadu Agricultural University, Coimbatore, India, over 10 years to study the efficacy of *Rhizobium* (local strain of *Rhizobium* – TNAU 14) on peanut indicated increased pod yield in the inoculated fields. The pod yield increase ranged from 14.66 to 30.80%(Table 1). Interestingly, in all the trials an increase in pod yield was recorded due to rhizobial inoculation, but slight variation was noticed which might be due to the some of the reasons we listed earlier. The trials were conducted under similar (though not identical) soil conditions, i.e., neutral soil, and confirmed.

Table 1. Demonstration trials on the use of *Rhizobium* in peanut by Tamil Nadu Agricultural University, Coimbatore, India (Thiyagarajan 2003)

Year	No. of trials conducted	Pod yield (kg ha^{-1})		Increase over control (%)
		Inoculated	Control	
1991-1992	6	3378	2825	19.57
1992-1993	2	3810	3100	22.90
1993-1994	3	1787	1480	20.74
1994-1995	2	1910	1460	30.80
1995-1996	4	2235	1811	23.41
1997-1998	7	1678	1403	19.60
1998-1999	11	1947	1698	14.66
1999-2000	8	1803	1533	17.61

Similar field trials conducted on soybean over 8 years also resulted in increased grain yield. (Table 2). The results clearly indicate that use of a suitable *Rhizobium* strain always results in a positive response.

Table 2. Demonstration trials on the use of *Rhizobium* in soybean by Tamil Nadu Agricultural University, Coimbatore, India (Thiyagarajan 2003)

Year	No. of trials	Grain yield (kg ha^{-1})	Increase over

	conducted	Inoculated	Control	control (%)
1991-1992	3	1688	1519	11.12
1992-1993	4	1860	1600	17.12
1993-1994	4	2091	1613	29.63
1994-1995	5	2003	1613	24.18
1995-1996	3	1907	1618	17.87
1997-1998	6	1296	1151	12.57

Rhizobial strains specific for various legume crops have been isolated and screened under laboratory conditions, and evaluated under greenhouse and field conditions over many years. Consequently, the following rhizobial strains were identified for inoculation to different leguminous crops in different states of India under the All India Coordinated Pulses Improvement Project (AICPIP) and Coordinated Biological Nitrogen Fixation Project (Table 3).

Table 3. Strains of *Rhizobium* and *Bradyrhizobium* developed in different parts of India

Host plant	Rhizobial strains	Location
Pigeon pea	BPR 9515, 9527	Badnapur (Maharashtra)
	RGR 10	Gulbarga (Karnataka)
	JARS 72	Sehore (Maharashtra)
	CRR 6, CC1	Coimbatore (Tamil Nadu)
	RA 30	Varanasi (Uttar Pradesh)
Soybean	COS-1, MTP-1	Coimbatore (Tamil Nadu)
Peanut	AH-6, TNAU-14	Coimbatore (Tamil Nadu)

The performance of the above strains in different locations was studied (Table 4) under the All India Coordinated Pulses Improvement Project.

Table 4. Performance of *Rhizobium* strains under different agroecological conditions on the grain yield of pigeon pea (Anonymous 2003)

Rhizobial strain (source)	Grain yield (kg ha^{-1}) at different locations				
	Banapur (Maharashtra)	Khagone (Madhya Pradesh)	Akola (Maharashtra)	Dholi (Bihar)	Gulbrga (Karnataka)
BPR-9527 (Badnapur)	1175	631	947	300	562
BPR-9515 (Badnapur)	1261	620	1096	-	615
RGR-10 (Gulbarga)	1435	828	1108	330	882
JARS-72 (Sehore)	1317	793	1214	270	552
CRR-6 (Coimbatore)	1519	752	1063	370	833
CC-1 (Coimbatore)	1276	768	1174	330	684
RA-30 (Varanasi)	1160	628	-	370	614
Local best	1231	835	1349	300	792
N-20 kg/ha	1079	756	1239	270	899
Uninoculated Control	1064	624	877	260	520
±SEM	46	41	23	14	22
CD at 5%	130	122	69	41	66

The results indicate the variation in the performance of the rhizobial strains at different locations. In Badnapur, the strains CRR 6, RGR 10, and JARS 72 – each is from a different agroecological condition – performed better than the bestlocal strains. Also, it should be noted that strains CRR 6 and CC1 differed in their performance at Badnapur though they are from the same location. However, the best performing strains in one location are likewise best at other locations tested, indicating their superiority. Anand and Dogra (1997) also reported the differential behavior of rhizobial and bradyrhizobial strains in pigeon pea.

17.5 How to Overcome the Hurdles

Earlier studies clearly indicated that great genetic diversity exists among rhizobial species/strains and this may be one of the reasons for the performance of the inoculated rhizobia. The particular species should have good compatibility with the host plant to achieve an effective symbiosis. The variation in the performance of strains of rhizobia also depends on the host's makeupHost physiology and genetics play an important role in selecting a symbiotic partner. We do not want to go into detail on those aspects. However, how the rhizobial strains behave with the varieties of host plants is important in selecting a strain of *Rhizobium* suitable for that legume. Experiments conducted at Tamil Nadu University and other places indicate that the above nitrogen-fixing traits varied with the host genotypes in various pulse crops (Chendrayan 2003). The results clearly showed that at a similar location the grain yield varies with the rhizobial strain employed. This preferential behavior may be exerted by the host variety or by the preference of the strain of *Rhizobium* for that particular variety. Hence, by selecting a specific rhizobial strain for a specific crop variety, we can achieve maximum productivity.

Inoculated rhizobia not only must compete for limited nutrients but also for interactions with indigenous heterotrophic microbes and predators. These reduce the capacity of inoculated rhizobia to maintain population densities at sufficient levels to ensure contact with the legume roots. This kind of competition with indigenous microorganisms can be manipulated, and thereby the rhizosphere colonization of rhizobia can be strengthened. In a study conducted on the survival and multiplication of introduced bacteria into soil, it was observed that the population of *R. leguminosarum* biovar *phaseoli*, *R. meliloti*, *Agrobacterium tumefaciens*, *Micrococcus flavus*, *Corynebacterium* sp., and *Pseudomonas* sp. declined because of their susceptibility to predation, starvation, or possibly antibiotic-producing or lytic microorganisms (Acea et al. 1988). Bacteriocin-producing *Rhizobium* strains have been reported to inhibit the growth of non-bacteriocinogenic strains (Ahlawatt and Dadarwal 1996).

The use of fungicides and antibiotics has been found to enhance the probability of rhizobial colonization. Hossein and Alexander (1984) augmented soybean rhizosphere colonization with *Bradyrhizobium japonicum* by using benomyl (a fungicide) and the antibiotics streptomycin and erythromycin. In this regard, an antagonistic bacterium isolated from the rhizosphere of pigeon pea on combined inoculation with rhizobia enhanced the root nodulation and crop yield (Chendrayan 2003). This antagonistic bacterium inhibits the growth of several soil bacterial and fungal isolates but not the rhizobia. Gunasekaran and Murugesan (2002) reported that combined inoculation of *Rhizobium* and antagonistic bacteria increased the growth, nodulation, and grain yield of green gram.

Soil temperature, moisture, and pH are the important physicochemical characters that influence the symbiotic nitrogen fixation in pulse crops. The soil temperature during summer months exceeds 54°C. This high temperature will certainly affect the efficiency of rhizobia. Since the soil temperature under field conditions cannot be controlled, temperature-tolerant strains have to be used. Similarly, soil moisture or water stress not only limits the survival of rhizobia, but also their symbiotic association with pulse crops (Saxena and Tilak 1999).

Soil reaction, salinity or acidity, has a great impact on rhizobia and their symbiotic activity. Salt stress decreases symbiotic efficiency to levels below the genetic potential of host-*Rhizobium* association, and thus decreases plant growth and grain yield (Singleton and Bohlool 1983). Likewise, soil acidity has been found to affect symbiotic nitrogen fixation, limiting *Rhizobium* survival and persistence in soils and reducing nodulation (Munns 1986). Hence rhizobial strains tolerant to soil acidity or alkalinity are to be explored.

Among the soil chemical factors that influence symbiotic nitrogen fixation in pulse crops, mineral nitrogen concentration is the most important one. In general, high soil nitrogen levels, applied or residual, reduce nodulation and N_2 fixation. Based on several studies, nodulation and/or nitrogen fixation were reduced by approximately 50% in different crops when nitrogen concentration in the root environment was between 20 and 90 mg/kg in the growth medium. The suppression in symbiotic nitrogen fixation is due particularly to the nitrate fraction in the root growth environment (Streeter 1988). Variations in the plant's ability to nodulate and fix N_2 in the presence of nitrate have been reported. Phosphorus deficiency is another factor that commonly restricts the realization of the potential of N_2 fixation by legumes. Only 30% of P is available to the crop if applied as chemical fertilizer due to fixation in soil. Experiments conducted at Tamil Nadu Agricultural University have indicated that dual inoculation of a pulse crop with rhizobia and phosphobacteria/arbuscular mycorrhizal fungi is found to enhance P uptake by plants, and consequently to improve nodulation, plant growth, and yield. Other than N_2 fixers and P solubilizers and mobilizers, many bacteria have the ability to produce plant growth-promoting substances. Inoculation of such bacteria (e.g., *Pseudomonas*) along with *Rhizobium* may further enhance the productivity of pulse crops, as has been evidenced by experiments conducted at Tamil Nadu.

17.6 Conclusion

Based on the above discussions, effective symbiosis between rhizobia and legume hosts can be said to be dependent upon various factors. , The diversity among rhizobia is wide, paralleling that of leguminous plants, as pointed out by Martínez-Romero and Caballero-Mellado (1996). Such diversity in this microsymbiont can be better utilized for crop production in two ways. First is the selection of suitable, efficient, situation-specific, location-specific, soil-specific, and crop-plant-genotype- specific strains of rhizobia; and second is the proper field evaluation of the strains. Genetic improvement of the selected strains is also necessary if maximum benefits are to be achieved.

Acknowledgements. The senior author is thankful to the Indian National Science Academy (INSA), New Delhi, to Tamil Nadu Agricultural University, Coimbatore, India, and to Deutsche Forschungsgemeinschaft (DFG), Germany, for providing financial support under the INSA-DFG bilateral exchange program in order to visit Philipps University of Marburg and to attend the conference.

17.7 References

Acea MJ, Moore CR, Alexander M (1988) Survival and growth of bacteria introduced into soil. Soil Biol Biochem 20:509-515

Ahlawat OP, Dadarwal KR (1996) Bacteriocin production by *Rhizobium* sp. Cicer and its role in nodule competence. Indian J Microbiol 36:17-23

Anand RC, Dogra RC (1997) Comparative efficiency of *Rhizobium* / *Bradyrhizobium* spp. strains in nodulating *Cajanus cajan* in relation to characteristic metabolic enzyme activities. Biol Fertil Soils 24:283-287

Anonymous (2003) All India Coordinated Pulses Improvement Project Annual report 2002 – 2003. ICAR New Delhi

Balasubramanian A, Prabahkaran J, Sundaram SP (1980) Influence of single and multi-strain rhizobial inoculant on cowpea. Madras Agric J 67:538-540

Balasundaram VR, Subba Rao NS (1977) A review of development of rhizobial inoculants for soybeans. Fert News 22:42-46

Chendrayan K (2003) Legume - *Rhizobium* interactions. In: Thangaraju M, Prasad G, Govindarajan K (eds) Inoculant production technology. Suri Associates, Coimbatore, India, pp 14-20

FAI (Fertilizer Association of India) (2001) Biofertilizer statistics 1999-2000. FAI New Delhi, India

Gunasekaran S, Murugesan R (2002) Effect of combined inoculation of *Rhizobium* and antagonistic bacteria on nodulation and growth of greengram. Madras Agric J 89:563-567

Hossain AKM, Alexander M (1984) Enhancing soybean rhizosphere colonization by *Rhizobium japonicum*. Appl Environ Microbiol 48:448-472

Kannaiyan S (2003) Inoculant production in developing countries - problems, potentials and success In: Hardarson,G Broughton,WJ (eds) Maximizing the use of biological nitrogen fixation in agriculture. Kluwer Academic Publ., Netherlands pp 187-198

Kannaiyan S, Kumar K, Thangaraju M, Govindarajan K (2001) Microbial inoculant production in developing countries - success, problems and potentials. Tamil Nadu Agricultural University Publication Series, Coimbatore, India

Lee KK, Wani SP (1989) Significance of biological nitrogen fixation and organic manures in soil fertility management. In: IFDC (ed) Soil fertility and fertilizer management - semi arid tropical India. Christianson Muscle Shoals, Alabama, USA, pp 89-108

Martínez-Romero E, Caballero-Mellado J (1996) *Rhizobium* phylogenies and bacterial genetic diversity. Crit Rev Plant Sci 15:113-140

Munns DN (1986) Acid soils tolerance in legumes and rhizobia. Adv Plant Nutr 2:63-91

Oblisami G, Balaraman K, Natarajan T (1976) Effect of composite cultures of *Rhizobium* on the pulse crops. Madras Agric J 63:587-589

Saxena AK, Tilak KVBR (1999) Potentials and prospects of *Rhizobium* biofertilizer. In: Jha MN, Sriram S, Venkataraman GS, Sharma SG (eds) Agromicrobes. Today and Tomorrow's printers and publishers, New Delhi, pp 51-78

Singleton PW, Bohlool BB (1983) Effect of salinity on the functional components of the soybean - *Rhizobium japonicum* symbiosis. Crop Sci 23:815-818

Streeter T (1988) Inhibition of legume nodule formation and N_2 fixation by nitrate. CRC Crit Rev Plant Sci 7:1-23

Subba Rao NS (1976) Field response of legumes in India to inoculation and fertilizer application. In: Nutman PS (ed) Symbiotic nitrogen fixation in plants. Cambridge Univ Press, Cambridge, pp 255-268

Subba Rao NS (1979) Biofertilizers in Indian agriculture: problems and prospects. Fert News 24:84-90

Subba Rao NS (2002) An appraisal of biofertilizers in India. In: Kannaiyan S (ed) Biotechnology of biofertilizers. Kluwer, Dordrecht, pp 1-9

Subba Rao NS, Balasundaram VR (1971) *Rhizobium* inoculants for soybean. Indian Farming 21:22-23

Subba Rao NS, Tilak KVBR (1977) Rhizobial cultures and their role in pulse production. Souvenir bulletin, directorate of pulses development, Lucknow, India, pp 31-34

Subba Rao NS, Venkataraman GS, Kannaiyan S (1993) Biological nitrogen fixation. Indian Council of Agricultural Research, New Delhi

Thiyagarajan TM (2003) Prospects of microbial inoculants in Indian agriculture. In: Thangaraju M, Prasad G, Govindarajan K (eds) Inoculant production technology. Suri Associates, Coimbatore, India, pp 1-8

18 Migration of Aquatic Invertebrates

L.W.G. Higler

Wageningen UR, ALTERRA, P.O. Box 46, 6700 AA Wageningen, The Netherlands, E-mail Bert.Higler@wur.nl

18.1 Abstract

Aquatic invertebrates are dependent on dispersion and migration for sustainable populations. Especially in running waters, these processes play a dominant role. Ephemeral water bodies have probably triggered dispersion phenomena in the evolutionary past. The life cycles of aquatic insects in particular show that dispersion and migration are essential elements for survival. The river Rhine case shows that recolonization of the main stream is only possible if the species still live in tributaries within the Rhine discharge area and, moreover, that introduction from other discharge areas (Ponto-Caspian species) is extraordinarily successful if natural predators and parasites are absent.

18.2 Introduction

Dispersion is an ecological phenomenon that is of fundamental importance for population biology. Dispersion is often mixed with migration. What is the difference? Migration (from one habitat to another) is firstly adaptive behavior and not haphazard. Migratory movements of aquatic insects occur at regular and specific times of the year. The flight is persistent, continuous, directional, and over long distances. Therefore, migration is also to be distinguished from more trivial flights within the area of reproduction. The area of reproduction is the habitat in which the eggs are laid, the larvae develop, and the adults emerge from pupae and nymphs. Most insects are distributed over larger areas than the places where they originated. Dispersion occurs when sustainable populations are being established in a very large area. The most common characteristic of most insect migrants is that migration begins immediately after hatching. The massive emergence and flying are parts of the life cycle of migrants and as inevitable as mating and egg-laying (Johnson 1969). Females predominantly migrate to new reproduction habitats, lay eggs and die, instead of returning to their original habitat. Descendants may recolonize the original habitats. There are other forms of migration, where individuals after visiting a winter habitat, return to their reproduction habitat.

Most insects do not have such a long life span; they have a short adult life.

Many insects are carried away by wind, and they have adapted to that by actively flying to higher air currents. Some insects are good fliers, but this does not automatically make them migrants. A migrant leaves its old habitat, either by flying with the wind or by choosing a certain direction..

In some species, all individuals of a generation migrate, in others only a restricted number of individuals of a generation migratet, and sometimes there are alternating generations of migrating and non-migrating individuals. Therefore, a precise estimation of migration on the basis of flying, the number of insects involved, flight distance, and dependency on season are difficult to determine.

Dragonflies (Odonata) are well-known migrators, but caddis flies (Trichoptera), stone flies (Plecoptera), may flies (Ephemeroptera), and alder flies (Megaloptera) are less efficient colonizers. In contrast, certain species of water bugs (Hemiptera), flies and mosquitoes (Diptera), and beetles (Coleoptera) are excellent colonizers, possessing invader qualities. Species from these orders disperse widely and they are often the first to colonize new habitats or temporary habitats.

18.3 Important Factors for Dispersion and Migration

Johnson (1969) distinguishes three types of strategies for migrating insects:
1. Emigration without return. Adults, living only one season, emerge in the reproduction habitat, disperse, lay eggs and die.
2. Emigration and return by the same individuals within one season. The adults migrate from reproduction habitats to nutrient-rich habitats, and return later in the season to lay eggs in the old habitats.
3. Emigration to hibernation or oversummering habitats and return by the same individuals after a diapause.

The flight range is the average distance covered by individuals of a species and can be estimated by regression of the density of the population at a certain distance from the reproduction habitat. The idea of flight range as a linear distance is, by the way, of restricted value. Individual insect differences in flight characteristics, weather, and landscape result in great differences in distances reached. The "effective flight range" is the distance where relatively large numbers are present, the maximum flight distance being where some individuals can be found.

18.4 Colonization Cycle Hypothesis

The primary goals in the adult phase of the typical aquatic insect's life cycle are to mate and to deposit eggs in habitats that are appropriate for the development of larvae. Dispersion in a stream of rheophilic aquatic insects has been hypothesized by Müller (1954). Adult insects fly upstream to lay eggs. The young larvae move downstream as a result of overpopulation or catastrophic drift and populate downstream habitats. When they are successful, they pupate and the emerged adults fly upstream to lay their eggs. Distances between the two habitats may vary between tens of meters to (sometimes) tens of kilometers.

An interesting feature is the necessity for mating. The females have to mate before they can oviposit. In many species, mating takes place in swarms in special places, for example, bushes or trees along the shore. Changes in riparian vegetation may inhibit mating and thus the size of populations of aquatic insects.

18.5 Inland Dispersion

Inland dispersion has been an under-investigated phenomenon for a long time, but it must also be considered as necessary for the colonization of new habitats, and as participation of aquatic insects in terrestrial food webs (Kovats et al. 1996).

Long-distance dispersion is a common feature in benthic invertebrates, and this has always been considered a passive but necessary phenomenon enabling some individuals to reach suitable habitats for colonization. The important role of physical transportation processes in regulating the number of colonizers in a new area has been emphasized for a long time. It is known as the recruitment limitation of supply-side ecology. Abundances may vary as a function of the time needed to reach new areas (settlement rate). Colonization is considered more important for the population structure than internal processes like predation and competition (Palmer et al. 1996).

Also, recolonization of formerly degraded rivers is dependent on dispersion of resting populations in the catchment, or on dispersion from inland areas. An example is the recolonization by certain species in the river Rhine. There is well-documented evidence of *Hydropsyche contubernalis* and *Epheron virgo* recolonizing the Rhine after the Sandoz fire that exterminated all life in the main stream. Gradual recovery of former habitats has been registered, whereby downstream recolonization took place within tens or hundreds of kilometers each year.

Generally speaking, restoration of streams is seldom followed by a rapid recovery of the original ecosystem. The reason for this is that egg-bearing females have to reach the restored habitat from faraway catchments, and this may take many years or even tens of years.

There are different factors influencing inland dispersion. Abiotic factors such as temperature, wind, clouds, and air moisture are directly responsible for departure, timing, and duration of flights. Properties of the habitat such as permanence, sustainability, frequency, and strength of disturbances may influence the dispersion behavior over the long term. Large rivers must be considered as permanent habitats with predictable discharge patterns and with substrate that will be disturbed less than in small rivers in case of high discharges. Species of large rivers are adapted to their specific conditions and show a restricted dispersion behavior. Moreover, large rivers tend to be far from each other and dispersing adults with a short lifetime are totally dependent on strong winds to reach other (large) catchments. Inland dispersion by adults from large rivers and lakes must be restricted and random, not spatially directed.

18.6 Power of Flying

Migration capacity is dependent on the ability to fly far in one effort, or frequently. In some populations of normally wingless water bugs, accidentally winged specimens occur (Brown 1951). In mosquitoes, non-blood-sucking species have wing muscles that degenerate by autolysis after the first flight. Their flight range is very restricted. Anderson (in de Moor 1992) showed that some Gerridae without wings or with reduced wings reproduce earlier and have a higher egg production rate than long-winged specimens of the same species. This shows that energy can be spent in one way or another, but seldom in both ways.

18.7 Habitat Properties

The resemblance or the difference between the source area and the colonized habitat can mean the difference between success and failure of colonizers. If individuals of a certain species are introduced into a new environment that is suitable for them, but so far from their natural distribution area that specialized predators and parasites are lacking, they have a selective advantage over the species present in that environment. A good example is the American crayfish *Orconectes limosus*, which has a stronger competitive power than the European *Astacus astacus. Orconectes* is a common inhabitant of many Dutch waters, and *Astacus* is now restricted to two small populations. Settlement is successful if there are few predators and enough shelters. *Orconectes* is not desirable to the parasite that has reduced *Astacus* all over Europe.

A migration or dispersion flight is ended if the animal is triggered to lay eggs, to search for food, or if the habitat is suitable. Migration behavior has developed in species that live in habitats temporarily unsuitable for reproduction or in habitats that totally disappear. It is accepted that abiotic factors such as temperature, precipitation, air pressure, insolation, and wind as well as botanic factors (perennial or annual vegetation) are the most important reasons for leaving a habitat.Other factors like crowded conditions, the presence of predators and a shortage of food are less important reasons.

Animal aggregations in temporary rivers are characterized by large numbers of substrate-generalists, creatures who are very mobile and capable of finding shelter in periods of drought (Brooks and Boulton 1991). Degraded or changed habitats are colonized by fast invaders instead of by former species. An example is the construction of a dam in a river. This causes a changed environment, where the habitats of the local species are destroyed or changed. They have to adapt to the new conditions, which brings them in the same position as new or introduced species to create a sustainable population. The new species have in general the advantage in that they have adapted to man-created circumstances (de Moor 1992). Colonizers are specialized in capturing disturbed areas. For example, in water bodies where natural enemies of the species have been killed by pesticides, or in aquatic biotopes where the heterogeneity of micro-habitats has disappeared, a homogeneous environment is left in which opportunistic species adapt.

The river Rhine demonstrates this. The success of introduced species from far away has been realized by the creation of the Danube/Rhine canal. This has caused many Ponto/Caspian species to become extraordinarily successful in colonizing the Rhine. Most of these species, probably, have not reached The Netherlands by their own efforts, but have been transported by ballast water. As a result, the largest part of the river ecosystem nowadays consists of neozoic species that never existed in the river Rhine. A balance in the ecosystem has not been reached. The small crustacean *Corophium curvispinum* has colonized all firm substrates and in this way hampers the possibilities for another, but much older, alien species, *Dreissena polymorpha*. And now, still another Ponto-Caspian neozoön, *Dikogammarus*, is reducing the vast numbers of *Corophium*.

There are theoretical models suggesting that organisms have to disperse, even under stable conditions, when the chance for survival is low in the reproduction habitat (Peckarsky et al. 2000). They have shown that short-living *Baetis* females after emerging in a dry environment have two alternatives: flying to a wet locality or dying. Flying, even over short distances, seems to be an important mechanism of dispersion with several local populations as result.

18.8 Disturbance and Dispersion

Frequency of disturbance and power of dispersion are two factors indicating the most probable control of aquatic communities (Palmer et al. 1996).

- Frequently or unpredictably disturbed running waters with well-dispersing organisms are under strong regional influence. The structure of the biocommunity is variable, has a high turnover, and is especially structured by "chance effects" of arriving and settling species (Moller Pillot 2003).

- A not too high disturbance frequency and a species with a low or average dispersion form a system that enables a competitively dominant species to develop a sustainable population. Invaders have the wherewithal to settle. Depending on the time between disturbances, local influences such as predator-prey relationships have a better chance to develop.

- Low to average disturbance in combination with high dispersion probably results in a balance between local and regional influences. Low disturbance frequency allows local interactions to play a substantial role, but a high input of colonizers results in the replacements of unsuccessful species. Such communities are generally rich in species.

- Finally, the combination of a low disturbance frequency and low dispersion results in the best possibility for local control and a low regional component.

18.9 Dispersion of Non-Flying Invertebrates

There are four categories of dispersion and migration by non-flying invertebrates or non-flying stages of insects:

Drift or downstream movement. As has already been described, downstream movement of insect larvae form an essential part of the life cycle. In advanced life forms, downstream movement or drift may occur actively or passively.

Upstream movement is always active. The animals have to overcome the forces of the current. They move through stretches with low current under the banks or sometimes overland.

Local or sideways movement seems trivial in the framework of this study, but it turns out to be of major importance for the survival of most running-water species.

Aerial movement is used in leaving an aquatic environment by clinging to birds or flying insects.

Drift is a well-studied phenomenon (Hynes 1970; Waters 1965; Pechlaner 1986; Brittain and Eikeland 1988).

There are different forms of drift:

Catastrophic drift is related to discharge conditions, in which the substrate is physically disturbed and animals are transported by the current (Minckley 1964). Catastrophic drift is caused by extreme spates (floods); but other causes may be drainage of effluents or toxic substances, inflow of warm water from cooling systems, or drying out of habitat. It is obvious that catastrophic drift occurs more often in the changed ecosystems of the agricultural or industrial landscape and is generally more disastrous than in natural systems.

Behavioral drift (Waters 1965) that is demonstrated during daily activities like feeding, mating, and moving around (mainly during the night; Brusven 1970; Statzner 1979), and actively seeking the water column (active drift) as happens to escape predators.

Distributional drift to colonize new habitats in the same system. This is seen especially in hatched young animals (Nishimura 1967; Ulffstrand et al. 1974). This form of drift may cover short distances (Brittain and Eikeland 1988) or larger stretches up to several kilometers. Distributional drift also takes place at night, probably to avoid predation by visual hunters like fish (Allan 1984).

Constant drift or background drift comprises low numbers of individuals that accidentally lose grip and come into the current.

Upstream movement takes place in many species, but seldom compensates for drift (except in the case of flying insects). Most macroinvertebrates are positively rheotactic and move under the banks in rows upstream. Hughes (1969) describes the relation between drift and upstream movement as two steps back (drift) to one step forward (upstream movement). Nonetheless, distances reached in an upstream direction may be considerable. Some *Leptophlebia* larvae have been found to migrate upstream 1-6 km (at a rate of 6 m per night; Neave 1930; Elliott 1971).

Gammarus species are still better: they have been noted to move 22 m/h (*G. pulex*: Hughes 1969), 40 m/h (*G. fossarum*: Meijering 1972), and 40-60 m/h (*G. zaddachi*: Dennert et al. 1969). Some individuals of the last species drifted 50-80 m/h, so there is hardly an overall downstream surplus. A surprising case contradicts the generally accepted hypothesis of Karl Müller. The larvae of a *Leptophlebia* species migrate upstream and the adult flies downstream to lay eggs (Hayden and Clifford 1974).

Upstream movement generally takes place during the night (Elliott 1971).

Some animals leave the water to go upstream. It is known that the river crayfish (Müller-Motzfeld et al. 1986) and several species of stone fly with wingless females creep upstream overland to deposit their eggs in the stream (Hynes 1970).

Local or sideways movement:

Under normal conditions, stream environment is always a mosaic of very different abiotic conditions in space and time. These are manifest by differences in:

Current velocity from one shore to the other (Chutter and Noble 1966)

Substrate (Tolkamp 1981)

Composition and magnitude of organic material (Egglishaw 1969)

Oxygen management (Mann 1961).

The inhabitants of streams choose the optimal places that differ during their life cycle and also between day and night. In streams with coarse substrate, many species live between the rocks and stones in the interstitial water (Williams 1977). In channelized streams with homogeneous bottoms most of these microhabitats are lacking, thus hampering the settlement and survival of running-water species.

Aerial distribution by hosts such as water birds and flying water insects occurs with Crustacea, Bivalvia, Ostracoda, Gastropoda, Hirudinea and water mites. This form of distribution is undirected, but as most of these hosts live in the same environment, there is a good chance that the hitchhikers will arrive in suitable habitats.

18.10 Climate Change

Hill and Fox (2003) have demonstrated that only about 20% of British butterflies benefit from a warming climate. These are good fliers and generalists. Those without these characteristics are decreasing through loss of habitat due to factors caused by climate change, i.e., northern species are losing their cool habitats, especially in fens. It is not improbable that aquatic invertebrates are similarly affected. Optimism about the results of global warming is not justified.

It is fascinating to see how these tiny organisms can change total ecosystems!

18.11 References

Allan JD (1984) The size composition of invertebrate drift in a Rocky Mountain stream. Oikos 42:68-76

Brittain JE, Eikeland TJ (1988) Invertebrate drift - a review. Hydrobiologia 166:77-93

Brooks SS, Boulton AJ (1991) Recolonization dynamics of benthic macroinvertebrates after artificial and natural disturbances in an Australian temporary stream. Aust J Mar Freshwater Res 42:295-308

Brown ES (1951) The relation between migration rate and type of. habitat in aquatic insects with special reference to certain species of Corixidae. Proc Zool Soc Lond 121:539-545

Brusven MA (1970) Drift periodicity and upstream dispersion of stream insects. J Entomol Soc Br Columbia 67:49-59

Chutter FM, Noble RG (1966) The reliability of a method of sampling stream invertebrates. Arch Hydrobiol 62:95-103

De Moor FC (1992) Factors influencing the establishment of aquatic insect invaders. Trans S R Soc S Afr 48:141-158

Dennert HG, Dennert AL, Kant P, Pinkster S, Stock JH (1969) Upstream and downstream migrations in relation to the reproduction cycle and to environmental factors in the amphipod *Gammarus zaddachi*. Bijdr Dierkunde 39:11-43

Egglishaw HJ (1969) The distribution of benthic invertebrates on substrata in fast flowing streams. J Anim Ecol 38:19-33

Elbersen-van der Straten JWH, Higler LWG (2002) Dispersion and migration of Aquatic insects in standing and running waters (in Dutch). ALTERRA report 572

Elliott JM (1971) Upstream movements of benthic invertebrates in a Lake District Stream. J Anim Ecol 40:235-252

Hayden W, Clifford HF (1974) Seasonal movements of the Mayfly Leptophlebia cupida (Say) in a Brown-water stream of Alberta, Canada. Am Midl Nat 91:990-102

Hill JK, Fox R (2003) Biologist, July 2003: Climate change and British butterfly distributions. Biologist 50:106-110

Hughes DA (1969) Some factors affecting drift and upstream movements of *Gammarus pulex*. Ecology 51:301-305

Hynes HBN (1970) The ecology of running waters. Liverpool Univ Press, Liverpool

Johnson CG (1969) Migration and dispersal of insects by flight. Methuen, London

Kovats ZE, Ciborowski JJH, Corkum LD (1996) Inland dispersal of adult aquatic insects. Freshwater Biol 36:265-276

Mann KH (1961) The life history of the leech *Erpobdella testacea* Sav. and its adaptive significance. Oikos 12:164-169

Meijering MPD (1972) Experimentelle Untersuchungen zur Drift und Aufwanderung von Gammariden in Fliessgewässern. Arch Hydrobiol 70:133-205

Minckley WL (1964) Upstream movements of *Gammarus* (Amphipoda) in Doe Run, Meade County, Kentucky. Ecology 45:195-197

Moller Pillot HKM (2003) The survival of aquatic animals in a dynamic world (in Dutch with an English summary). Stichting het Noordbrabants Landschap

Müller K (1954) Investigations on the organic drift in North Swedish streams. Rep Inst Freshwater Res Drotningholm 35:133-148

Müller-Motzfeld G, Duty J, Strunk P (1986) "Krebs-Sterben" im Herta-See (Rügen). Naturschutzarb in Mecklenburg 29:93-97

Neave N (1930) Migratory habits of the mayfly *Blasturus cupidus* Say. Ecology 49:75-81

Nishimura N (1967) Ecological studies on net-spinning caddisfly *Stenopsyche griseipennis* McLachlan. II. Upstream migration and determination of flight distance. Mushi 40:39-46

Palmer MA, Allan JD, Butman CA (1996) Dispersal as a regional process affecting the local dynamics of marine and stream benthic invertebrates. Tree 11:322-326

Pechlaner R (1986) Traps for drift and barriers for the upward migration of invertebrates in the rhithral zone of running waters (G.e.). Wass Abwass 30:421-463

Peckarsky BL, Taylor BW, Caudill CC (2000) Hydrologic and behavioral constraints on oviposition of stream insects: implication for adult dispersal. Oecologia 125:186-200

Statzner B (1979) Der Obere und Untere Schierenseebach (Schleswig-Holstein), Strukturen und Funktionen in zwei norddeutschen See-Ausfluss-systemen, unter besonderer Berucksichtigung der Makroinvertebraten. PhD Thesis, Kiel University

Tolkamp HH (1981) Organism-substrate relationships in lowland streams. PhD Thesis, Wageningen University

Ulfstrand S, Nilson LM, Stergar A (1974) Composition and diversity of benthic species collectives colonizing implanted substrates in a south Swedish stream. Entomol Scand 5:115-122

Waters TF (1965) Interpretation of invertebrate drift in streams. Ecology 46:327-334

Williams DD (1977) Movements of benthos during recolonization of temporary streams. Oikos 29:306-312

19 Migration of Fishery Resources in the World's Oceans

Paul G. K. Rodhouse

British Antarctic Survey, High Cross, Madingley Road, Cambridge CB3 0ET, UK, E-mail pgkr@bas.ac.uk

19.1 Abstract

All fish, and other motile fishery resources, make migrations. These range from tens or hundreds of metres to large-scale oceanic migrations over thousands of kilometres. Migrations are usually cyclical over time scales from diurnal to annual or longer. Diurnal migrations are driven by the light/dark cycle and seem to have evolved to balance the requirement to feed against the risk of predation. The major seasonal migrations have evolved because of the differing habitat requirements for breeding and feeding, and in the oceans these migrations are linked to prevailing current patterns. There is considerable concern in world fisheries about the top-down effects of exploitation of marine resources because approximately one third of the world's fishery resources are either fully or over exploited. However, there are also bottom-up effects, driven by pollution and manifestations of global climate change, which are increasingly recognized to be important drivers of change in the world's fisheries. Changing ocean current systems at the surface which might be caused by global climate change have the potential to cause disruption of migration patterns and prevent the successful completion of life cycles. Bottom-up effects of variable environments on populations of migratory species such as Atlantic cod, Antarctic krill and shortfin squid have been identified and provide different kinds of examples of how changing oceanographic conditions can drive change in exploited stocks. Examples of this kind of variability provide valuable insights into the likely effects of large-scale ecological change on the world's fisheries.

19.2 Introduction

Migration - moving from one place to another - is a fundamental feature of the life style of many, if not all, motile aquatic organisms and has been a subject of research since the early days of fishery science (Meek 1915). Migratory movements include daily changes, often from one depth layer to another, linked to the light-dark cycle, and these seem to have evolved to balance the requirement to feed against the risk of predation. Over longer time scales there are local and seasonal movements, dispersals and substantial movements between widely separated and well-defined areas. It is the last of these, the true migrations defined by Harden Jones (1968), that have most relevance to the management of marine fishery resources. It is these that form the focus of this review.

The major migrations of fishery resources generally follow the pattern in Fig. 1. Feeding grounds and spawning grounds are typically separated geographically. Outside the spawning season, adults on the feeding grounds grow and ripen and then, prior to spawning, commence the migration to the spawning grounds. This is usually against the flow of the prevailing current system. At the spawning ground eggs are released and fertilized and the spent adults then return downstream to the feeding grounds. The young stages, which are frequently planktonic and unable to make headway against the prevailing current, are carried downstream to the nursery grounds. From there the juveniles move downstream again, and recruit to the adult stock on the feeding grounds.

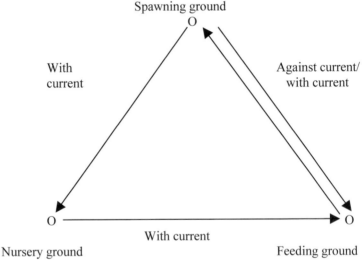

Fig. 1. Generalised migration pattern of marine fish and other fishery resources. (After Harden Jones 1968)

This basic migration pattern has evolved because of the differing habitat requirements of adults, the young stages and juveniles. Larvae are released into the marine production cycle at an appropriate time and place so they grow and avoid predation on the nursery grounds. Species with pelagic eggs and larvae maintain themselves within a particular oceanographic regime because the adults make active, compensatory movements in the opposite direction to which the pelagic stages are carried passively in the prevailing current. The whole process is an adaptation to the growth pattern of fish, which increase in mass by several orders of magnitude over the life cycle (Cushing 1982).

There are contrasting migration patterns in high and low latitude seas (Cushing 1975). Poleward of 40° in both hemispheres the marine production system is highly seasonal and migration circuits have evolved so that the release of larvae is timed to coincide with the spring peak of production. The spawning and nursery grounds are therefore fixed spatially and temporally in these high latitude systems. In tropical and subtropical regions, the annual production cycle is much less pronounced and has resulted in the evolution of more diffuse, less precise, migration patterns of fish and other resources in space and time.

19.3 Current Status of Marine Fishery Resources

Global fishery production from capture fisheries and aquaculture is currently at the highest level recorded and provides more than 15% of the total animal protein consumed by humans (FAO 2002). There is, however, a trend of world population increasing faster than the total fish supply, and per capita consumption of fish decreased by more than 10% between 1987 and 2000.

Total production of marine capture fisheries was 86.0 million tonnes in 2000 following a period of instability in total production after the 1997-1998 El Niño, which especially affected the fisheries of the eastern Pacific. About 47% of the major exploited stocks are fully exploited and delivering catches that are at, or close to, their maximum sustainable limits. Another 18% of stocks are overexploited with no prospect of increasing production, and the strong possibility that catches will decrease unless overfishing is prevented. A further 10% of stocks are substantially depleted or recovering and much less productive than in the past. Finally, about 25% of the stocks are underexploited or only moderately exploited. As fishing pressure increases worldwide, the trend is for the number of underexploited and moderately exploited stocks to decrease, the number of fully exploited stocks to remain relatively stable and the number of overexploited, depleted and recovering stocks to increase.

Global fishing pressure appears to be bringing about large-scale changes in marine ecosystems. This is reflected by an apparent downward shift in the mean level in the food web exploited by fisheries -- the so-called process of fishing down the food web (Pauly et al. 1998). There are also apparent changes in the community structure where groundfish stocks have been overexploited (Caddy and Rodhouse 1998).Worldwide depletion of top predatory fish in various marine ecosystems has occurred (Myers and Worm 2003).

Fishery science has historically been concerned with providing the basis for management of the top-down effects of exploitation on stocks caused by exploitation. There is now an increasing realisation that bottom-up effects, caused by changes in the marine environment, especially changes caused by human activities, are also important in driving change in the fisheries. These changes, which range from pollution of the seas to changes in oceanography driven by global climate change, affect fundamental ecological processes that inevitably impact the stocks of exploited fishery resources. Global warming will cause change in sea temperature, winds, freshwater discharge into the oceans, mixing of the ocean surface and sea_ice extent in the polar regions. Warming will also cause change in large scale oceanic advection processes which, in turn, will affect transport of the planktonic egg and larval phases of fish life cycles.

Given the links between migrations and ocean currents, questions about climate and oceanographic variability and change have emerged as key issues in fisheries science. The next part of this review will examine three examples of fishery resources in which biomass has been shown to respond to climate/oceanographic variability over a range of time scales. These provide valuable insights into the likely effects of large-scale ecological change on the world's fisheries.

19.4 Atlantic Cod – Response to Decadal Change

Atlantic cod, *Gadus morhua*, has been one of the most heavily exploited species, and some stocks have effectively been reduced to commercial extinction. In the North and Irish Seas, stocks have declined by 90% since the 1960s, whilst in Canada some stocks are down by more than 99%. Whilst the general scientific consensus is that overfishing is a major factor in this decline, there is also strong evidence that environmental factors also drive change. In some stocks disruption of the migration triangle by oceanographic variability is a clear factor, whilst in others, perhaps because of complexities in the environment and differing interactions between top-down and bottom-up effects, the relationship between environment and stock biomass is harder to establish. Atlantic cod are relatively long-lived with a maximum lifespan of about 25 years. Populations therefore contain numerous year classes of different ages which buffer population biomass from the short-term effects of the environment on recruitment. They are nevertheless subject to the effects of decadal scale variability.

An apparently clear example of the interaction between environment and the migration triangle is illustrated by the period of cod abundance at Greenland over a 50-year period from about 1920 to 1970 (Brander 1996). The Greenland cod stock is the most northerly in the northwest Atlantic and it may be that at the extremity of the species' geographical range, biological changes driven by the environment are better defined.

The stock spawns off the coasts of southwest and east Greenland, both inshore in the fjords and offshore. In addition, substantial numbers of developing larvae drift into the area from Iceland when oceanographic conditions are right and there is a return migration of spawners from Greenland to Iceland. There is an approximately 350-year record of cod catches from the area. In the 20th century the first decade or so was marked by low catches off west Greenland, but after 1917 cod spread gradually northward and in response the Greenland Administration set up a number of fishing stations, and catches rose to 460,000 tonnes/year. These events took place at the start of a period of relatively warm sea temperatures. By the late 1960s, the stock had again retreated south, the local fish had all but disappeared, and the remaining stock was sustained by recruitment from Iceland. The decline was accompanied by the onset of cool conditions after 1968.

The collapse of the Greenland fishery was undoubtedly a function of heavy fishing pressure. Regional cooling since the late 1960s is anomalous seen against the global warming trend, but it there is a clear correlation between regional warming and the increase in biomass of the local Greenland stock. Because at least some spawners from Greenland migrate to Iceland to spawn, the decline of the west Greenland stock probably also contributed to a decline in the Iceland stock.

19.5 Antarctic Krill – Response to Sub-Decadal Change

Antarctic krill, *Euphausia superba*, are planktonic, shrimp-like crustaceans that are unable to swim against the prevailing current. They do, however, make vertical migrations and it is generally thought that by moving between depth layers, where prevailing current speed and direction differ, they are able to exert some control over their drift. Krill undoubtedly move over long distances in the circumpolar current of the Southern Ocean, but their distribution is not even throughout the system. Instead, they are concentrated in particular areas of the Southern Ocean (Marr 1962), so there are clearly processes by which these concentrations are maintained although they are not understood.

Because of their remote location and low market value Antarctic krill are one of the few large fishery resources that are still lightly exploited, so changes in population biomass can largely be attributed to bottom-up effects of the environment. Krill are relatively short-lived with a maximum lifespan of about 5 years. Populations therefore contain fewer year classes of different ages than longer-lived species. Population biomass is therefore less buffered from annual and sub-decadal effects of the environment on recruitment.

In the Atlantic sector of the Southern Ocean the main breeding area for krill is in the region of the Antarctic Peninsula. Large numbers drift from there across the Scotia Sea to the vicinity of South Georgia where they provide a food resource for large populations of seabirds, seals and, in the past, baleen whales. Krill biomass at the Antarctic Peninsula fluctuates with year class success and this is positively correlated with extensive sea ice cover the previous winter (Loeb et al. 1997). This seems to be because krill can exploit the algal blooms beneath sea ice during years of extensive sea ice. Conversely, salps, which compete with krill for planktonic food, benefit from open water conditions and so, under condition of poor sea-ice development, reduce the phytoplankton available for the krill, thus reducing their spawning success.

At South Georgia krill biomass also fluctuates widely over sub-decadal time scales. The pattern of variability is marked by occasional poor years of recruitment when low biomass causes breeding failure in the krill-dependent species of higher predators. These years of low recruitment are undoubtedly due to variability in spawning success at the Antarctic Peninsula, but the effects may be magnified downstream at South Georgia because the main region in which krill drift from the Antarctic Peninsula is also subject to variable sea-ice cover from year to year. Variation in sea-ice extent affects the success of overwintering krill larvae drifting downstream from the spawning area, and this is a further source of variability of recruitment into the seas around South Georgia (Murphy et al. 1998).

In the 1990s there was a marked reduction in the number of years in which there was a strong recruitment of krill at South Georgia compared with the 1980s, and this was probably linked to regional warming and fewer years of extensive sea ice development (Reid and Croxall 2001). If, as generally assumed, regional warming at the Antarctic Peninsula and in the Scotia Sea is linked to global climate change, then we may be witnessing the onset of ecological change in this high latitude system driven, bottom up, by the activities of humans.

19.6 Shortfin Squid – Response to Annual Change

The Argentine shortfin squid, *Illex argentinus,* is a southwest Atlantic species that has a lifespan of 1 year. It has been commercially exploited since the early 1980s and is now probably close to being fully exploited. Whilst there has been no general trend of increasing or decreasing catch rate over that time, the biomass of the population fluctuates dramatically between years. In common with other squid species the short lifespan is associated with rapid growth and a "spawn once and die" reproductive strategy. These squid are ecological opportunists that have evolved to increase population size rapidly when environmental conditions are favourable, but are equally prone to collapse when conditions are unfavourable.

The winter spawning population, which comprises the great majority of the catch, spawns off the continental shelf break in the latitude of the Plate River where there is considerable meandering in the current system so that eggs and larvae are retained in the region. The juveniles move in over the continental shelf as they grow and migrate south to the main feeding grounds around the Falkland Islands where they grow into adults and attain sexual maturity. Towards the end of the life cycle they then make the return migration to the spawning grounds where they spawn and die. This species therefore migrates between the low-latitude, low-primary-production system of the sub-tropical South Atlantic, where it spawns, and the high-latitude, seasonally high-production system where it feeds, grows and attains sexual maturity.

Using remotely sensed sea surface temperature data, it has been shown that about 55% of variability in strength of recruitment into the fishery around the Falkland Islands can be explained by variation on the spawning grounds of the area of ocean surface of optimum temperature for spawning (Waluda et al. 2001). These conditions, in turn, are associated with differences in the amount of horizontal temperature gradients at the ocean surface, presumably due to changes in mesoscale meandering in the surface layers. In the spawning season prior to poor recruitment the area of optimum temperature is reduced and at the same time conditions become less suitable for retention of eggs and juveniles close to the continental shelf. The environmental variability in the south Atlantic that drives variability in the squid population has been shown to have teleconnections with the El Niño-Southern Oscillation (ENSO) cycle in the Pacific, albeit with a time lag (Waluda et al. 1999), so it is reasonable to conclude that variability in the squid population has links back to ENSO.

ENSO events appear to be increasing in frequency and severity in response to global warming. This highly opportunistic, short-lived, squid species can be expected to respond very quickly to changes in this large-scale ocean/climate phenomenon.

19.7 Conclusion

Atlantic cod, Antarctic krill and shortfin squid provide different examples of the way the environment can drive change in standing stock biomass in fisheries by acting on different parts of the migration cycle. In the case of Atlantic cod it was the adult phase, feeding off west Greenland, that appears to have been most influenced by shifts in the temperature regime during the 20th century. Warm conditions starting in the 1920s allowed an expansion of the feeding grounds of the adult fish northwards along the west coast of Greenland which increased the size of the spawning stock, not only off Greenland but also off Iceland where some of the west Greenland cod migrated to spawn. With the end of the warm period in the 1970s the west Greenland fishery declined and, because of a reduction in the scale of the spawning migration from west Greenland, this probably also caused a reduction in the size of the Iceland stock.

In the case of Antarctic krill, environmental conditions associated with the extent of annual sea ice influence both the spawning condition of the adults in the region of the Antarctic Peninsula, and hence breeding success, and the early life stages of the krill as they drift in the Antarctic Polar Current system towards South Georgia from the region of the Peninsula.

In the longfin squid, environmental variability on the spawning grounds seems to have most influence on recruitment and hence biomass. In this case, it appears to be the early life phase which is influenced by the environment, being affected both by temperature and by ocean currents which transport the planktonic eggs and larvae. We therefore see from these three examples that a variable environment can act on each limb of the migration triangle, shown in Fig. 1, in ways that can increase or decrease stock biomass.

The time scale over which biomass responds to environmental variability is in large part a function of the life span of the species. In long-lived Atlantic cod variability is on a time scale of decades. In this species the effects of short-term variations in spawning success on biomass of the stock are damped by the large numbers of year classes present in the population. Antarctic krill, with a small number of year classes in the population, are subject greater variation so that to poor years of low stock biomass are caused by the failure of a single year class to recruit. The ecologically opportunistic squid, with their 1-year life span, have the most labile populations and biomass fluctuates dramatically from year to year. Each year biomass is determined to a major extent by the breeding success of a single year class which dies after spawning and which, depending on the success of the vulnerable early-life stages, gives rise to the recruitment of a single new cohort in the population.

As fishing pressure on the world's stocks continues to increase, it is important to understand how environmental variability exerts bottom-up effects on the populations. These effects can have the effect of either increasing or ameliorating stress on stocks. Either way as stocks come under increasing exploitation pressure it will become more important for managers of fisheries to incorporate ecosystem thinking into their management strategies.

Understanding the effects of environmental variability also enables us to better understand the likely effects of global climate change on the fisheries. The long-lived species such as Atlantic cod have evolved to fill their ecological niche in the environment and then maintain their position by virtue of their size and dominance as competitors for food and other resources. The shorter-lived, fast-growing species have evolved to respond rapidly to favourable environmental conditions, increasing stock size quickly when opportunities are presented. They are analogous to weed, or plague, species in agricultural systems. Under conditions of environmental change and deteriorating environmental conditions, it is the stocks of species with the former strategy that we might expect to be most affected, especially when they are simultaneously exposed to high levels of fishing pressure.

19.8 References

Brander K (1996) Effects of climate change on cod (*Gadus morhua*) stocks. In: Wood CM, McDonald DG (eds) Global warming: implications for freshwater and marine fish. Cambridge Univ Press, Cambridge, pp 255-278

Caddy JF, Rodhouse PG (1998) Do recent trends in cephalopod and groundfish landings indicate widespread ecological change in global fisheries. Rev Fish Biol Fish 8:431-444

Cushing DH (1975) Marine ecology and fisheries. Cambridge Univ Press, Cambridge, 278 pp

Cushing DH (1982) Climate and fisheries. Academic Press, London, 373 pp

FAO (2002) The state of world fisheries and aquaculture. www.fao.org

Harden Jones FR (1968) Fish migration. Arnold, London, 325 pp

Loeb V, Siegel V, Holm-Hansen O, Hewitt R, Fraser W, Trivelpiece W, Trivelpiece S (1997) Effects of sea-ice extent and krill or salp dominance on the Antarctic food web. Nature 387:897-900

Marr JWS (1962) The natural history and geography of the Antarctic krill (*Euphausia superba* Dana). Discove Rep 32:33-464

Meek A (1915) Migrations in the sea. Nature 95:231

Murphy EJ, Watkins JL, Reid K, Trathan PN, Everson I, Croxall JP, Priddle J, Brandon MA, Brierley AS, Hofmann E (1998) Interannual variability of the South Georgia Marine Ecosystem: biological and physical sources of variation in the abundance of Antarctic krill. Fish Oceanogr 7:381-390

Myers RA, Worm B (2003) Rapid worldwide depletion of predatory fish communities. Nature 423:280-283

Pauly D, Christensen V, Dalsgaard J, Froese R, Torres F Jr (1998) Fishing down marine food webs. Science 279:60-863

Reid K, Croxall JP (2001) Environmental response of upper trophic level predators reveals a system change in an Antarctic marine ecosystem. Proc R Soc B 268:377-384

Waluda CM, Trathan PN, Rodhouse PG (1999) Influence of oceanographic variability on recruitment in the genus *Illex argentinus* (Cephalopoda: Ommastrephidae) fishery in the South Atlantic. Mar Ecol Prog Ser 183:159-167

Waluda CM, Rodhouse PG, Podestá GP, Trathan PN, Pierce GP (2001) Oceanography of the *Illex argentinus* (Cephalopoda: Ommatrephidae) hatching grounds and influences on recruitment variability. Mar Biol 139:671-679

20 Migration of Marine Mammals

Ian L. Boyd

Sea Mammal Research Unit, Gatty Marine Laboratory, University of St Andrews, St Andrews KY16 8LB, UK, E-mail ilb@st-andrews.ac.uk

20.1 Abstract

Marine mammals utilize marine resources throughout the world's oceans, but the location of food is often spatially and temporally separated from environments that are required for reproduction. There is an increasing understanding of migratory behaviour because of the use of new techniques to track marine mammals. Many of the mysticete cetaceans exploit seasonally rich food supplies in the polar summer but migrate to sub-tropical waters during winter when mating and birth take place. In some cases, these migrations follow predictable routes but those cetacean species with the largest body size are, in general, those that migrate over the longest distances. The greatest understanding of migratory patterns comes from pinnipeds that are restricted to giving birth on land or ice. Again, body size appears to have co-evolved with migration behaviour. Those animals that have the largest body size also have the largest absolute energy requirements. Consequently, large marine mammals must forage in regions of relatively abundant prey in order to be able to feed profitably. Owing to differences in the allometric scaling of metabolic rate and energy stores with body mass, large body size confers greater fasting capabilities. This means that large marine mammals can migrate further in search of richer food patches. The range of body size amongst the pinnipeds and cetaceans may have evolved as a result of selection to exploit the high degree of heterogeneity of food supply in the oceans and to allow animals to exploit food remote from where they are constrained to reproduce.

20.2 Introduction

Migration is an important part of the life histories of most marine mammals. Relatively little is known in detail about the patterns of migration in many species, but new methods including non-invasive mark resighting (Clapham et al. 1993; Smith et al. 1999; Stevick et al. 2002), tracking using vocalizations (Clapham and Mattila 1990; Clark 1995) and the tracking of individuals using satellite tags (e.g., McConnell et al. 1992; Martin et al. 1993; Mate et al. 1997) are beginning to add greatly to our knowledge of the frequency and extent of movements of individuals. Historically, evidence of migration has come from the observations by whalers and the traditional hunting cycles of native peoples such as those associated with the annual migration of grey whales (Swartz 1986) along the Pacific rim of North America.

A recent brief review of migration patterns in marine mammals has been provided by Stevick et al. (2002). This includes consideration of migration routes distance, speeds and intraspecific variation in migratory behaviour. It is not the intention of this paper to reiterate this review. Here, I will examine the underlying biological, geographical and physical drivers of migration in order to identify principles that may apply to all marine mammals and, therefore, to help develop a predictive framework for migration.

I define migration as the repeated movement of individuals between different environments that have a clear geographical separation. Of the suite of critical co-evolved life history characteristics of marine mammals, migration is perhaps the one that is most susceptible to the effects of human disturbance. Increasing use of the marine environment is introducing hazards to marine mammals that could affect migration patterns. These hazards include physical strikes of these mammals in increasingly busy shipping lanes, noise pollution in the marine environment in the frequency ranges used for communications by many cetaceans, and physical barriers to migration involving new sites for renewable energy which are increasingly being established offshore. The extent to which each of these might affect marine mammals is difficult to assess at present and will vary between species, populations and circumstances, but it is important to understand the underlying biology of migration in order to understand the potential effects of disruption.

Migration may occur as a result of a range of co-evolving factors. In polar regions migration may occur because of the need for animals to move out of ice-bound areas (Martin et al. 1993). Corkeron and Conner (1999) hypothesized that the migration of mysticete whales to tropical regions was a response to relatively high predation risk for offspring due to killer whales at high latitudes, but it seems most likely that a general principle underlying migration in marine mammals will turn out to be the need to sustain a neutral or positive net energy balance. Two factors compete against each other on the income and expenditure sides of the energy balance sheet. These are the distribution and abundance of energy sources in the oceans, and the thermal characteristics of the oceans which act as a sink for heat energy from animals that are endothermic and that maintain a body temperature up to 40°C above that of the surrounding water.

Because these are general constraints, they should apply equally to all marine mammals, although species may have found different solutions to maximization of their genetic fitness. These solutions are likely to have resulted in the co-evolution of many of the important phenotypic characteristics of marine mammals, including body size, migratory behaviour, longevity and reproductive rate -- and each of these will be modified by individual circumstances such as predation pressure. Cetaceans may have some of the most extreme examples of these processes, but the difficulties of studying cetaceans means that the general principals of life-history evolution in marine mammals more often come from studies in pinnipeds (seal, fur seal, sea lion and walrus) as examples of more tractable species.

This chapter is in three sections concerning: (1) physical constraints that lead to marine mammals having to occupy different environments at different times in their lives; (2) geographical constraints that represent the distance between the different environments they occupy; and (3) phenotypic constraints that dictate the extent to which allometric considerations are important to patterns of migratory behaviour.

20.3 Physical Constraints: Thermal Challenges

Unlike cetaceans, pinnipeds have to give birth on land and they have therefore retained a terrestrial phase in their life histories. For some species, such as the hooded seal (*Cystophora cristata*), this phase can be as short as 4 days each year but this alone has tied seals to a narrow range of life-history options because food is, more or less, remote from the location where mothers must give birth. Certain factors, such as the risk of predation while ashore as well as social evolution (Bartholomew 1970), have had a secondary influence on the selection of the birth site, but the primary consideration is likely to have been proximity to food.

The pinniped life cycle has been thought to have been constrained to a phase on land because the small newborn seal is unlikely to be able to thermoregulate in water during the early stages of life (Boyd 2002). Most pinnipeds inhabit cold, polar and temperate oceans and the relatively small body size and thin layers of insulating blubber at birth mean that the newborn is vulnerable to hypothermia. The energetic demands required to counterbalance this challenge can only be met by increasing the rate of energy turnover in order to sustain heat balance, by increasing body size and insulation to reduce heat flux, or by migration to warmer habitats. Since air is 24 times less conductive than water, giving birth on land has provided pinnipeds with the opportunity to reduce heat flux in a less conductive medium.

Thus, the life history of pinnipeds appears to have been driven by a need to reconcile the constraints of foraging in the sea and having to give birth in a medium that is thermally tractable for young offspring. This is a physical constraint that should apply to cetaceans as much as it does to pinnipeds. Both experience the same constraints associated with being endothermic in a cold, conductive medium. However, cetaceans are entirely aquatic and, rather than migrating to land, cetaceans migrate to warmer oceanic regions to give birth and nurse offspring (Swartz 1986; Clapham et al. 1993; Kasamatsu et al. 1995). Thermal constraints may also have contributed to the adaptive importance of generally large body size of cetaceans because of the thermal advantages this provides as a means of reducing total body heat flux. Cetaceans of small body size, characterized by the porpoises and dolphins, are species of tropical and temperate latitudes; the high latitudes are almost exclusively occupied by the cetacean species of large body size.

This may be partly reflective of different thermal challenges in these regions and the advantages brought by different body sizes.

20.4 Geographical Constraints: Heterogeneous Food Distribution

Food for marine mammals is not evenly distributed in the sea. In fact, there is a highly heterogenous distribution of potential food, and this heterogeneity occurs in both space and time (Steele 1985). Food availability may vary unpredictably on a multi-annual and ocean-basin scale (Trillmich and Ono 1991) or seasonally with a relatively high level of predictability. At smaller scales there is also likely to be a high variance associated with prey availability, but marine mammals may learn to reduce the variance associated with finding food by using oceanographic features to indicate the likely presence of prey. The extent to which marine mammals tend to do this is the subject of current research.

Not all potential prey items in the oceans are suitable as food for marine mammals. Marine mammals must feed on patches of prey that provide a nett energy gain as a result of foraging. This means that, depending upon the energetic costs of foraging on a particular prey type, marine mammals may not be able to forage profitably on a particular patch of prey. Therefore, the profitability of a particular patch will depend upon the rate of energy expenditure of the marine mammal, the specific cost of catching the prey type and the distance to the next profitable prey patch. Recent studies of Antarctic fur seals (*Arctocephalus gazella*) suggest that marine mammals are capable of optimizing their decisions across these three factors in order to maximize their overall rate of energy gain while foraging (Mori and Boyd 2003).

The distribution of exploitable patches of food in the sea is such that there is an abundance of small patches while large patches are rare (Fig. 1). The problem for marine mammals is that, as the energy cost of foraging increases, fewer of the abundant but small prey patches are exploitable because they will not return a net energy gain (Fig. 1). This means that, as energy costs increase (e.g. with increasing body size), marine mammals must search further in order to find suitable areas where they can feed. Marine mammals with large foraging costs must travel further from suitable areas to give birth than those with small costs.

20.5 Phenotypic Constraints: the Consequences of Size

The absolute energy costs of maintenance increase with increasing body size (Boyd 2002). Therefore, large marine mammals will only be able to forage profitably on the large, infrequent patches of prey (Fig. 1) and this will result in travel of greater distance, on average, in order to find them.

Large body size, however, brings the advantage of increased fasting capability. This occurs because the metabolic costs of larger species are a smaller proportion of their energy storage capacity than that of smaller species. Therefore, the time that large species can spend searching for profitable food patches is greater than for small species.

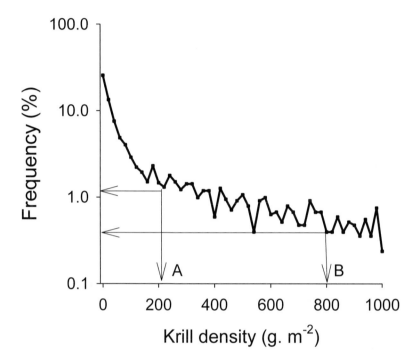

Fig. 1. An example of food distribution in the sea. In this case, the distribution of food patch size for Antarctic fur seals feeding on krill in the Southern Ocean is shown. Frequency is given as a log scale and shows the relatively high abundance of low density krill patches and the relatively low abundance of high density krill patches. In case A, a marine mammal is assumed to require a minimum krill density of 200 g m^2 in order to forage profitably. This shows that only just over 1% of patches encountered by such a marine mammal would be worth exploiting. In case B, a marine mammal is assumed to require a minimum krill density four times that of case A. This individual would find that only about 1 in 400 patches encountered would be worth exploiting.

20.6 Migratory Patterns

The geographical and phenotypic constraints described in the previous sections should combine to give a set of migratory behaviours in marine mammals. I will now assess the degree of support that exists for this theoretical framework from observations of marine mammals.

Amongst the pinnipeds, the fur seals are a species with relatively small body size (at least amongst females). These species migrate between the breeding grounds, where pups are reared, and foraging areas offshore. Each migration takes place over periods of days to weeks but is limited in duration by the ability of the pup to survive starvation during the absence of the mother. Due to their relatively small body size, the absolute costs of foraging are relatively small so they can exploit relatively low quality patches of prey (Fig. 1). This appears to be the reason why these small pinnipeds can sustain this form of short-term migratory behaviour (Boyd 1998).

Increasing body size results in a lower frequency of exploitable prey patches (Fig. 1) and, in pinnipeds, this appears to result in a threshold body mass of about 70-90 kg above which the type of short-term migratory behaviour shown by the small fur seals cannot be sustained. This happens because there is no longer a sufficient supply of profitable prey patches available in the vicinity of the pup to allow the mother to return to the pup with food before it dies of starvation. Consequently, we see a quantum change in the migratory behaviour of pinnipeds as body size increases and this is accompanied by a change in the life-histories of the seals. The harbour seal (*Phoca vitulina*) appears to have a body size that is close to the point at which this change takes place (Boness et al. 1994). Larger species tend to begin lactation with most of the resources needed to sustain both the mother and pup through the pup-rearing period stored in the body of the mother in the form of adipose tissue reserves. Thus, they remain with the pup through lactation, only then to migrate to foraging grounds (e.g. Stewart and DeLong 1995). In the most extreme cases, the hooded seals remain with the pup for only 4 days and then return to distant foraging areas (Bowen et al. 1987). Consequently, the capital breeders are the phocid pinnipeds (true seals) with their larger body size than the otariid pinnipeds (eared seals, including the fur seals), which are income breeders, meaning that they raise their offspring on energy gained during lactation. This contrast leads to differing migratory behaviour: in the case of the income breeders there are frequent short migrations between the offspring and the foraging ground, whereas in the capital breeder there is only one migration (Boness and Bowen 1996). The elephant seals provide the most extreme example of the capital breeder model (Stewart and DeLong 1995).

Therefore, within the pinnipeds, there appears to be a general pattern of co-evolution of body size and migration behaviour. Exceptions to this general pattern can be found in many species, but two particular exceptions are the Steller sea lion (*Eumetopius jubatus*) and ringed seal (*Phoca hispida*). The Steller sea lion is an otariid with the phenotype (in terms of adipose tissue reserves) of an income breeder (relatively small adipose tissue reserves) but with the body mass of a capital breeder. Consequently, rather than migrating long distances to find profitable prey patches, it has developed behaviour that involves moving the pup to sites on land that are closer to the location of the food. It appears that the Steller sea lion has adapted its migration behaviour to adjust for its specific circumstances but its behaviour is still consistent with the general model for marine mammals, assuming that these species are constrained by energetics.

In contrast, the ringed seal is a phocid pinniped which has the phenotype of a capital breeder but the body mass of an income breeder. These characteristics appear only to be compatible because ringed seals occupy ice-bound habitats where there is no need for migration because food is available in the immediate vicinity of the breeding site. The examples of ringed seals and Steller sea lions both suggest how the trade-off between the need to occupy suitable habitat for the offspring and the need to obtain energy can affect marine mammal distributions as well as migration patterns.

Most of the well-known migratory cetaceans, such as right whales (Mate et al. 1997) and humpback whales (Stevick et al. 2002), appear to follow the general pattern of the capital breeding pinnipeds. In these species, large blubber reserves are accumulated during intensive feeding in polar regions during summer. Their large body size means that they are only able to feed profitably on the concentrated resources occurring in polar regions during summer. During winter, these species are often found in very specific wintering grounds where births take place but where relatively little feeding takes place.

Other cetaceans, mainly those with relatively small body sizes, inhabiting tropical and temperate regions where food is available within habitat that is appropriate for reproduction at that location may have relatively little need to migrate except as a means of following the availability of food. In these cases migratory pathways may simply follow the movement of marine secondary and tertiary production.

20.7 References

Bartholomew GA (1970) A model for the evolution of pinniped polygyny. Evolution 24:546-559

Boness DJ, Bowen WD (1996) The evolution of maternal care in pinnipeds. BioScience 46:645-654

Boness DJ, Bowen WD, Oftedal OT (1994) Evidence of a maternal foraging cycle resembling that of otariid seals in a small phocid, the harbour seal. Behav Ecol Sociobiol 34:95-104

Bowen WD, Boness DJ, Oftedal OT (1987) Mass transfer from mother to pup and subsequent mass loss by the weaned pup in the hooded seal, *Cystophora cristata*. Can J Zool 65:1-8

Boyd IL (1998) Time and energy constraints in pinniped lactation. Am Nat 152:717-728

Boyd IL (2002) Energetics: consequences for fitness. In: Hoelzel R (ed) Marine mammal biology: an evolutionary approach. Blackwell, Oxford, pp 247-277

Clapham PJ, Mattila DK (1990) Humpback whale songs as indicators of migration routes. Mar Mamm Sci 6:155-160

Clapham PJ, Barff LS, Carlson CA et al (1993) Seasonal occurrence and annual return of humpback whales, *Magaptera novaengliae*, in the southern Gulf of Maine. Can J Zool 71:440-443

Clark CW (1995) Application of US Navy underwater hydrophone arrays for scientific research on whales. Rep Int Whaling Comm 45:210-212

Corkeron PJ, Connor RC (1999) Why do baleen whales migrate? Mar Mamm Sci 15:1228-1245

Kasamatsu F, Nishiwaki S, Sakuramoto K (1995) Breeding areas and southbound migrations of southern mink whales *Balaenoptera acutorostrata*. Mar Ecol Prog Ser 119:1-10

McConnell BJ, Chambers C, Nicholas KS, Fedak MA (1992) Satellite tracking of grey seals (*Halichoerus grypus*). J Zool Lond 226:271-282

Martin AR, Smith TG, Cox OP (1993) Studying the behaviour and movement of high Arctic belugas with satellite telemetry. Symp Zool Soc Lond 66:195-210

Mate BR, Nieukirk SL, Kraus SD (1997) Satellite-monitored movements of the northern right whales. J Wildl Manage 61:1393-1405

Mori Y, Boyd IL (2004) The behavioural basis for non-linear functional responses: the case of the Antarctic fur seal. Ecology

Smith TD, Allen J, Clapham PJ et al (1999) Ocean-basin-wide mark-recapture study of the North Atlantic humpback whale (*Megaptera novaeangliae*). Mar Mam Sc 15:1-32

Steele JH (1985) A comparison of terrestrial and marine ecological systems. Nature 313:355-358

Stevick PT, McConnell BJ, Hammond PS (2002) Patterns of movement. In: Hoelzel R (ed) Marine mammal biology: an evolutionary approach. Blackwell, Oxford, pp 185-216

Stewart BS, DeLong RL (1995) Double migrations of northern elephant seal, *Mirounga angustirostris*. J Mammal 76:196-205

Swartz SL (1986) Gray whale migratory, social and breeding behaviour. Rep Int Whaling Comm, Spec Issue 8:207-229

Trillmich F, Ono KA (1991) Pinnipeds and El Niño: responses to environmental stress. Springer, Berlin Heidelberg New York

21 The "Global Register of Migratory Species" – First Results of Global GIS Analysis

Klaus Riede

Zoologisches Forschungsinstitut und Museum Alexander Koenig (ZFMK), Adenauerallee 160, 53113 Bonn, Germany, E-mail k.riede.zfmk@uni-bonn.de

21.1 Abstract

The advantages of using Geographical Information Systems (GIS) for the study of animal distribution ranges and movements are evident. Projections of GIS maps can be easily changed, maps can be exported into other software applications, and they can be combined with a wide variety of other relevant GIS data sets, containing environmental, ecological, or economical data. GIS maps contain underlying tables, connecting geographic information with attributes stored in complex databases. This is of particular importance when mapping a migratory species, because a time code has to be data-based for different areas of its range, reflecting its changing seasonal activities, such as breeding, wintering, spawning, and feeding.

The Global Register of Migratory Species (GROMS) information system has been designed to store these complexities of migration within one geo-database. Up to now, 1,000 migratory vertebrate species have been GIS-mapped. Though still on a global scale, by merging or intersecting maps, this comprehensive data set allows a wide variety of overall analyses, such as:

1. Diversity gradients of migrants (species numbers per territory or grid cell)
2. Intersection with GIS data sets on ecoregions, providing a list of migrants per ecoregion
3. Identification of "migratory footprints" for certain regions or ecosystems, by overlaying all distribution maps for species visiting this region
4. Range state calculation by intersection with political boundaries, to be used in conservation policy and multilateral environmental treaties, such as the "Convention on the Conservation of Migratory Species of Wild Animals".

All results are re-imported into the GROMS database, allowing easy query for geodata with standardized and fast relational database query tools.

The GROMS project is funded by the UNEP Secretariat, Bonn, for the "Convention on the Conservation of Migratory Species of Wild Animals" (also known as CMS or the Bonn Convention, http://www.cms.int/). It is hosted by the ZFMK (Zoologisches Forschungsinstitut und Museum Alexander Koenig;

http://www.museumkoenig.uni-bonn.de/). Details and data sets are available at the GROMS website http://www.groms.de, or off-line on CD-ROM (Riede 2004).

21.2 Introduction

Because migratory species cross borders, their efficient protection requires international cooperation. Endangered migrants are protected by the Convention on the Conservation of Migratory Species of Wild Animals (also known as CMS or the Bonn Convention), an international treaty bringing together 84 member states (as by October 2003). However, knowledge about animal migration is insufficient and widely scattered. The Global Register of Migratory Species (GROMS) supports the Bonn Convention by summarizing the present state of knowledge in a relational database connected to a geographical information system (Riede 2001, 2004 and http://www.groms.de).

Among zoologists, there are several definitions of migration in use. Some scientists prefer a rather broad definition, such as "*Migration: the act of moving from one spatial unit to another*" (Baker 1978, p. 23).

In contrast, "migratory species" as defined by the CMS means:

"... the entire population or any geographically separate part of the population of any species or lower taxon of wild animals, a significant proportion of whose members cyclically and predictably cross one or more national jurisdictional boundaries" (CMS 1979, Article 1)

An important component of the definition is its emphasis on the return component of migrations ("*cyclically and predictably*"). This is close to the biological concept of:

"'true migration': ... a seasonal movement that implies some element of a return to some initial starting point – the traveller needs a 'return ticket'" (Dingle 1996).

The definition of "true migration" was adopted by GROMS. As a biological concept, it is independent of national boundaries, and therefore includes species migrating within large-range states. However, for reasons of practicability, the minimum migration distance was set at 100 km, which means that small-scale migrants such as amphibians were not included within the GROMS.

In temperate regions, migrants occur in high numbers, and their presence during summer is an important component of biodiversity at higher latitudes. Birds come to mind first, but migration is widespread throughout the animal kingdom. Among the less well-known wanderers are bats, marine mammals, turtles, fishes, and insects (see Boyd, Higler, Placher, Rodhouse, this Vol.). The ideal tool to study their distribution and movements is a Geographical Information System (GIS), for evident reasons: projections of GIS maps can be easily changed, maps can be exported into other software applications, and they can be combined with a wide variety of other relevant GIS data sets, containing environmental, ecological or socioeconomical data. GIS maps contain underlying tables, connecting geographic information with attributes stored in complex databases (Longley et al. 1999). This is of particular importance when mapping a migratory species, because a time code has to be data-based for different areas of its range, reflecting its changing seasonal activities, such as breeding, wintering, spawning, and feeding.

The GROMS information system has been designed to store these complexities of migration within one geo-database. First results from a meta-analysis of migratory vertebrate species distribution on global scale are presented here.

21.3 Methods

The aim of GROMS is to produce an overview of the distribution of migratory species on a global scale. The first step was to identify migrants, to generate a reference list of migratory vertebrates. At present, the reference list contains 4,357 species (295 mammals, including bats, terrestrial mammals, seals and sirenia, whales, and dolphins; 2,145 birds; 10 reptiles, including marine turtles; 1,895 fishes; and a few invertebrate species). The second step was to map species in GIS format. A set of approximately 1,000 GIS maps, mainly based on handbooks, reviews, and digital sources, was generated (for map sources and species groups covered, see Riede 2001, Riede 2004, and www.groms.de). This GIS data set has been used for a variety of purposes, such as publication of maps on an OpenGIS server or within species fact sheets (www.groms.de), or intersection with other GIS layers, for instance, with the political boundaries of states and provinces (administrative units). The basic concept of GIS intersection is outlined in textbooks and GIS user guides (ESRI 1998), and examples are described in detail in Riede (2001, p. 40). The following additional steps were necessary for an intersection of the entire set of GROMS distribution maps:

1. Merging of all distribution maps to one GIS data file (allpolygons.shp),
2. Intersecting of "allpolygons.shp" with administrative boundaries (admin98.shp by ESRI)

Operation 2 results in a geo-table, containing presence-absence data for all administrative units, and was re-imported into the GROMS relational database (an MS-ACCESS version on CD-ROM is published in Riede 2004). The geo-table contains basically three fields: species_name, province_name and literature_source, which is the reference for the map source. Once integrated into the database, geographic information can be retrieved within the GROMS database, by standardized Standard Query Language (SQL) database tools. Species lists for each country or province can easily be generated by clicking on the "species lists" button within the GROMS database main selection panel. In addition, it is possible to list range states or provinces for each of the mapped species, as part of the "species report" option.

21.4 Results

Based on the GIS operations outlined above, a meta-analysis on global level was performed, resulting in synoptic maps covering the following themes:

- Diversity gradients of migrants (species numbers per province or grid cell)
- Intersection with GIS datasets on ecoregions, providing a list of migrants per ecoregion
- Identification of "migratory footprints" for certain regions or ecosystems, by overlaying all distribution maps for species occurring within the respective region or ecosystem (Fig. 1)

a)

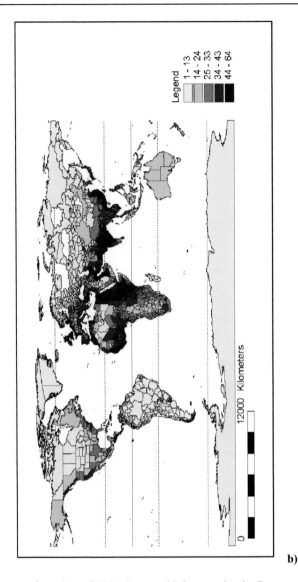

b)

Fig. 1. Migratory footprints of 215 migratory birds occurring in Germany. The number of these species in any other province of the world has been calculated, based on GIS inter-section of bird distribution maps with province borders (administrative units). See text for further interpretations. **a)** Number of migratory birds occurring in Germany, breeding else-where. **b)** Number of migratory birds occurring in Germany, wintering elsewhere

21.4.1 Migratory Species Diversity Maps

Intersection of maps of migratory mammals and birds allows one to calculate the number of migratory species per province or grid cell. The resulting distributions of diversity are similar and have been published in Riede (2001, Figs. A2.86 and A2.87). A striking result was the high diversity of migrants in temperate regions and coastal areas, while there is no increase in diversity toward the tropics, as frequently noted for other diversity distribution maps on a global scale (see for example Barthlott et al. 1999 for plant diversity). Besides coastal areas, the highest diversity is observed within areas used by different species for breeding and wintering, often from different biogeographical regions (e.g., western Asia, northwestern Europe). These maps will certainly change when more species are added, in particular the lesser known inner-tropical migrants and passerine birds. However, it is evident that a high diversity of migrants can be found within the highly industrialized countries, which may call into question the current focus of conservation efforts on "biodiversity hotspots" in the tropics (cf. Myers et al. 2000). Therefore, industrialized and overdeveloped nations should take a lead in demonstrating that sustainable coexistence of man and animals is possible by developing or redeveloping ecologically sound agricultural practices, and minimizing the ecological footprints of their citizens.

21.4.2 Migratory Species and Ecosystem Diversity

Using the GIS-based map of ecoregions provided by ArcView/ESRI [890 ecoregions based on the World Wildlife Fund (WWF) classification: www-eco.shp], the number of migratory species occurring in each ecoregion could be calculated by GIS intersection with 846 GROMS species distribution maps. Only terrestrial species have been evaluated. In accordance with the data presented above, there is a low diversity of migrants in ecoregions otherwise considered as "diversity hotspots" such as tropical rainforests. A surprisingly high diversity is observed in ecoregions severely stressed by human civilization (South American grasslands, European broad-leaved forest, Sahel belt, Himalayan subtropical pine forest). The number of species has then been calculated for each of the 890 ecoregions, and plotted as a color-coded map (Riede 2004, Fig. 3.16), to be used by the "Millennium ecosystem assessment" (see Alcamo et al. 2003 and http://www.millenniumassessment.org/en/).

21.4.3 Range State Calculation

The maintenance of rangestate lists is an important and time-consuming task for conservationists. Our example shows that range states or even provinces for each species can be calculated by using GIS maps based on a well-defined data source (in this case, the reference for the distribution map). Thereby, rangestate information can be made more transparent. Different sources can be compared, and in case of contradictory results, further investigations can be initiated. As outlined in Section 21.3, the geo-table generated by intersection with political boundaries contains columns referring to species name, province, and literature source. The respective countries inhabited by each species (the so-called range states) can be easily listed. Such rangestate lists are important tools in conservation policy and multilateral environmental treaties, such as the "Convention on the Conservation of Migratory Species of Wild Animals." However, the number of elements resulting from a combination of 1,000 widely distributed migratory species with 2,522 administrative units (provinces or "states" such as Hesse) is huge, resulting in a table of 340,000 entries. This indicates the huge task ahead for monitoring, even if the task is restricted to a reliable update of presence/absence within a certain province.

21.4.4 Migratory Footprints

Once integrated within the database, the geo-table can be data-mined to reveal patterns of distribution of migrants. For example, it is possible to map the "migratory footprint" of a certain region and its species. Figure 1 shows the migratory footprints of those migratory species occurring in Germany, and mapped by the GROMS. The "breeding" footprint shows that most German breeding birds are also distributed all over Eurasia. The "feeding, wintering" footprint shows the strong connection with Africa and parts of Asia, which are the main wintering areas for most central European bird species. Therefore, the "migratory footprint" is an instructive way to visualize interrelations between continents and ecosystems, as generated by migratory animals.

21.5 Conclusions

The integration of geographical information within databases is a particularly complex task, and migratory species are an even greater challenge, as they are on the move. Any information system dealing with migratory species has to cover:
− movements of migrants, requiring a GIS with time-code,
− a higher taxonomic resolution including subspecies and/or populations, which often show different migratory routes.

GROMS is unique because it has developed a data model covering these aspects.

It can be used for any biodiversity information system dealing with geographical changes over time, such as immigrations of alien viruses, species, or genes, or even migration of alleles.

Efficient conservation has to be based on knowledge, including input from other disciplines. Agricultural practice, fishery policy, and forestry have tremendous impacts on conservation, especially on animals migrating through different habitats. GIS technology is the ideal integrative approach to deal with these challenges (see Feoli, this Vol.).

For their protection across borders, the "Convention on the Conservation of Migratory Species of Wild Animals" (or "Bonn Convention") was established as a result of a recommendation of the United Nations Conference on the Human Environment, held in Stockholm, 1972. Today, the Bonn Convention has become an effective legal instrument, providing recommendations for the Convention on Biological Diversity, which resulted from the Rio Conference on the Human Environment, 1992.

21.6 References

Alcamo J et al. (2003) Ecosystems and human well-being: a framework for assessment/millennium ecosystem assessment. World Resources Institute, Washington, DC

Baker RR (1978) The evolutionary ecology of animal migration. Hodder and Stoughton, London

Barthlott W, Biedinger N, Braun G, Feid F, Kier G, Mutke J (1999) Terminological and methodological aspects of the mapping and analysis of global biodiversity. Acta Bot Fenn 162:103-110

CMS (1979) Convention on the conservation of migratory species of wild animals. http://www.wcmc.org.uk/cms/cms_conv.htm

Dingle H (1996) Migration - The biology of life on the move. Oxford University Press, New York

ESRI (Environmental Systems Research Institute) (1998) Getting to know ArcView GIS (for Version 3.1). ESRI, Redlands, California

Longley PA, Goodchild MF, Maguire DJ, Rhind DW (eds) (1999) Geographical information systems: principles, techniques, applications and management, 2 volume set, 2nd edn. Wiley, New York

Myers N, Mittermeier RA, Mittermeier CG, da Fonseca GAB, Kent J (2000) Biodiversity hotspots for conservation priorities. Nature 403:853-858

Riede K (2001) The global register of migratory species - database, GIS maps and threat Analysis. Landwirtschaftsverlag, Münster

Riede K (2004) The global register of migratory species - from global to regional scales. (with CD-ROM). Landwirtschaftsverlag, Münster

22 Coupled Dynamics of Lemmings and Long-Distance Migratory Birds

Noél Holmgren

School of Life Sciences, University of Skövde, P.O. Box 408, 541 28 Skövde, Sweden, Email noel.holmgren@inv.his.se

22.1 Abstract

The reproductive success of tundra-nesting birds exhibits considerable variation. It has been proposed that this variation is the result of predators switching from lemmings to bird eggs and nestlings in years after the lemming population has crashed. In a recent study (Blomqvist et al. 2002) that included a long time series of data of tundra-breeding birds, two species of geese and two species of waders, were analyzed. Several results support the "bird-lemming" hypothesis: (1) measurements of reproductive output correlates with lemming abundance, (2) measurements of reproductive output exhibit 3-year cycles as does the lemming abundance index, (3) adult waders migrate earlier in years when nest predation is expected to be most intense, (4) timing of migration of adult waders exhibits 3-year cycles. An attempt to test the impact of climate on reproductive success or survival of juveniles was unsuccessful.

22.2 Introduction

Large numbers of arctic-breeding geese spend the winter in western Europe (Cramp and Simmons 1977). Historically, very little was known of where these birds breed and of their living conditions during summer. The long tradition in counting these waders and geese in winter enabled the tracking of population changes and changes in reproductive success. This revealed a considerable variation in the number of juvenile birds across years, first reported in the brant goose (*Branta bernicla*; Summers 1986; Summers and Underhill 1987; Fig. 1E) and the curlew sandpiper (*Calidris ferruginea*; Roselaar 1979).

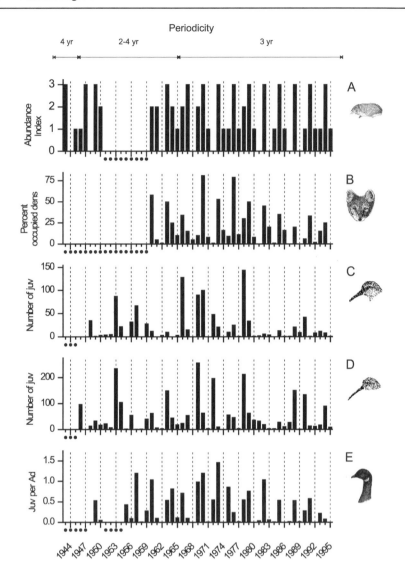

Fig. 1. Comparison of time series on population fluctuations in tundra-nesting birds and arctic fox with a presumed association with lemming cyclicity. **A** Abundance index (0–3) of lemmings on the Taimyr peninsula; **B** proportion of dens occupied by breeding arctic fox from western Taimyr; **C** number of ringed juvenile red knots during autumn migration at Ottenby, Sweden; **D** number of ringed juvenile curlew sandpipers during autumn migration at Ottenby, Sweden; **E** percentage of juvenile dark-bellied brant geese at wintering sites in northwestern Europe. *Dots* under the time axis indicate missing data. At *top* we have indicated the inferred periodicity of the lemming population numbers. Reprinted with permission from Springer-Verlag ©. Appears as figure 2 in Blomqvist *et al.* (2002)

A number of hypotheses have been proposed to explain the interannual variation in breeding success in Arctic breeding birds: (1) variation in weather conditions during breeding; (2) variation in feeding conditions on wintering grounds in
spring before migration, which assumes that the birds' breeding success to some
extent depends on the reserves built up before departure; (3) variation in wind and
weather conditions during spring migration. This hypothesis, as well as the previous one, assumes that energy balance prior to breeding affects breeding success;
(4) variation in predators' disturbance of birds on the nest, which assumes that in
years with many predators, those seeking food scare birds off nests and increase
the risk of hatching failure; (5) variation in egg and nestling predation, which was
more explicitly defined as the bird-lemming hypothesis (Roselaar 1979; Summers
1986). This final hypothesis proposes that the interannual variation in predation
pressure is driven by lemming cycles. The breeding of Arctic waders and geese is
successful when lemming populations increase or stay high, but poor in years
when lemmings are scarce and the predators switch from feeding on lemmings to
bird eggs or nestlings. None of the hypotheses mentioned above are mutually ex-
clusive, they may in fact all be true to a greater or lesser extent.

The lemmings of the Russian Arctic exhibit regular population cycles of 3-year
periodicity (Kokorev and Kuksov 2002; Fig. 1A). If the bird-lemming hypothesis
is true, two implications can be drawn: (1) lemming density and bird reproductive
success should correlate positively; (2) bird reproductive success might be expected to exhibit regular interannual cycles as does variation in lemming populations.

22.3 Material

In a recent paper, Blomqvist et al. (2002) analyzed data from two species of geese
and two species of waders. Forty-one years of brant goose, and 48 years of white-
fronted goose data of winter counts in western Europe have been compiled from
the literature. In addition, for 50 years, juvenile red knots and curlew sandpipers
were ringed at the autumn stopover site, Ottenby, SE Sweden. The breeding
ranges of these birds are concentrated to the Taimyr peninsula, Siberia. In contrast, the winter grounds are widely separated. The geese winter in the temperate
zone of western Europe, whereas the waders have tropical and subtropical winter
quarters in, for example, western and southern Africa. If all four species exhibit
similar variation in number of juveniles, it would strengthen the case for hypotheses pertinent to the breeding grounds rather than those involving wintering
grounds.

22.4 Test of the Bird-Lemming Hypothesis

The bird-lemming hypothesis predicts that the predators who normally feed on
lemmings, such as the arctic fox, switch to bird eggs and nestlings in years when

lemmings are scarce. There is evidence from 36 years of data from Taimyr for a positive correlation between the abundance of lemmings and percentage of artic fox dens being occupied (Kendall τ=0.663, n=36, p<0.001, one-tailed test; Fig. 1A,B; Blomqvist et al. 2002). When lemming populations crash, arctic foxes have been observed to still be abundant (Underhill et al. 1993) and to put an intense predation pressure on bird eggs and nestlings. Hence, reproduction of arctic-breeding birds is expected to be least successful in years after a lemming peak. Predation intensity was estimated by an index, calculated as the difference in lemming abundance between the focal year and the previous year (Blomqvist et al. 2002).

The predation index showed a hypothesis-predicted negative correlation with the number of juvenile red knots (Kendall τ=-0.225, n=40, p<0.05, one-tailed test; Fig. 1A vs. C) and curlew sandpipers ringed at Ottenby (Kendall τ=-0.402, n=40, p<0.001, one-tailed test; Fig. 1A vs. D; Blomqvist et al. 2002). In other words, when predation was expected to be most intense, reproduction appeared least successful. The number of ringed adults showed no correlation with the predation index for any of the two wader species (Blomqvist et al. 2002).

The brant goose and the white-fronted goose also breed on Taimyr, but winter in western Europe as opposed to the red knot and the curlew sandpiper that winter in tropical and subtropical Africa. The average proportion of juveniles in winter flocks was negatively correlated with the predation index, both for the brant goose (Kendall τ=-0.600, n=39, p<0.001, one-tailed test; Fig. 1A vs. E), and for the white-fronted goose (Kendall τ=-0.305, n=38, p<0.01, one-tailed test; Blomqvist et al. 2002).

In years when breeding fails due to nest predation, the adults are assumed to initiate fall migration earlier than otherwise. In an analysis of the median dates for different years, Blomqvist et al. (2002) showed that passage of migratory adult birds were earlier in years with high predation index than in years with low. This was true for both the red knot (Kendall τ=-0.204, n=35, p<0.05, one-tailed test) and the curlew sandpiper (Kendall τ=-0.215, n=39, p<0.05, one-tailed test).

If the predation on nestlings and eggs is ultimately driven by lemming numbers, and lemming numbers change in a 3-year cycle, reproductive success of waders and geese on the tundra can be expected also to exhibit a 3-year cycle. Fourier analysis did indeed reveal significant periodicity in number of juvenile red knots (Fig. 2C), number of juvenile curlew sandpipers (Fig. 2D), median passage date of adult red knots (Fig. 2E), median passage date of adult curlew sandpipers (Fig. 2F), number of juvenile brant geese per reproductive adult (Fig. 2G), and number of juvenile white-fronted geese per reproductive adult (Fig. 2H; Blomqvist et al. 2002). In all cases but one, the dominating periodicity was 3 years. The passage of adult curlew sandpipers showed a stronger periodicity of 5 to 6 years than of 3 years. The results based on ringing data are obtained without controlling for trapping conditions. Variation in trapping conditions, between and within years, can easily bias data.

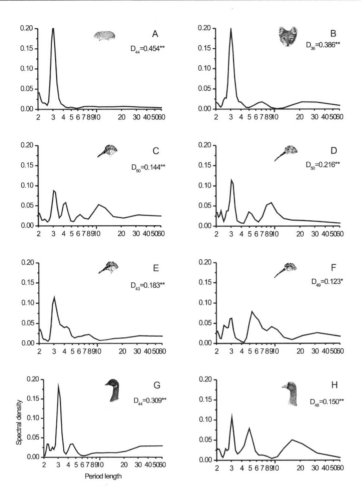

Fig. 2. Spectral analyses using Fourier transformation of different population fluctuation measures: **A** abundance index of the lemmings on Taimyr; **B** percentage of dens occupied by breeding arctic fox on western Taimyr; **C** annual ringing numbers of juvenile red knots during autumn migration at Ottenby, Sweden; **D** annual ringing numbers of juvenile curlew sandpipers during autumn migration at Ottenby, Sweden; **E** median ringing date of adult red knots during autumn migration at Ottenby, Sweden; **F** median ringing date of adult curlew sandpipers during autumn migration at Ottenby, Sweden; **G** inferred breeding success (number of juveniles per breeding adult) in the dark-bellied brant goose from winter counts in northwestern Europe; **H** inferred breeding success (number of juveniles per breeding adult) in the white-fronted goose from winter counts in Britain and the Netherlands. The D-statistic of the Kolmogorov-Smirnov test is presented, here indicating deviation from random noise. The symbol * indicates $p<0.05$ and ** indicates $p<0.01$. Reprinted with permission from Springer-Verlag ©. Appears as figure 3 in Blomqvist *et al.* (2002)

22.5 Test of the Spring Progression Hypothesis

To test two alternative weather hypotheses, the bird data were compared with two climate indexes. Blomqvist et al. (2002) used the May index of the Eurasian pattern (EAP; also called the Scandinavian pattern, Fig. 3) as a proxy for spring progression on Taimyr. The positive phase of this pattern is associated with positive height anomalies, sometimes reflecting major blocking anti-cyclones (NOAA 2001). Hence, if interannual variation in spring progression affects breeding conditions, one can expect a positive correlation between EAP and the breeding success of waders and geese. Neither the number of juveniles per reproductive adult of the two goose species, the number of ringed juveniles of the two wader species, nor the passage of the adults of the two wader species correlated with the EAP (Table 1).

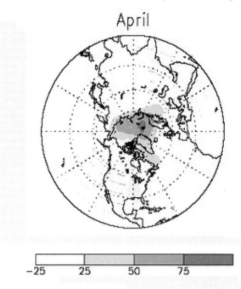

Fig. 3. Map of the positive phase of the SCAND pattern in April. For details see NOAA (2001)

Table 1. Kendall rank-order correlation coefficient (τ) for the recorded population variables of lemmings, arctic fox, geese, and waders *versus* the EAP climate index for the month of May, 1946–1995. The climate impact hypothesis predicts positive correlations, therefore, *p*-values refer to one-tailed tests. Table from Blomqvist et al. (2002)

Species	Measure	Area	n	Kendall–τ	p-Level
Lemming	Abundance index	Taimyr	37	0.11	>0.05
Lemming	Predation index	Taimyr	36	-0.15	>0.05
Arctic fox	Occupied dens (%)	Taimyr	36	0.22	>0.05
Brant goose	Juv./reprod.adult (%)	NW Europe	42	0.10	>0.05
White-fronted goose	Juv./reprod.adult (%)	NW Europe	45	-0.01	>0.05
Red knot	N:o. juv.	Ottenby	46	0.16	>0.05
Curlew sandpiper	N:o. juv.	Ottenby	46	0.19	>0.05
Red knot	Passage of adults	Ottenby	41	0.05	>0.05
Curlew sandpiper	Passage of adults	Ottenby	46	-0.05	>0.05

22.6 Test of the Winter Survival Hypothesis

The monthly indexes from October to March of the North Atlantic oscillation (NAO, Fig. 4) were used as proxies of autumn and winter conditions in western Europe. Strong positive phases of the NAO tend to be associated with above-normal temperatures and precipitation across northern Europe (NOAA 2001). If winter survival of juvenile geese varies between years due to variation in harshness of the fall and winter, NAO would exhibit a positive correlation with geese reproductive success as it appears from winter counts. The number of juveniles per reproductive adult brant goose or of white-fronted goose also did not correlate with any of the monthly NAO indexes (Table 2).

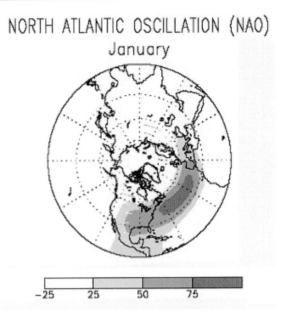

Fig. 4. Map of the positive phase of the NAO pattern in January. For details see NOAA (2001)

Table 2. Kendall rank-order correlation coefficient (τ) for the annual percentage of juvenile birds per reproductive adult in the dark-bellied brant goose and white-fronted goose versus the NAO climate index for the months October–March, 1950–1995. The climate impact hypothesis predicts positive correlations, therefore, p-values refer to one-tailed tests. Table from Blomqvist et al. (2002)

Month	Brant goose			White-fronted goose		
	n	Kendall τ	p-Level	n	Kendall τ	p-Level
October	42	0.02	>0.05	45	-0.04	>0.05
November	42	-0.05	>0.05	45	-0.24	>0.05
December	42	0.09	>0.05	45	-0.07	>0.05
January	42	-0.06	>0.05	45	0.10	>0.05
February	42	-0.01	>0.05	45	0.08	>0.05
March	42	-0.03	>0.05	45	-0.08	>0.05

22.7 Conclusions

Studies on the interannual variation in reproductive output of tundra-breeding birds suggest that the variation is mainly driven by the prey switch of predators

(Roselaar 1979; Summers 1986; Greenwood 1987; Summers and Underhill 1987; Underhill et al. 1989; Underhill and Summers 1990; Spaans et al. 1993; Summers et al. 1998; Blomqvist et al. 2002). The predators switch from lemmings to birds when lemming populations crash. Since the lemming abundance changes in a regular three-year cycle on the Russian tundra, the three-year periodicity is transferred to the populations of juvenile tundra-nesting birds. The periodicity itself is by some ecologists believed to be generated from the interaction between the lemmings and their food plants (Seldal et al. 1994; Turchin et al. 2000). The three-year periodicity signal may be transmitted from the plants of the Taimyr Tundra, through the migration of Tundra breeding birds, to the benthos fauna of the estuaries in South Africa on which the birds feed (Fig. 5). Future studies will reveal what significance the chemical defense mechanisms of plants on Taimyr has on the abundance of benthos fauna in South Africa.

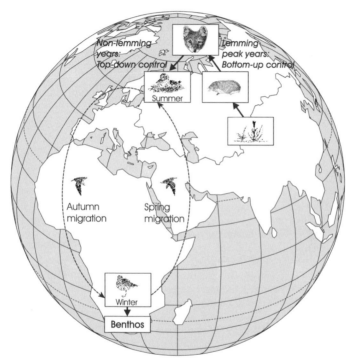

Fig. 5. Suggested ecological interactions between the plant-lemming-predator system on the Siberian tundra and migrating tundra-nesting shorebirds, which may result in variations in the predation pressure on macrobenthic invertebrates of shallow wetlands in southern wintering areas of the shorebirds, for instance, in the western and southern parts of Africa. Predation by rodentivores (e.g., arctic fox, skuas) is a key factor controlling the breeding success of the red knot and the curlew sandpiper. This top-down control is particularly strong in years immediately following a lemming peak year. Hence, there is a time lag of 1 year between the lemming decline and the numerical response of the waders, resulting in reduced predation rate on the macrobenthic invertebrate fauna

22.8 References

Blomqvist S, Holmgren N, Åkesson S, Hedenström A, Pettersson J (2002) Indirect effects of lemming cycles on sandpiper dynamics: 50 years of counts from southern Sweden. Oecologia 133:146-158

Cramp S, Simmons KEL (1977) The birds of the western Palearctic, vol I. Oxford Univ Press, Oxford

Greenwood JJD (1987) Three-year cycles of lemming and arctic geese explained. Nature 328:577

Kokorev YI, Kuksov VA (2002) Population dynamics of lemmings, *Lemmus sibirica* and *Dicrostonyx torquatus,* and arctic fox *Alopex lagopus* on the Taimyr peninsula, Siberia, 1960-2001. Ornis Svecica 12:139-143

NOAA (2001) http://www.cpc.ncep.noaa.gov/data/teledoc/telecontents.html

Roselaar CS (1979) Variation in numbers of curlew sandpipers (*Calidris ferruginea*) [in Dutch with English summary]. Watervogels 4:202-210

Seldal T, Andersen KJ, Högstedt G (1994) Grazing-induced proteinase inhibitors: a possible cause for lemming population cycles. Oikos 70:3-11

Spaans B, Stock M, St Joseph A, Bergmann H-H, Ebbinge BS (1993) Breeding biology of dark-bellied brent geese *Branta b. bernicla* in Taimyr in 1990 in the absence of arctic foxes and under favourable weather conditions. Polar Res. 12: 117-130

Summers RW (1986) Breeding production of dark-bellied brent geese *Branta bernicla bernicla* in relation to lemming cycles. Bird Study 33: 105-108

Summers RW, Underhill LG (1987) Factors related to breeding production of brent geese *Branta b. bernicla* and waders (Charadrii) on the Taimyr peninsula. Bird Study 34: 161-171

Summers RW, Underhill LG, Syroechkovski EE Jr (1998) The breeding productivity of dark-bellied brent geese and curlew sandpipers in relation to changes in the numbers of arctic foxes and lemmings on the Taimyr Peninsula, Siberia. Ecography 21: 573-580

Turchin P, Oksanen L, Ekerholm P, Oksanen T, Henttonen H (2000) Are lemmings prey or predators? Nature 405: 562-565

Underhill LG, Prys-Jones RP, Syroechkovski EE Jr, Groen NM, Karpov V, Lappo HG, van Roomen MWJ, Rybkin A, Schekkerman H, Spiekman H, Summers RW (1993) Breeding of waders (Charadrii) and brent geese *Branta bernicla bernicla* at Pronchishcheva Lake, northeastern Taimyr, Russia, in a peak and a decreasing lemming year. Ibis 135: 277-292

Underhill LG, Summers RW (1990) Multivariate analyses of breeding performance in dark-bellied brent geese *Branta b. bernicla.* Ibis 132: 477-482

Underhill LG, Waltner M, Summers RW (1989) Three-year cycles in breeding productivity of Knots *Calidris canutus* wintering in southern Africa suggest Taimyr Peninsula provenance. Bird Study 36: 83-87

23 Israel - an Intercontinental Highway for Migrating Birds

Martin Kraft

Allgemeine Ökologie und Tierökologie, Fachbereich Biologie, Philipps-Universität Marburg, Karl-von-Frischstrasse, 35032 Marburg, Germany, E-mail kraftm@mailer.uni-marburg.de

23.1 Abstract

Israel is a small country, yet it blessed with an extremely varied wildlife. Its avifauna is especially plentiful, with well over 485 species of birds recorded during various seasons of the year. Lying at the crossroads of one of the main migration routes from Africa to the Palearctic, Israel attracts many species of migrant birds, often in huge numbers. It is situated at the meeting point of three continents - Europe, Asia and Africa. Israel's avifauna includes representatives from all three of them, as well as species of this region alone. It is one of the best places in the world for watching a great variety of migrating birds within a small geographical area. Several excursions have been made by the author to Israel during spring migration between March and April. Depictions of the highest seasonal totals of the six most abundant raptors over Israel in autumn and spring are shown, as well as figures of migration routes and major watch points. Different habitats in southern and middle Israel are shown, as well as typical migrating and resident bird species.

23.2 Introduction

Israel is located along one of the principal migration routes of Eurasian birds. The region forms almost the only land bridge between the Mediterranean and the Arabian deserts, and is therefore extensively used by many birds on their way south to Africa in the autumn and back north again to Europe and Asia in the spring. Israel can be called an intercontinental highway for migrating birds, especially pelicans, storks, raptors, and song birds, as for example the different races of the yellow wagtail *Motacilla flava* (e.g., Lachman 1983; Paz and Eshbol 1987; Shirihai and Bahat 1993; Kraft 1996; Shirihai 1996). The major reason for the richness, both in species and in numbers, of migrants passing through Israel is the country's location at an intercontinental junction. This wealth is also due to the country's latitudinal location, which is such that a maximal number of species use Israeli air. There are strictly diurnal and nocturnal migrants, as well as some species that adopt both strategies. In 1996, 283 species were known to pass over Israel during the migration seasons (Shirihai 1996). Most of these (177) species do not breed in

Israel, 127 species are winter visitors as well as passage migrants and more than 100 species are also summer visitors or residents. Spring migration normally takes place between January and mid-June, the largest numbers from March through May. The huge migration over Israel is an outcome of the wide geographical distribution of the many species.

The breeding ranges together covering a vast area encompassing a broad spectrum of climates from Eurasia to the northern Arctic. Most of the species that pass via Israel during the first half of spring are short-distance migrants; some of these also overwinter in southern Europe, northern Africa, and the Middle East (e.g., bluethroat, chiffchaff, chaffinch) and many are widespread breeders in Eurasia with a broad food diversity. Long-distance migrants which winter chiefly south of the Sahara have longer periods of appearance on passage through Israel, especially in the form of successive waves of different populations or subspecies (e.g., yellow wagtail). Species such as Orphean and Bonelli's warblers pass through Israel very early in spring, as soon as their food resources on the breeding grounds allow them to return. Numerous species pass within three or four peak days, and counts have shown that over 50% of the total were recorded in one or two days of passage (e.g., 20,000 white-winged black terns and 23,000 levant sparrowhawks in one day). Six abundant, fully migratory species of raptor leave the western half of the Palearctic entirely and fly with a score of other, less numerous raptor species to winter in Africa. The six are honey buzzard *Pernis apivorus*, steppe buzzard *Buteo buteo vulpinus*, steppe eagle *Aquila nipalensis*, lesser spotted eagle *Aquila pomarina*, black kite *Milvus migrans*, and levant sparrowhawk *Accipiter brevipes*. In autumn, of these six abundant Palearctic species that migrate through the Middle East to Africa, only four cross Israel in really huge numbers: honey buzzard (max. counted per season 440,000), levant sparrowhawk (max. 50,000) and lesser spotted eagle (max. 141,000), which migrate along the eastern Mediterranean bypass route, and steppe eagle (max. 24,000), which follows the northern Red Sea bypass route (Dovrat 1991; Shirihai and Christie 1992; Shirihai 1996; see Fig. 1a, after Shirihai 1996).

Spring. Unlike the circumstances during autumn migration, the returning raptors avoid crossing the southern end of the Red Sea, and all six "major species" cross its northern end into Sinai, then into Israel (Shirihai and Christie 1992; Shirihai 1996). Lesser spotted eagle again passes over the northern and central Negev, and thus is not recorded in large numbers at Eilat. Most of the other species move along eastern Sinai and over Eilat and that portion of the Rift Valley known as the Arava in the southern Negev, then they cross into Jordan. The five most abundant species at Eilat are steppe eagle (max. 75,000), black kite (36,000), steppe buzzard (460,000), levant sparrowhawk (49,000), and honey buzzard (850,000), all between mid February and the end of May. The Eilat route in spring is spread over a front of 100 km or more. Migration begins in the mountains to the west in the morning (the start varying among species), becoming stronger, higher, and denser as thermal conditions develop, and drifts eastward. In the early morning raptors cross the Arava eastward to the Jordan mountains at a point up to several kilometers north of Eilat, cross nearer to Eilat as midday approaches and air heats so that thermal currents allow them to rise to 1,500 m and glide very rapidly.

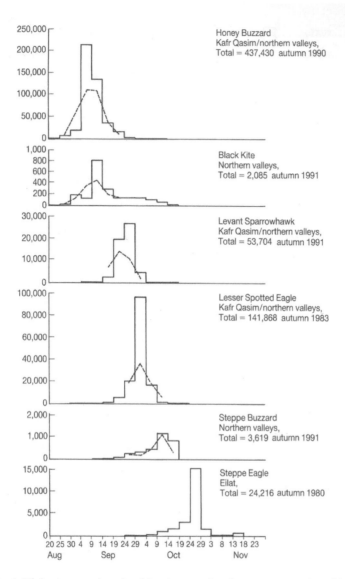

Fig. 1. Highest seasonal totals of the six most abundant raptors over Israel in autumn during 5-day periods. *Broken lines* show average 5-day totals in the autumns 1982-1991 of birds passing on the "north-central-western route" (counts made at Kafr Qasim and the northern valleys except for the steppe eagle, which was counted from Eilat and in one autumn only). Note that species sequence follows temporal pattern of species' passages. Autumn figures for black kite and steppe buzzard are relatively small, since these species mostly pass along a more easterly corridor in autumn (after Shirihai 1996)

For about 4 h beginning at noon, northerly winds blow the passage off course and the valley crossing occurs south of Eilat or over the southernmost Negev (depending on species and time of season), while later in the afternoon the raptors lose height and return to migrate above the Eilat mountains, where they also roost. Thus, under suitable conditions raptors cross the Rift Valley near the northern part of the Bay of Eilat, but when crossing conditions there are unfavorable, or when southerly or easterly winds deflect the passage, they may pass on a broad front over the southern or central Negev, thus crossing the valley between its central part and the Dead Sea area (Shirihai and Christie 1992; Shirihai 1996; See Figs. 2 and 3 (after Shirihai 1996).

The Alpine Swift *Apus melba* is one of the fastest birds in the world and can be observed during two distinct longish influxes that represent the bulk of passage: third week in August to mid-October (mostly *tuneti*), peaking end of September and beginning of October, and mid-October to end of November (mostly *melba*), peaking the last days in October. This is a broad-front passage usually at great altitudes in flocks of tens to hundreds, with mass movement of thousands in a continuous stream on peak days through northern, central, and western Israel, and the central Negev mountains. Spring migration takes place mainly in mid-February to the end of March, but generally has slightly smaller numbers than in autumn (but still thousands on peak days) and is over more eastern areas.

In the lecture, the following photographs of typical habitats, birds and bird-watchers were used:

Short-toed eagle *Circaetus gallicus* (Y. Eshbol)

Common passage migrant over most parts, mainly along migration routes, and very rare winter visitor; also quite common summer visitor in most parts, especially in north and center (Shirihai 1996).

Steppe buzzard *Buteo buteo vulpinus* (H. Shirihai)

Fairly common autumn and abundant spring passage migrant over all parts. Sometimes more than 400,000 individuals during spring migration (Shirihai 1996).

Honey buzzard *Pernis apivorus* (Y. Eshbol)

Abundant passage migrant over much of the country, mostly along specific routes and during a very brief period; in autumn mainly end August and during September; in spring mainly mid-April to end of May. At Eilat, max. spring total 851,598, from 11 April to 25 March 1985 (Shirihai 1996).

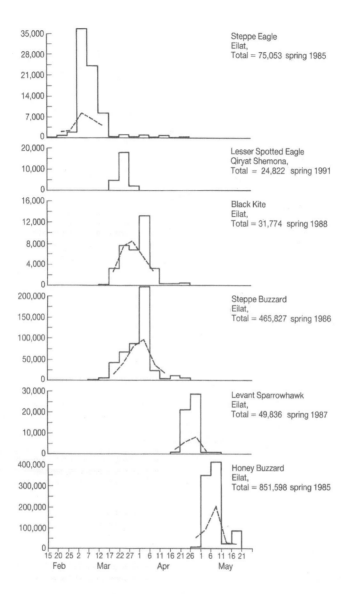

Fig. 2. Highest seasonal totals of the six most abundant raptors over Israel in autumn during 5-day periods. *Broken lines* show average 5-day totals in springs of 1977, 1983, and 1985-1988 (counts all made on "Eilat route" except for lesser spotted eagle, data for which are from only one partial count made in one spring only at Qiryat Shemona). Note that species sequence follows temporal pattern of species' passages (after Shirihai 1996)

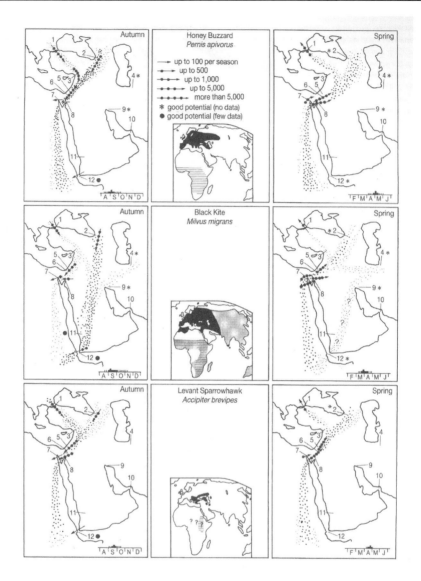

Fig. 3. Migration routes of the abundant raptor species through the Middle East region in autumn and spring, with (*inset*) approximate breeding range (*black* populations which migrate to Africa) and African wintering range. *Stippled area* shows apparent routes followed (*larger dots* indicate where vast majority of passage occurs). *1* Bosporus, *2* East Pontics, *3* Iskenderun, *4* southeastern Caspian, *5* Kafr Qasim/northern valleys, *6* northern Negev and Dead Sea, *7* Suez, *8* Eilat, *9* Kuwait, *10* Strait of Hormuz, *11* North Yemen, *12* Bab-el-Mandeb (after Shirihai 1996)

Imperial eagle *Aquila heliaca* (P. Doherty)

Quite rare to scarce passage migrant over most parts, and winter visitor mainly in low-lying areas of northern and western Israel (Shirihai 1996).

Black bush robin *Cercotrichas podobe* (D. Cottridge)

Occasional to extremely rare local spring and summer visitor from Africa between March and July and apparently casual breeder in southern Arava (Shirihai 1996).

Several excursions led the author and his students to Israel where photographs of typical habitats in southern and central Israel were taken, such as "High Mountain" near Eilat, "Red Sea" south of Eilat, "Timna Valley" north of Eilat, "Ein Gedi," and "Dead Sea," "habitats near Jerusalem," and "Jordan Valley near Jericho". All these areas are good for watching migrating, resting and roosting birds, as well as resident bird species.

The town of Eilat and its surroundings are most famous for the enormous numbers of birds passing through during migration seasons. Especially noteworthy: the spring migration of raptors is the heaviest recorded in the world. In March and April many birds of prey, e.g., steppe eagles *Aquila nipalensis* and other soaring birds such as black storks *Ciconia nigra* can be seen in quite large numbers. The Eilat area is also a resting site for many species and there is much to be seen in the daytime. It is possible to see, for example, Baillon's crake *Porzana pusilla*, black-winged stilt *Himantopus himantopus*, collared pratincole *Glareola pratincola*, white-tailed lapwing *Vanellus leucurus*, Caspian plover *Charadrius asiaticus*, greater sand plover *C. leschenaultii*, little ringed plover *C. dubius*, little stint *Calidris minuta*, ruff *Philomachus pugnax*, green and wood sandpipers *Tringa glareola and T. ochropus*, white-eyed gull *Larus leucophthalmus*, black-headed gull *L. ridibundus*, slender-billed gull *Larus genei*, Caspian tern *Sterna caspia*, turtle dove *Streptopelia turtur*, European bee-eater *Merops apiaster*, barn swallow *Hirundo rustica*, tawny and red-throated pipits *Anthus campestris and A. cervinus*, yellow and citrine wagtails *Motacilla flava and M. citreola*, bluethroat *Luscinia svecica*, Crezschmar's bunting *Emberiza caesia*, and Spanish sparrow *Passer hispaniolensis*. The last occurs in very large numbers in the "North Fields" north of Eilat in the Arava. The "Arava" is the name given to that part of the Rift Valley running from the southern edge of the Dead Sea to Eilat. The region is flat with only low hills breaking the landscape, and extensive areas are dotted with acacia trees, and often Rüppell's warbler *Sylvia rueppelli* and the beautiful masked shrike *Lanius nubicus* can be found here. During the winter months, the colorful Sinai rosefinch *Carpodacus synoicus* often appears in some wadis, e.g., the wadi "Shlomo" near Eilat. In the breeding season this bird is an uncommon local (partial) resident, chiefly above 500 m in arid desert regions of the Eilat mountains, southeastern Negev, the central Negev highlands, and the southern and central Judean Desert -- with local roaming also . It inhabits rocky highland deserts broken by cliffs, ravines, and stony wadis with scattered bushes and some acacias, often at edges of small upland human habitations. Another speciality for Israel is the Hoopoe lark *Alaemon alaudipes*, which is a scarce local resident in northeastern, central, and southern Arava, eastern and southern Negev, and Nizzana area. This bird has a beautiful melodious and melancholy song that starts slowly, accelerates and ascends in tone, then drops in speed and tone and slowly dies away (see Hollom et al. 1988; Porter et al. 1996). This human-like song can often be heard early in the morning in the southernmost area of the Arava. Two other stunning residents are the blackstart *Cercomela melanura* and the white-crowned black wheatear *Oenanthe leucopyga*. These birds are common around the Timna Valley north of Eilat and they are often very tame, offering good views and easily taken photographs (e.g., Fig. 4; Kraft 1996; Kraft and Frede 1996).

Fig. 4. Flock of migrating black-winged stilts *Himantopus himantopus* near Eilat 1988 (after Shirihai 1996)

2.3 Conclusions

Only a few birds are described, but they illustrate a small part of the richness of Israel's avifauna. Especially storks and raptors pass over Israel in hundreds of thousands and can easily be seen at strategic observation points along their traditional flight paths. Other species which migrate on a broader front, such as passerines, can also be seen en masse, attracted to the green oases of the new Israeli kibbutz cultivations scattered throughout the desert landscapes of the southern part of the country. In addition to the migrant birds, Israel also holds a fascinating resident avifauna, including many of the Middle East's desert species. It offers excellent opportunities for studying bird migration, migration routes, behavior at resting and roosting sites, as well as the ecology and behavior of resident bird species in their special habitats. It is therefore no surprise that in the last 25 years Israel has become increasingly popular with European birdwatchers, who now flock there as regularly as the avian migrants. When the author and his students first came to Israel in spring 1987, it sometimes took their breath away when thousands of these interesting birds flew overhead within just a few minutes.

Never before has wildlife conservation been more important. The rapid pace of development in Israel, as elsewhere, puts many of the country's bird habitats at ever greater risk. Despite this, it is hoped that the recent Middle East peace process may help to promote many conservation issues in this unique part of the world and may political developments also lead to peace in such a beautiful country.

2.4 References

Dovrat E. (1991) The Kafr Quasim (Kefar Kassem) raptor migration survey, autumns 1977-1987: a brief summary. Raptors in Israel. Passage and wintering populations. IBCE, IRCI, pp 13-30

Hollom PAD, Porter RF, Christensen, Willis I (1988) Birds of the Middle East and North Africa. Calton, Staffordshire

Kraft M (1996) Israel - Zentrum des Vogelzugs. MagNatur 1:66-71

Kraft M, Frede M (1996) Das Vogelportrait - der Saharasteinschmätzer Oenanthe leucopyga. In: Die Pracht unserer Vögel. Sonderbroschüre mit Leica. Solms

Lachman E (1983) Birdwatching in Israel. Israel Economist, Jerusalem, Tel Aviv

Paz U, Eshbol Y (1987) The birds of Israel. Helm, London

Porter RF, Christensen S, Schiermacker-Hansen (1996) Field guide to the birds of the Middle East. Poyser, London

Shirihai H (1996) The birds of Israel. Academic Press, London

Shirihai H, Bahat O (1993) Birdwatching in the deserts of Israel. International Bird Watching Center Eilat

Shirihai H, Christie DA (1992) Raptor migration at Eilat. Br Birds 85:141-186

24 Contrasting Molecular Markers Reveal: Gene Flow via Pollen Is Much More Effective Than Gene Flow Via Seeds

Birgit Ziegenhagen, Ronald Bialozyt and Sascha Liepelt

Philipps-University, Faculty of Biology, Nature Conservation Division, Working Group of Conservation Biology and Conservation Genetics, Karl-von-Frisch-Strasse, 35032 Marburg, Germany, E-mail ziegenha@staff.uni-marburg.de

24.1 Abstract

Today's distribution of genetic diversity in plants is to a large extent shaped by historical and recent gene flow mediated by seeds and pollen. It is only recently that in certain plant species DNA markers with contrasting modes of inheritance have been explored. Such markers are desired to differentiate between seed- and pollen-mediated gene flow. The present chapter is a review of our studies in the tree genus fir (*Abies* sp.). The genus served as a model to study gene flow by contrasting markers, since as a member of Pinaceae its mitochondrial DNA (mtDNA) is uniparentally maternally inherited, whereas its chloroplast DNA (cpDNA) is uniparentally paternally inherited.

In the species silver fir (*Abies alba* Mill.), a large, range-wide study using contrasting markers was performed including more than 1,000 individuals from about 100 European populations. The markers used were a maternally inherited mtDNA marker (*nad*5-intron4 polymorphism) and a paternally inherited cpDNA marker (*psb*C polymorphism). In each case, two variants were detected which indicated the existence of at least two glacial refugia in the western and eastern Mediterranean area. Whereas the distribution of the maternal lineages remained mainly separated and displayed main postglacial migration routes, the efficiency of pollen-mediated gene flow became evident through the paternally inherited marker. It was suggested that range-wide gene flow through pollen may have confounded the genetic imprints of the ice ages during interglacial periods. The efficiency of pollen-mediated gene flow could even be traced at the next higher taxonomic level, including seven more *Abies* species around the Mediterranean Basin. There is first evidence for past natural hybridization events, the central European species *A. alba* harboring paternal lineages of almost all other *Abies* species studied. The evolutionary implications are discussed with special regard to conservation management of a species gene pool.

24.2 Introduction

The earth's orbital cycles have frequently caused climate changes that tremendously impacted the terrestrial biota. This led to permanent changes in biological diversity, including ongoing evolutionary changes above and within the species level. European biota and biological resources as they present themselves today have been mainly shaped by the effects of the last Quaternary ice age, followed by range expansion of the species that had survived in southern or eastern European refugia. At the same time, during the Holocene, man appeared and started to increasingly interfere with the natural distribution processes of biological resources and the inherent genetic resources. Before any efficient management of the remaining resources both at the organismic and genetic level can be taken, profound knowledge is needed of past processes. In particular, knowledge is needed of the large-scale distribution processes, which implies knowledge of the number and locations of refugia and the routes taken for range expansion. For the latter process, we will use the term "invasion" in agreement with a recent review paper by Petit et al. (2004a). According to these authors, invasion means an irreversible process of range expansion, at least under the climatic conditions of the current postglacial period. A question of increasing interest is that of the genetic history of the invading organisms. What are the genetic consequences of dispersal along the migration routes and in areas of secondary contact between different refugial lineages? We present recent studies including our investigations on trees, the keystone species of the European ecosystems. Postglacial invasion is mainly mediated by seed dispersal along migration routes. The term "migration" has long been used legitimately in regard to these sessile and long-lived organisms if large spatio-temporal scales are addressed.

24.2.1 Postglacial History of European Trees

Much knowledge of the postglacial history of European tree species has been accumulated in the due course of palynological studies. By means of fossil pollen records it has been to some extent possible to determine the location of refugia and to date the arrival of the species at selected locations of the present range (e.g., Huntley and Birks 1983). Furthermore, using the fossil pollen records, assumptions were made about the average speed of seed migration or the range of long-distance dispersal events (summary in Bonn and Poschlod 1998). Since the 1990s, DNA markers have been increasingly introduced in range-wide studies of numerous forest trees, mainly in angiosperm species. These studies were facilitated by systematically exploring complete chloroplast DNA sequences for intraspecific variation (review in Petit and Vendramin 2004). Moderate mutation rates in combination with uniparentally maternal inheritance made the later-detected variants ideal markers in range-wide studies. At that time, studies using cpDNA markers in combination with methods of phylogeography were introduced. Pioneered by studies in oaks, it became possible for European tree species to distinctly delineate refugial maternal lineages and the putative routes they took during postglacial invasion (Dumolin-Lapègue et al. 1997; Petit et al. 2003). With accumulation of knowledge, not only of tree species, a comparative phylogeography was approached in order to elaborate putative generalizations on the postglacial invasions of European biota (Taberlet et al. 1998; Hewitt 1999; Petit et al. 2003). Three general types of recolonization patterns were defined according to different speeds and routes of migration from refugia south of the Alps and the Pyrenees, and from the Balkan Peninsula. Analysis of sequence divergence between maternal lineages originating from different refugia led to the conclusion that they have been isolated for several glacial periods (Hewitt 1999, 2000). Therefore, Hewitt (2000) postulated a strong "genetic legacy" of the ice ages and suggested that the climatic oscillations of the Quaternary had an impact on speciation due to long-term isolation in refugial areas.

24.2.2 "Genetic Legacy" of Ice Ages?

Hewitt (2000) also suggested that the "genetic legacy" referred to data that were mainly obtained on the basis of the maternal gene pools of species. The question arises whether long-term isolation in a genetic context holds true for the whole gene pool defined by the (re)combination of maternal and paternal lineages. In particular for tree species, this is an important question since wind-dispersed pollen may mediate a highly efficient long-distance gene flow. First evidence for pollen as a main agent of gene flow arose from studies using genetic markers with contrasting modes of inheritance (e.g. Richardson et al. 2002). Another question may be of importance with special regard to forest trees: Is interspecific hybridization, such as speciation, a significant evolutionary driver for generating adaptively relevant genetic diversity? The present chapter is on *Abies,* which for the reasons given below, is an interesting model to give preliminary answers to these questions.

24.3 The Tree Model *Abies* (sp.)

The following sections review our results obtained on the forest tree fir (*Abies* sp.) as a putative model for evolutionary implications of the last ice age. The within-species study is presented in Sections 24.3.1 and 24.3.2, the study at the genus level in Section 24.3.3

Why is *Abies* an appropriate model to study the effect of postglacial range expansion and gene flow through seeds versus gene flow through pollen? First, *Abies* is a member of the Pinaceae, a family with frequently found contrasting modes of inheritance of their two organelle genomes (review in Petit and Vendramin 2004; for *Abies* in particular see our studies in Sects. 24.3.1 and 24.3.2). Further traits favor *Abies*: The genus belongs to the majority of the temperate forest tree species which spread their pollen grains through wind. Thus, we may expect a considerable capacity for long-distance pollen dispersal. Regarding the

Mediterranean *Abies* species, evidence exists for interspecific gene flow in controlled pollination experiments revealing sexual compatibility among them (Kormuták, pers. comm.). These findings challenged us to study the geographic ranges at which interspecific gene flow occurs or has occurred among the species. The opportunity is comparatively better than in other tree species since nowadays *Abies* species are mainly allopatric (see Fig. 1), with the one exception that *A. alba* and *A. cephalonica* have contact in the Balkans.

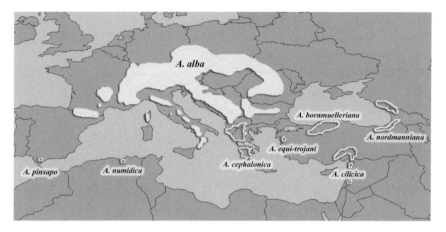

Fig. 1. Natural ranges of the Mediterranean species analyzed. The natural range of the central European species silver fir (*A. alba*) is indicated by *white background*

At the species level, for some extraordinary reasons, *A. alba* may serve as a well-suited model for a whole class of wind-pollinated plant species: the life history traits of *A. alba* are bound to encumber more than to facilitate a highly efficient pollen-mediated gene flow. *A. alba* has very large pollen grains with a comparatively high velocity of fall (Stanley and Linskens 1974), long generation times (~40 years), and a very long lifespan (>300 years). Generalizations from our model will therefore rather underestimate than overestimate the effect of pollen-mediated gene flow in other wind-pollinated species (for *Abies* seed and pollen morphology and sizes, see Fig. 2A, B). Another advantage of our within-species model was observed during marker development. Besides contrasting modes of inheritance, the two detected markers have ideal properties because they are neutral, highly conserved, and not prone to homoplasty, as for instance microsatellite markers (Liepelt et al. 2001). There are only two alleles for each marker, and each allele corresponds to a single lineage, which can therefore be identified without further phylogenetic methods.

A

B

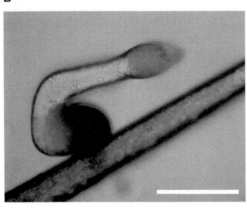

Fig. 2. A Seed of silver fir (*Abies alba* Mill.) with membranous wing, *scale bar* 1 cm. **B** Germinated pollen grain of *Abies* sp. sticking to a human eyelash, *scale bar* 100 μm

24.3.1 Geographic Distribution of Maternal Lineages in *Abies alba*

Silver fir (*Abies alba* Mill.) started its invasion about 11,000 years ago and reached the limits of its present range 6,000 years B.P. (Huntley and Birks 1983). Nowadays, it covers a broad mountainous area across central Europe, ranging from approximately 0° to 27°E and 38° to 52°N (Meusel et al. 1978). Early hypotheses on the distribution of glacial refugia were based on fossil pollen data (Langer 1963; Kral 1980), but were substantially revised by a study of Konnert and Bergmann (1995). Using traditional isozyme gene markers, these authors inferred the existence of five glacial refugia. They argued that *Abies alba* most likely recolonized Europe from three of the five, namely, from the Apennine and Balkan peninsulas and less probably from the Massif Central in southern France.

Since isozyme gene markers are biparentally inherited, it is not possible to explicitly reconstruct the routes of seed migration or the explicit locations and expansions of suture zones where the refugial lineages met or mixed.

Therefore we were encouraged to develop purely maternally inherited markers in order to apply them in a range-wide study including more than 1,000 individuals originating from about 100 populations. As described in Liepelt et al. (2002), two size variants (allele 1 and allele 2) were detected within intron No. 4 of the mitochondrial *nad5* gene. Sequence analysis revealed an insertion/deletion of about 80 bp (GenBank accession nos. AY147793 and AY147794). Before we applied the newly developed marker we needed to verify its mode of inheritance. The marker locus n*ad*5-4 also exhibited interspecific length differences that were used to determine the mode of inheritance in numerous controlled interspecific crosses. We were able to confirm the uniparentally maternal inheritance of the marker. Figure 3 gives an example of two controlled crossing combinations where the maternal transmission of the variation is obvious. The distribution of the two alleles throughout Europe showed a very strong subdivision of the natural range. Populations from the western part of the range exclusively contained allele 1 and populations from the eastern part contained allele 2. Mixed populations were observed in Croatia, Slovenia, and northeastern Italy (Liepelt et al. 2002) and to a small extent also in the northern Carpathians (Gomöry et al. 2004). East and west of these locations populations were fixed for one allele, except for three populations of the western part, where one individual each exhibited allele 2. Those populations were located in the eastern Pyrenees, in the Black Forest, and in central Italy. In total, as an unexpected result compared with the former isozyme gene studies, a greater part of the range was recolonized by the western rather than by the eastern lineage. It seems likely that, following the last Ice Age, the maternal lineage carrying allele 1 originated from a western Mediterranean refugium, while the other lineage carrying allele 2 originated from an eastern Mediterranean refugium. This is in accordance with former allozyme studies indicating putative glacial refugia in the Apennine and the Balkan peninsulas (Konnert and Bergmann 1995). Regarding the newly identified introgression zones, pollen maps give evidence for an early contact in Croatia, Slovenia, and northeastern Italy about 7,000 to 7,500 years ago (Huntley and Birks 1983). The contact and/or introgression zone in the northern Carpathians is evident from pollen maps as well (Huntley and Birks 1983). This contact was only established between 4,000 B.P. and 1,000 B.P., which may be the reason for the only narrow zone of introgression.

The low amount of introgression between the maternal lineages indicates a small extent of gene flow via seeds after populations became established. This is probably not a function of selection, but rather one of colonization time and population density. After colonization was completed and the populations reached their optimum density, the latter acted as a barrier to consequent gene flow via seeds.

To summarize: As in many other European taxa (see Sect. 24.2), in Abies alba maternal lineages originating from different glacial refugia remained mostly separated during postglacial range expansion.

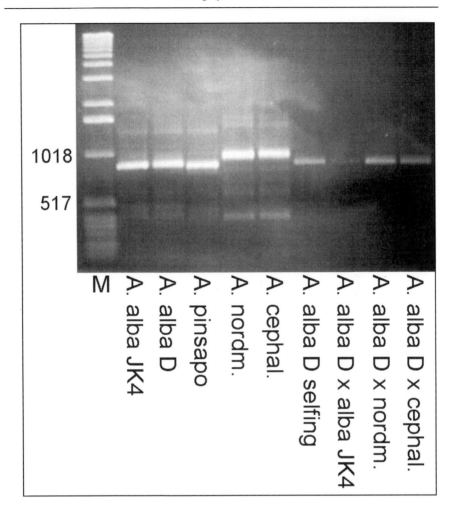

Fig. 3. Agarose gel electrophoresis of size variants at mtDNA locus *nad5* – intron4 for five parental trees and progeny of four of their combinations after controlled pollination experiments. *Two right lanes* The progeny of a controlled cross between *A. alba* (mother) and either *A. nordmanniana* or *A. cephalonica* (fathers) carry the maternal size variant of *A alba*. *M* Molecular size standard (1 KB Ladder, Gibco BRL, Invitrogen Karlsruhe)

24.3.2 Geographic Distribution of Paternal Lineages in *Abies alba*

As outlined in the Section 24.2, the question arises whether gene flow through wind-dispersed pollen is confounding the long-term effect of contraction and expansion of isolated maternal refugial lineages and thereby of maternal gene pools.

We applied a paternally inherited cpDNA marker (Ziegenhagen et al. 1995) to all sampled silver fir populations. Sequence analysis of the PCR-RFLP marker revealed a silent mutation, which caused no change in the amino acid sequence of the translated protein (GenBank accession no. AY147792; for details of material and methods, see Liepelt et al. 2002). This silent mutation represents two alleles (allele A and allele B), which were present throughout most of the analyzed populations of *A. alba*, although they were not evenly distributed. A geographical cline was visible, with allele A being more frequent to the east and allele B more frequent to the west of the distribution range. Populations that were fixed for one cpDNA variant were found in northwestern Italy (allele B) and in Bulgaria (allele A), thus marking the putative western and eastern refugial areas as confirmed by the maternally inherited *nad5*-intron4 marker (Liepelt et al. 2002). In order to statistically analyze the function of geographic distances of gene flow through seeds versus wind-dispersed pollen, we calculated clines of allele frequencies (Liepelt et al. 2002). As revealed in Fig. 4A,B, the cline of chloroplast *psbC* alleles is as wide as the natural range of *A. alba*, while the cline of mitochondrial *nad5*-4 alleles spreads over less than 2° of longitude. The centers of both clines were close together. A Kolmogorov-Smirnov test confirmed that the differences in allele frequency distributions of both markers were highly significant ($p<0.001$, Liepelt et al. 2002). The observed range-wide cline for the pollen-mediated gene flow spanned a much greater distance than suggested from the distribution of biparentally inherited allozyme variation (Konnert and Bergmann 1995). Thus, using uniparentally inherited markers, new light could be thrown on the contribution of seed-mediated gene flow versus the contribution of pollen-mediated gene flow across the whole analyzed region.

To summarize: Our results provide first striking evidence that even a species with very long generation times and heavy pollen grains was able to establish a highly efficient pollen-mediated gene flow between refugia. Therefore, we postulate that an exchange of genetic information between refugia by range-wide paternal introgression is possible in wind-pollinated plant species. In the case of A. alba, this exchange did not need more time than the duration of the current postglacial period.

A

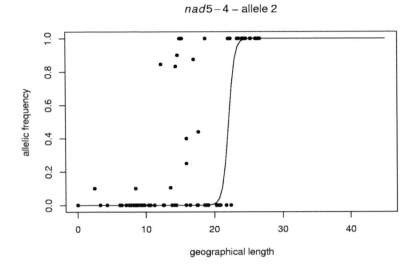

B

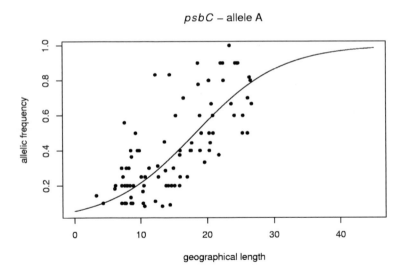

Fig. 4. Clines of allele frequencies for the two contrasting markers in range wide studies of silver fir (*Abies alba* MILL). **A)** Cline of frequencies of the allele 2 of the mtDNA locus *nad*4-intron5. **B)** Cline of frequencies of the allele A of the cpDNA locus *psbC*. Figures from Liepelt et al. (2002).

24.3.3 Geographic Distribution of Paternal Lineages in the Genus *Abies*

Given a sexual compatibility of the Mediterranean *Abies* species, we were interested in the question whether a putative past or recent pollen-mediated gene flow among species can be made visible in a geographic context using paternally inherited cpDNA markers. Currently, we are developing cpDNA markers with variants that are to a certain extent fixed for different species or species groups. It seems that the variable intergenic spacer region between the chloroplast genes *trn*C and *trn*D may be a potential source of the appropriate markers (Parducci and Szmidt 1999; Liepelt et al., in prep.). So far, in analyzing representative individuals of the eight Mediterranean *Abies* species under study (see Fig. 1), we detected 13 cpDNA variants. Preliminary results have indicated that gene flow via pollen occurred among certain species, preferentially among today's geographically neighboring species. Most interesting is the preliminary finding that the central European species *A. alba* harbors a large portion of the total paternal gene pool of the Mediterranean *Abies* species (data not shown, Liepelt et al., in prep.).

To summarize: There is preliminary evidence that gene flow via wind-dispersed pollen took place among the Mediterranean Abies species. However, uncertainties still exist about the scales relating to both time and locations of the hybridization events.

24.4 Discussion and Conclusions

Our studies have revealed that in judging the efficiency of intraspecific gene flow during postglacial range expansion one needs to differentiate between flow via seeds and flow via pollen. In the case of species with wind-dispersed pollen, it has become evident that, due to highly efficient gene flow, even geographically far distant refugial lineages may exchange genetic information with each other. Thus, the long-term effect of refugial isolation as observed with maternal lineages may be considerably confounded. Our studies have further revealed that interspecific hybridization and subsequent gene flow have to be discussed as evolutionary mechanisms, and must be considered among today's allopatric tree species such as *Abies*.

24.4.1 Evolutionary Implications

It has been debated whether gene flow is a restrictive or creative force in evolution (Slatkin 1987). Indeed, in the case of *A. alba*, both forces may act. On the one hand, high rates of pollen-mediated gene flow may slow down or prevent speciation. On the other hand, there could also be a creative aspect (Slatkin 1987), because superior alleles or allele combinations can spread throughout the species enhancing its adaptive potential. Such processes may follow the Wrights' "shifting balance theory" (Futuyma 1998). For this, new advantageous adaptation to be spread for the benefit of the entire species gene flow is essential (Slatkin 1987). In the light of repeated range shifts during the glacial periods, gene flow might be the key to maintain diversity and adaptability of forest trees. Yet, not all genetic differences may be erased that have accumulated between refugia, as can be observed in studies using nuclear markers (e.g., Konnert and Bergmann 1995; Zanetto and Kremer 1995; Comps et al. 2001). The diversification effect through pollen-mediated genetic admixture might hold true, even if nuclear introgression is not significant. Nucleus and chloroplast contain genes that code for different subunits of proteins (e.g., ribulose biphosphate decarboxylase). If these evolved independently for a longer period, new nuclear-organelle genetic combinations could result.

Regarding the strong driving effect of pollen-meditated gene flow within a species, we may assume an even stronger enhancement of the adaptive potential through interspecific pollen-mediated gene flow as suggested for *Abies*. This is not the only evidence for hybridization. In the white oak species complex many species hybridize (Dumolin-Lapègue et al. 1997) and thus form a huge gene pool, where genes can be exchanged widely via paternal gene flow. In a recent review, Petit et al. (2004b) argued that *Quercus robur* L. was the one to invade via seeds and *Q. petraea* Matt. Liebl. the one that followed through "pollen swamping" into the established central European oak populations. By means of thus far unknown processes, *Q. petraea* has been relatively recently "resurrected" as a species. This is how Petit et al. (2004b) call such a process of species re-establishment. They interpret the two processes "pollen swamping" and "resurrection" as evolutionary mechanisms to facilitate the invasion of a species with less efficient seed dispersal and/or capacity to establish as a pioneer.

In conclusion, trees that are known to evolve slowly in terms of speciation obviously have adopted intra- and interspecific genetic strategies to enlarge or maintain diversity in relatively short time spans. This at least holds true for the climatic oscillatory conditions in Europe during the last ice ages.

Conservation management of a species gene pool focuses on the distribution of diversity within a species (Young et al. 2000). Commonly, areas of high genetic diversity are delineated as a putative source in gene resource programs. Furthermore, regional restoration activities are devoted to make use of autochthonous populations, if possible. With the new knowledge of the efficiency of pollen-mediated gene flow within and among related species, we need to critically revise the basic assumptions for common programs. However, more new questions than answers arise, for instance: where is one to delineate adaptively relevant zones for a sustainable genetic management of a species? Is it still appropriate to focus on species as "units of diversity" (also see critical comments by Bachmann 1998), when we consider an efficient gene flow among related species of a genus? In this context and referring to *Abies,* we also need to ask: When did interspecific hybridization take place among today's allopatric species; and is there current gene flow of adaptive relevance? Answers may be expected to come from ongoing efforts in functional and population genomics; and in due course adaptively relevant genomic regions may be identified, genomically mapped, and related to a geographical context. Our efforts are aimed at understanding the spatio-temporal dynamics of intra- and interspecific genetic diversity by analyzing fossil material, particularly of *Abies*, an entertainment which has become feasible (Petit and Parducci 2004; Ziegenhagen et al., in prep.). Furthermore, in our ongoing studies, seed dispersal mechanisms for maintaining diversity are being explored with computer simulations.

Acknowledgements. We are grateful to S. Jelkmann, V. Kuhlenkamp, Christina Mengel, and I. Schulze for excellent work in the laboratory. Further we would like to thank I. Barbu, H. Braun, W. Eder, B. Fady, A. Franke, L. Llamas Gómez, J. Gracan, E. Hussendörfer, M. Konnert, A. Kormuták, L. Paule, D. Peev, E. Ritter, P. Rotach, I. Strohschneider, W. van der Knaap, G.G. Vendramin, G.J. Wilhelm for kindly providing us with sample material. We are grateful to R.J. Petit and G.G. Vendramin for sending us their manuscripts or page proofs prior to publication. This work was funded by the EU project FOSSILVA (CT-1999-00036) within the EU Program ENVIRONMENT (FP5).

24.5 References

Bachmann K (1998) Species as units of diversity: an outdated concept. Theor Biosci 117:213-230

Bonn S, Poschlod P (1998) Ausbreitungsbiologie der Pflanzen Mitteleuropas. UTB, Quelle and Meyer, Wiesbaden

Comps B, Gomöry D, Letouzey J, Thiebaut B, Petit RJ (2001) Diverging trends between heterozygosity and allelic richness during postglacial colonization in the European beech. Genetics 157:389-397

Dumolin-Lapegue S, Demesure B, Fineschi S, Le Corre V, Petit RJ (1997) Phylogeographic structure of white oaks throughout the European Continent. Genetics 146:1475-1487

Futuyma D (1998) Evolutionary biology, 3rd edn. Sinauer Ass, Sunderland, Mass, USA

Gömöry D, Longauer R, Ballian B, Brus R, Kraigher H, Parpan VI, Parpan TV, Stupar VI, Ladislav P, Liepelt S, Ziegenhagen B (2004) Variation patterns of mitochondrial DNA of Abies alba Mill. in suture zones of postglacial migration in Europe. Acta Soc Bot Pol (submitted)

Hewitt GM (1999) Post-glacial recolonization of European biota. Biol J Linn Soc 68:87-112

Hewitt GM (2000) The genetic legacy of the Quaternary ice ages. Nature 405:907-913

Huntley B, Birks HJB (1983) An atlas of past and present pollen maps for Europe: 0-13,000 years ago. Cambridge Univ Press, Cambridge

Konnert M, Bergmann F (1995) The geographical distribution of genetic variation of silver fir (*Abies alba*, Pinaceae) in relation to its migration history. Plant Syst Evol 196:19-30

Kral F (1980) Waldgeschichtliche Grundlagen für die Ausscheidung von Ökotypen bei *Abies alba*. Proceedings 3. IUFRO Tannensymposium Wien. Österreichischer Agrar Verlag, Wien, pp 158-168

Langer H (1963) Einwanderung und Ausbreitung der Weisstanne in Süddeutschland. Forstwiss Centralbl 83:33-52

Liepelt S, Kuhlenkamp V, Anzidei M, Vendramin GG, Ziegenhagen B (2001) Pitfalls in determining size homoplasy of microsatellite loci. Mol Ecol Notes 1:332-335

Liepelt S, Bialozyt R, Ziegenhagen B (2002) Wind-dispersed pollen mediates postglacial gene flow among refugia. Proc Natl Acad Sci USA 99:14590-14594

Meusel H, Jäger E, Rauschert, S, Weinert T (1978) Vergleichende Chorologie der zentraleuropäischen Flora, Bd 2. Text und Karten. Fischer, Jena

Parducci L, Szmidt AE (1999) PCR-RFLP analysis of cpDNA in the genus *Abies*. Theor Appl Genet 98:802-808

Petit RJ, Parducci L (2004) Ancient DNA – unlocking plants' fossil secrets Introducing genetic and palaeogenetic approaches in plant palaeoecology and archaeology, Bordeaux, France, September 2003. New Phytol Forum (in press)

Petit RJ, Vendramin GG (2004) Plant phylogeography based on organelle genes: an introduction (in press)

Petit RJ, Aguinagalde I, de Beaulieu JL, Bittkau C et al (2003) Glacial refugia: hotspots but not melting pots of genetic diversity. Science 300:1563-1565

Petit RJ, Bialozyt R, Garnier-Gère P, Hampe A (2004a) Ecology and genetics of tree invasions from recent introductions to Quaternary migrations. For Ecol Manage (in press)

Petit RJ, Bodénès C, Ducousso A, Roussel G, Kremer A (2004b) Hybridization as a mechanism of invasion in oak. New Phytol doi 10.1046/j.1469-8137.2003.00944.x

Richardson BA, Brunsfeld SJ, Klopfenstein NB (2002) DNA from bird-dispersed seed and wind-disseminated pollen provides insights into postglacial colonization and population genetic structure of whitebark pine (*Pinus albicaulis*). Molecular Ecology 11:215-227

Slatkin M (1987) Gene flow and the geographic structure of natural populations. Science 236:787-792

Stanley RG, Linskens HF (1974) Pollen. Springer, Heidelberg, Berlin, New York

Taberlet P, Fumagalli L, Wust-Saucy A-G, Cosson J-F (1998) Comparative phylogeography and postglacial colonization routes in Europe. Mol Ecol 7:453-464

Vendramin GG, Degen B, Petit RJ, Anzidei M, Madaghiele A, Ziegenhagen B (1999) High level of variation at *Abies alba* chloroplast microsatellite loci in Europe. Mol Ecol 8:1117-1126

Young A, Boshier D, Boyle T (eds) (2000) Forest conservation genetics. Principles and practice. CSIRO Publishing, Collingwood, Australia

Zanetto A, Kremer A (1995) Geographical structure of gene diversity in *Quercus petraea* (Matt.) Liebl. I. Monolocus patterns of variation. Heredity 75:506-517

Ziegenhagen B, Kormutak A, Schauerte M, Scholz F (1995) Restriction site polymorphism in chloroplast DNA of silver fir (*Abies alba* Mill.). For Genet 2:99-107

25 Principal Changes of Ecological Migration and Dispersal: Nature Conservation Consequences

Harald Plachter

Philipps-University, Faculty of Biology, Nature Conservation Division, Karl-von-Frisch-Strasse, 35032 Marburg, Germany, E-mail h.plachter@staff.uni-marburg.de

25.1 Abstract

Migration of individuals is one of the major ecological processes for the survival of species. During the past 100 years human impacts fundamentally changed the amount and qualities of migration capabilities. This not only relates to structural phenomena like habitat fragmentation and population isolation but also to the process itself. Migrating individuals are impacted e.g. by collisions with vehicles, electromagnetic waves, light during night. Many vectors for passive dispersal, which may exceed active mobility by magnitude in many species, vanished (like nomadic livestock) or are disrupted by technical measures (like flowing waters). On the other hand globalized mobility of humans, including international trade, supports the extension of the area of distribution even trans-continentally (Neophytes, Neozoans). Due to changes in migration patterns and capabilities, we therefore face fundamental changes of the species sets all over the world in near future. Current human migration tendencies will also have severe effects on the regional biodiversity by the erosion and loss of sophisticated knowledge on the natural properties of rural areas and of related, traditional techniques of land use.

25.2 Migration and the Survival of Species

During the past two decades, mankind has impacted nature fundamentally. One effect is the extinction of species which substantially exceeds the natural mass extinctions in the earth's previous history. About 6,800 plant and 5,500 animal species and subspecies are named in the global Red List of the World Conservation Union (IUCN) to be extinct or threatened (www.redlist.org 2004). However, this is only a vague indicator since our knowledge of most invertebrate species is inadequate.

Besides individual impacts such as hunting and fishing impose, habitat loss is cited to be the most prominent factor in human-induced extinction. Many studies have demonstrated the loss in area and the increasing fragmentation of natural and semi-natural ecosystems.. Corridors on different spatial levels of landscapes are planned, and some already realized, to reconnect those fragmented ecosystems. However, it is doubtful whether such a structurally based approach may sufficiently address the basic problems, for instance, whether preservation or creation of structural landscape elements such as hedges or field margins can compensate fragmentation.

All modern models of population ecology are based on the dynamics of two contrasting effects, colonization and local extinction, and they include stochastic factors to explain what happens in nature. Thus, kind, frequency, and spatio-temporal pattern of migration of individuals from one habitat to another are features as important as changes within the subpopulations themselves, or in the structural environment of the species. It is much more difficult to analyze human influence on ecological processes related to individual migration than to demonstrate visible structural changes. Consequently, our knowledge is still fragmentary. We must assume from the available data that the human effects on migration are at least as profound as on structural elements.

25.3 What Is Migration?

In the literature there is some terminological confusion concerning the dislocation of individuals or groups of individuals from one place to another. The most common terms are "migration" and "dispersal", but they are applied to very different ecological phenomena. Dislocation of individuals occurs on very different spatial levels of nature. [I do not understand, for native speakers over years had no problems to cope with this term; see comment in email to Prof. Werner; Is "spatial levels" more appropriate?] for very different purposes, and thus with very different ecological effects. Many vertebrates seasonally migrate between summer and winter habitats, some of them, e.g., the Arctic tern (*Sterna pardisaea*), the white stork (*Ciconia ciconia*), and the grey whale *(Eschrichtius robustus)* travel between continents. Others, including many bat species, the Saiga antelope (*Saiga tatarica*), or the monarch butterfly *(Danaus plexippus),* migrate on a more regional level. Quite different from these are stochastic social migrations such as those of the lemming *(Lemus lemus),* or of some caterpillars that are often induced by overpopulation. The latter can, but need not, result in colonization of new habitats. Beyond this, there is a continuous level of dislocation of individuals from existing populations, actively or passively, directed or undirected, that can end with the death of the individual (the most common effect in many cases), the integration into another population, the colonization of a new habitat, or even with the extension of the area of distribution of the species. Stochastic events apparently play an important role in the result of such dislocations. This is obvious in the spread of species, such as the collared dove *(Streptopelia decaocto)* or the cormorant (*Phalacrocorax carbo*), where sudden transcontinental extension of the area occurred after a long period of stasis.

In the following, only the word "migration" is used for all these phenomena, whereby case-by-case explanations are added as to type of dislocation..

25.4 Functional Changes in Migration Patterns

Discussion of the human impact on migration patterns of animals and plants living in the wild often focuses on linear technical structures such as roads, railroads, or channels. Obviously, they have significant effects on the migration of animals and not rarely also on the migration chances of plants. However, these elements are only a narrow fragment of a broad spectrum of features related to migration that humans have changed over the past several centuries.

25.4.1 Technical Features

There are, in principle, five different types of impact from technical installations on actively migrating animals:
1. Individuals killed by vehicles, land machines, turbines (as in the dams of hydroelectric power plants), high voltage transmission lines, air-conditioning tubes, and so on;
2. Blocking of migration routes by invincible, mostly linear technical facilities, such as fences, channels with steep artificial banks, aboveground pipelines, dams;
3. Blocking of migration routes by unfavorable environmental factors that cause animals to change their direction of movement, such as strips of open land along roads or beneath high voltage lines;
4. Setting up of territories, which are avoided by migrating animals, such as human settlements or clusters of wind power facilities;
5. Pollution by acoustic signals and electromagnetic waves.

From these five types of impacts, the last two are perhaps the most effective, especially with respect to their presence almost everywhere on earth, however, our knowledge of these impacts is very poor. We know that the effects of wind power plants on migrating birds can be very significant if they are placed in prominent locations of traditional flyways. Many animals perceive acoustic or visual signals beyond the human range. While only one century ago all these frequency windows remained fairly natural, they are now filled with an enormous number of human signals. It is rather likely that human echo-sounder signals disturb the orientation of whales, as may be true for bats. Huge areas of Europe and North America are lit up at night by electric lights, as can be impressively seen in satellite images. Many animals migrate at night in order to minimize the danger of predation and they are able to perceive these new signals. We are far from understanding what effects this substantial change of the environment may have.

25.4.2 Passive Migration by Livestock

Livestock keeping has been a very common type of human use of nature since the Neolithic revolution. In most early cultures the dependence on this kind of resource was so strong that human activities had to adapt themselves to the habits of the domesticated animals. Nomadic life habits was one such way.. Other, more stationary variants are transhumant livestock keeping systems (i.e., seasonal movement of livestock, especially sheep between mountain and lowland pastures, either under the care of herders or in company with the owners), which became common especially in the European region millennia before our time. The summer and the winter pastures were often several hundred kilometers apart. The need for adequate pastures was one of the causes of colonization, of even unfavorable regions such as the high mountain areas of Europe. As large cities developed, their meat supply came to be regularly transported to them alive and on foot.

Over a very long period of time these migrating livestock flocks served as an efficient vector for the passive migration of wild animal and plant species. Because of the consistency of their fur and the morphological structure of their hooves, sheep have come to be recognized as the best vehicle for the live transport of small units of plant or animal life from one location to another. Sheep carry plant diaspores of many species for weeks in their fur (Fischer et al. 1996). Grasshoppers may ride on their bodies for more than 700 m, farther than can be covered by flightless females of some species (Warkus et al. 1997). Young live snails have been found between the nails of sheep and even lizards ride on their backs. Taking the experimental data and comparing them with the extent of former nomadic and transhumant livestock keeping, the magnitude of this passive transport of potential propagation units must have been enormous.

All data support the theory that migrating livestock has been an efficient vector for the passive connectivity between populations of wild animals and plants and has served as an important factor in the propagation of species.

Nomadic life behavior has been largely abandoned all over today's world for it is no longer compatible with modern lifestyle and supply systems. Lifestyle is stationary today, including the meat production afforded by livestock. Livestock migrations have ceased to exist in most parts of the world. It is true, product transport from one location to another has been extended tremendously during the past century as a consequence of higher technical mobility and international markets. And there is has been substantial ecological change: while the transport of occasional "passengers" formerly often ended in equally suitable habitats – and thus successfully - it is now almost always a one-way road. Sheep transported by a truck still carry the same number of diaspores as in former times, but they now end up in a slaughterhouse, not in a grassland.

The same type of outcome applies to many other kinds of passive transport. By historical means of transport, many snail species were moved from one place to another, thus facilitating population establishment in new regions (Dörge et al. 1999). Modern transport and production schemes no longer support such ecological processes.

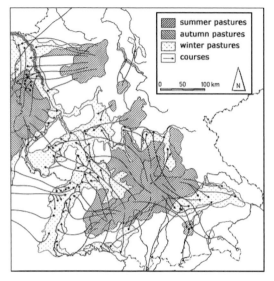

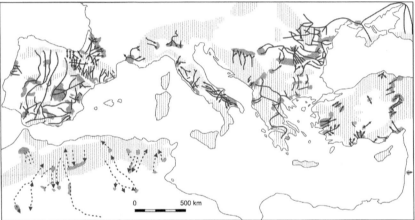

Fig. 1. Transhumant livestock keeping in former times in Germany (after Hornberger 1959, redrawn) and in the Mediterranean countries (from Schultz 2002, after Grigg 1974; *hatched* mountain areas, redrawn)

25.4.3 Passive Transport by Natural Forces

Basic natural forces such as wind, landslides, oceanic currents, and inland streams also support passive dispersal and colonization by wild species. Regularly, huge numbers of propagation units are carried from one location to another by flowing waters (Bill et al. 1999). However, information is poor as to the extent to which this last phenomenon really supports the stability of populations on a landscape level or the colonization of new habitats.

The discharge of natural streams changes significantly over time. Low water situations alternate with floods. During high discharge, significant terrestrial areas within the floodplain, where terrestrial species live, are inundated. It is most probable that during such events potential propagation units are taken up by the flowing water and taken to other locations downstream. A successful "arrival" of the transported units should have significant effects on the connectivity between populations and thus on the survival probabilities of certain species on a landscape level.

During average discharge, about 10^6 animal organisms and molts are transported per meter of stream width per day on the surface of an alpine wild river (Hering and Plachter 1997). Many species in floodplains are adapted to sudden inundations. In such a case, ants of the species *Formica selysii* form living rafts of adhering individuals, eggs, and one queen. These are able to float on the water, regain land downstream, and initiate new populations (Lude et al. 1999). Experimental designs with marked snails, carabid beetles, and spiders prove that these animals are not only able to survive swimming on the water for days, but that they are also capable of landing successfully on the banks and of actively searching for suitable habitats. Often plant debris such as pieces of wood or grass bundles are used for rafting. The number of individuals decreases with distance, but living snails could be traced for more than 20 km (Tenzer and Plachter 2003; Fig. 2).

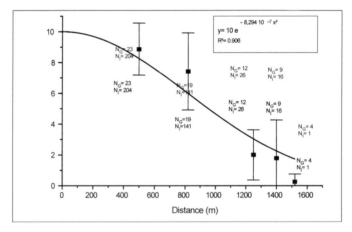

Fig. 2. Average numbers of marked individuals of the snail *Arianta arbustorum* by distance from the starting point on artificial bundles of plant debris in swimming experiments on the river Lahn (Germany). N_G Number of caught bundles, N_i total number of snails remaining on the bundles, ±SD. (Tenzer 2004)

According to water engineering measures, humans have changed this ecological process fundamentally on most rivers of the world. Dikes significantly lower the area where living animals and diaspores can be taken up by floods; dams form an invincible barrier for the transport on the water table. Steep man-made banks diminish the chance that "traveling" propagation units may land, and not rarely there are no longer suitable habitats in the vicinity of landing places.

25.5 Extension of the Area of Distribution

During the past decades, the amount of globally transported goods increased enormously. In general, human mobility is much greater than only 50 years ago, the means of transport much broader. One important feature of globalization is that continental borders are increasingly bridged. Of course, transcontinental trade and human mobility have existed for centuries, but never before in history have such huge distances been covered in such a small span of time. This fact offers propagation chances for species that are able to survive travel for only some hours or days. In sum, modern international trade and human mobility may indeed be classified as a new feature of evolution. Human globalization is a starting point for globalization of the evolution of nature.

Already now the multiple invasion of neophytes and neozoans is a major issue for conservation and agriculture in many regions of the world, e.g., in Australia and in South Africa. Although management and economic problems are fewer in the temperate zones of the world, even there a never seen invasion of new species is recognized (Kowarik 2003). In Europe, new channels have broken natural fresh water borders between major river catchments, resulting, e.g., in a significant invasion of freshwater organisms from the Black Sea basin (Fig. 3). In the Rhine River the number of neozoans has increased fivefold since 1900, many of the species originating from eastern Europe, but also from other, far distant places of the world (Kinzelbach 1995).

One can argue that the loss of natural migration routes and passive transport systems is ecologically compensated by these modern phenomena. However, this is not the case. The species profiting from natural and modern technical transport systems are quite different, not only in taxonomy, but also in life-form types. While the earlier migrators were often specialized forms, the current transport systems mainly favor aggressive forms with a broad environment and habitat adaptability . Moreover, natural barriers, which are a major feature for speciation, are increasingly broken down.

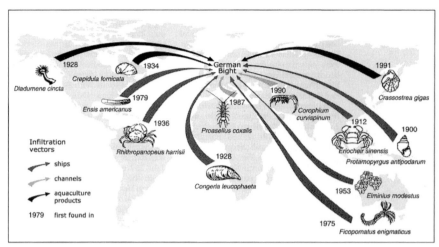

Fig. 3. Introduction of neozoans from different parts of the world to the German Bight, classified according to vectors. (Kowarik 2003, after Nehring und Leuchs 1999)

25.6 Human Migration and Conservation

Social and economic disparities are increasing in many modern societies. Famine and undersupply of food are still the major problems of mankind. Increasingly more people do not even have access to clean water. This results in growing streams of migrating people, from rural areas to the cities on a regional level, and from continent to continent, hoping for a better life. Those who stay often change their lifestyle fundamentally. Traditional farmers become traders or service people for another kind of modern "migration", the seasonal migration of tourists.

Many landscapes of the world are clearly shaped by humans over millennia. giving way to "cultural landscapes." Even many tropical rainforest areas with indigenous cultures have been so affected because of shifting agriculture, use of forest products, and hunting. Nature has adapted to this situation over time. In European cultural landscapes, where such developments have become quite obvious, natural values including biodiversity are closely linked to the traditional landscapes.

Due to limited technical capabilities, people have had to adapt their kind of land use to the local constraints of nature. Indeed, in many cultural landscapes, a co-evolution between nature and local cultures has taken place (Plachter and Roessler 1995). This in no way means that historically people have lived "in harmony" with nature. Many old land-use practices were degrading and detrimental to nature, especially concerning forests and soils. However, over centuries, by an ongoing process of trial and error, both culture and nature "learned" from each other. Earlier cultures "respected" nature, and therefore attributed values to "their" surrounding nature. Much of this sophisticated knowledge was never written down. It was passed on orally from generation to generation.

Nowadays, predominantly young people leave rural areas all over the world to migrate to more favorable places. The land is now kept in use by older people, but what will be the perspective in a longer period of time?. Emigrating people take with them all their knowledge of nature where they were born, such as it is, and of how to make use of it in a sustainable way. Thus, to a considerable extent, recent human migration also results in a permanent erosion of adapted knowledge of use and management techniques as developed in more remote cultural landscapes.

This effect is limited not only to underdeveloped countries. Even in central Europe there is quite a similar tendency resulting in the abandonment of agriculture in unfavorable areas. This need not mean that those landscapes will be free of human settlements. The rural exodus is partially compensated by individuals seeking a more "natural" home environment. This is already the case around many European towns, and it even includes traditional agricultural regions, e.g., the Tuscany in Italy. Tourists, as another sociological group, especially invade cultural landscapes still rich in nature.

In any case, these new "users" of the land have no knowledge of the specificities of nature, and they have totally different attitudes from the former farmers toward nature and its requirements. There are two major accompanying problems. First, who will manage these landscapes in the future? They are far from self-sustaining and need permanent human care in order to survive. Second, what will be the consequences of these new "functions" of traditional cultural landscapes on biodiversity?

Both questions need answers soon, for this process of erosion of adapted knowledge and respect for nature is still ongoing all over the world.

25.7 References

Bill H-J, Poschlod P, Reich M, Plachter H (1999) Experiments and observations on seed dispersal by running water in an Alpine floodplain. Bull Geobot Inst ETH Zuerich 65:13–28

Dörge N, Walther C, Beinlich B, Plachter H (1999) The significance of passive transport for dispersal in terrestrial snails (Gastropoda, Pulmonata). Z Ökol Naturschutz 8:1–10

Fischer S, Poschlod P, Beinlich B (1996) Experimental studies on the dispersal of plants and animals on sheep in calcareous grasslands. J Appl Ecol 33:1206–1222

Grigg DB (1974) The agricultural systems of the world. Cambridge Univ Press, Cambridge

Hering D, Plachter H (1997) Riparian ground beetles (Coleoptera, Carabidae) preying on aquatic invertebrates: a feeding strategy in alpine floodplains. Oecologia 111:261–270

Hornberger T (1959) Die kulturgeographische Bedeutung der Wanderschäferei in Süddeutschland. Süddeutsche Transhumanz. Bundesanstalt für Landeskunde, Remagen

Kinzelbach R (1995) Neozoans in European water. Exemplifying the worldwide process of invasion and species mixing. Experientia 51(5):526–538

Korwarik I (2003) Biologische Invasionen: Neophyten und Neozonen in Mitteleuropa. Ulmer, Stuttgart

Lude A, Plachter H, Reich M (1999) Life strategies of ants (Hymenoptera, Formicidae) in unpredictable floodplain habitats of alpine rivers. Entomol Gen 24:75–91

Nehring S, Leuchs H (1999) Neozoa (Makrozoobenthos) An der deutschen Nordseeküste: eine Übersicht. Bericht des Bundesamt für Gewässerkunde, Bd 1200, Koblenz, Berlin

Plachter, H, Rössler, M (1995) Cultural landscapes: reconnecting culture and nature. In: von Droste B, Plachter H, Rössler M (eds) Cultural landscapes of universal value. Components of a global strategy. Fischer, Jena, pp 15–18

Schultz J (2002) Die Ökozonen der Erde. Ulmer, Stuttgart

Tenzer C (2004) Ausbreitung terrestrischer Wirbelloser durch Fließgewässer. Dissertation Univ. Marburg, Fac. Biology.

Tenzer C, Plachter H (2003) Dispersal of terrestrial invertebrates by rivers – an important ecological process. Verh Ges Ökol 33. Fischer, Stuttgart, p 327

Warkus E, Beinlich B, Plachter H (1997) Dispersal of grasshoppers (Orthoptera: Saltatoria) by wandering flocks of sheep on calcareous grassland in southwest Germany. Verh Ges Ökol 27. Fischer, Stuttgart, pp 71–78

26 Agronomical Practices Maximizing Water Use

Michele Rinaldi

Istituto Sperimentale Agronomico, via Celso Ulpiani 5, 70125 Bari, Italy, E-mail michele_rinaldi@libero.it

26.1 Abstract

Water management must be the focus of every management decision and practice used in dryland farming systems in semiarid regions. Unless water is available, technologies such as improved varieties, fertilizers, and pest control will do little or nothing to improve yields. A small increase in seasonal evapotranspiration in semiarid regions can result in a significant increase in grain yield when the threshold amount of evapotranspiration required for producing grain has been met. The best way to increase seasonal evapotranspiration is by reducing runoff and reducing evaporation from the soil surface. Both of these losses can be reduced in most situations by using conservation agriculture systems that leave more crop residues on the soil surface to serve as a mulch. This also reduces tillage that in turn enhances soil organic matter content to increase the rate that water infiltrates into the soil to increase the amount of plant-available soil water that the soil can hold. Finally, a structure of an interdisciplinary project to study crop management systems to maximize water use is proposed.

26.2 Introduction

The competition among agricultural, industrial, and urban entities for water resources is high, and agriculture has been the weakest player. Lack of water, associated with the lowered soil fertility, causes desertification, which is becoming more and more important in all over the world. The areas with a major risk of desertification are increasing in the last years, especially in Mediterranean countries. The Mediterranean ecosystem is one of the most beautiful and richest in the world, but it is also one of the most fragile and vulnerable.

The definition of instruments, norms, and indicators suitable to identify the degree of desertification risk, and then the application of common policies for protecting fragile ecosystems are the aims of concluded and on-going projects. The United Nations Convention to Combat Desertification, held in Rio de Janeiro in 1992, took into account the complexity of a phenomenon that reflects physical, biological, and climatic issues, as well as social, economic, and strategic ones. Highlighting the adverse natural conditions, very often worsened by unsustainable economic development, the Convention adopted a definition of desertification:

"Land degradation in arid, semiarid and dry sub-humid areas resulting from various factors, including climatic variations and human activities."

Land degradation means loss of biological and/or economic productivity and biodiversity in croplands, pastures, forests, and woodlands due to several processes among which are: water and wind erosion; modification of the physical, chemical, and biological soil properties; reduction or modification of the vegetation cover. Among the causes related to human activities: the cultivation of inappropriate areas, overgrazing, deforestation, and inadequate irrigation and tillage practices.

26.3 The Role of Conservative and Dryland Agriculture

As a consequence, agriculture plays an important role in preventing water loss and desertification processes. This objective can be achieved using the principles of conservative agriculture (CA) linked to those of dryland agriculture (DA).

The four key principles of CA are:
1. cover soil with residues;
2. reduce mechanical soil disturbance;
3. control traffic;
4. alternate commercial and cover crops to increase organic matter, lessen erosion, and increase the water use efficiency.

For many authors DA is the "growing of crops in areas where every management decision is based on knowing the soil water status and knowing how every practice is likely to affect the soil water."

Dryland agriculture is practiced mostly in semiarid regions. Semiarid regions occur in many parts of the world and make up a large part of the world's soil resources. There are differences, however, in semiarid regions. Some areas have limited precipitation during all months of the year and the potential evaporation greatly exceeds the rainfall. One of the methods used to classify climates is the aridity index (AI), which is the ratio of the annual precipitation to the annual potential evapotranspiration. When the AI is between 0.2 and 0.5, the region is classified as "semiarid."

There are three keys of DA applicable to all semiarid regions to define them in a specific classification system. These are:

Key 1. No growing season is or will be nearly the same in precipitation amount, kind, or range, or in temperature average, range, or extreme as the previous growing season and because of this, crop management requires an adjustment every year.

Key 2. The soil and moisture resources do not remain the same for any long period of time. The competition for moisture and nutrients to produce crops requires removal of the protective grass cover, and severe drought often leaves the soil highly vulnerable to wind erosion.

Key 3. There is abundant sunshine due to many cloud-free days that induces rapid growth that cannot be sustained. Grain crops can desiccate just days before ripening due to sun-drenched, rainless conditions. It is equally possible for a few mm of precipitation to occur at almost the last moment and produce a good grain crop.

DA systems are always risky because of the limited amounts and variability of precipitation. Perhaps even more importantly, key 2 points out that the soil and moisture resource changes rapidly in semiarid regions once these lands are cultivated. This means that the already limited water resources become even more limited. Therefore, it is critical that producers and managers in semiarid regions clearly understand the consequences of carrying out certain practices. Practices that are very successful and sustainable in humid regions can be disastrous in semiarid regions. Also, some practices such as intensive tillage can produce very positive results in the short term, but can lead to very negative consequences in the long term.

Dryland agriculture is highly dependent on precipitation, both snow and rainfall. The water budget shows how water received as rain or snow can be easily lost before it has an opportunity to be used by a crop. *Water use by weeds* and *evaporation from the soil surface* are the two most important pathways of water loss that must be avoided if improvements in precipitation-use efficiency are to be accomplished.

Soil tillage and *weed controls* are important crop management practices affecting soil moisture, fertility conservation, and economic yield in arid and semi-arid countries.

Several studies have been carried out on the old practice of "*fallow*", where the soil is left without crop, or with another crop usually incorporated into the soil as green manure before maturity. Fallow is considered essential in arid environment to ensure an economic crop yield. In low-rainfall areas, the cropping system "fallow wheat" is commonly adopted. The reason is to guarantee a water storage during the fallow period to ensure an adequate water supply for the following wheat.

Both conservative and dryland agriculture consider a list of crop management techniques that influence the soil-plant-atmosphere relationships in semiarid environments:

Crop. High water use efficiency (C3 vs. C4 plants, deep root, good physiological adaptation), drought resistant species

Tillage. Crop residue management, water balance, weed control, soil physical aspect, infiltration rate, soil evaporation, mulching, green manure

Sowing. Timing, intercropping, sod seeding

Irrigation. Critical phase for water, deficit controlled irrigation, low quality water, water use efficiency of irrigation water

Fertilization. Organic fertilization to increase soil fertility

Cropping Systems. All the above practices, rotation or crop sequence including fallow period.

The main aim of this contribution is to show the effect of crop management following the principles of conservative and dryland agriculture on water infiltration and storage in the soil.

26.4 Reducing Soil Evaporation

26.4.1 Soil Cracks

In many dryland areas, significantly more water evaporates from the soil surface than is used by the growing crops. Soil surface evaporation can be reduced with minimum tillage, mulches, and subsurface irrigation. One of the most important ways to reduce water evaporation from the soil surface is to get the water as deep as possible into the profile. Understanding crack development is important for managing soils characterized by this phenomenon. In soils that contain montmorillonitic clay, cracks often develop with low soil water content and to depths of 30 to 60 cm or more. If precipitation occurs when these cracks are open, some of the water will move quickly to the bottom of the cracks, bypassing the soil matrix.

26.4.2 Biological Mulches

The use of a mulch, vegetal or not, is in most cases the most effective way to minimize evaporation of soil water in dryland regions. There are two reasons for this. The first is that the mulch on the soil surface breaks the "pores continuum" near the soil surface where the water evaporates. The second is that maintaining a mulch on the surface minimizes tillage that dries the soil. In dryland regions, soil quickly dries, and it dries to the depth that it has been tilled.

26.4.3 Increase Plant Extraction of Soil Water

Plants extract water at a soil water content above that of the permanent wilting point. At field capacity or above, some of the water can be extracted by plants relatively easily because it is contained within rather large soil pores and it is held at fairly low potential. As plants extract water, the more easily extractable water is taken first, and then the water becomes increasingly difficult for plants to extract because it is held increasingly tightly by the soil. This is mainly because water is held in smaller and smaller pores as soil water is depleted. A lot of water is still present in the soil at the permanent wilting point, particularly in fine-textured soils, but it is held so tightly by the soil particles that the plants cannot extract it.

The Fig. 1 shows the photosynthetic rate of wheat as a function of water potential. For many soils the field capacity is approximately -0.033 MPa and permanent wilting point is approximately -1.5 MPa. The rate of growth decreased rapidly as the soil water potential increased showing that the plants simply could not extract water from the soil at a rate fast enough to meet the needs of the plants even though the soil water content was considerably higher than the permanent wilting point. The shape of this curve, however, is climate-dependent. In a humid area where the water demands are much lower, the photosynthetic rate would remain relatively high even as the soil dried because the demand for water by the plants would be much lower and the plants could extract water at a rate fast enough to meet the plant needs even though the soil was becoming relatively dry.

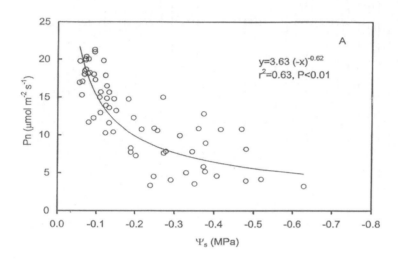

Fig. 1. Relationship of net photosynthesis vs. plant water potential

The Fig. 1 shows, however, that for maximum crop growth, the soil water potential needs to be maintained at relatively high values. This seldom occurs in dryland agriculture except for very short time periods.

26.4.4 Increase Crop Water Use Efficiency

Since water is the limiting factor in dryland production, it is essential to utilize it as efficiently as possible. This is very difficult, however, in semiarid regions because of high evaporation demands of the climate. Figure 2 shows the theoretical relationships between transpiration and biomass production, and between seasonal evapotranspiration and grain yield. The first important point is that there is a straight-line relationship between biomass production and transpiration and the line passes through the origin. Although this relationship holds for all climates, the slope is very much dependent on climatic conditions. As the climate becomes hotter and drier, the slope of the line decreases, showing that more water must be transpired to produce one unit of biomass. Likewise, there is a theoretical straight-line relationship between the yield of grain and seasonal evapotranspiration. In this case, the line does not pass through the origin, meaning that a substantial amount of water must undergo evapotranspiration before the first unit of grain is produced. Under drought conditions, a common occurrence is the production of some biomass without any grain production. Again, the point that the line intercepts the *x*-axis and the slope of the line are climate-dependent. In dryland regions, there is often only enough precipitation to allow seasonal evapotranspiration to barely exceed the amount required to begin grain production, and so grain yields are very low.

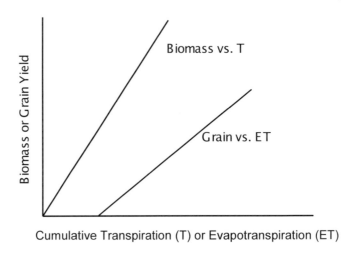

Fig. 2. Relationship of biomass or grain yield vs. cumulated transpiration or evapotranspiration

Another very important point is that once grain production starts, each additional unit of water will produce a proportional amount of added grain yield. This is why dryland grain yields can increase very significantly with a relatively small addition of water. Any practices that can reduce evaporation, runoff, or other water losses so that more of the limited precipitation can be used for transpiration will greatly increase grain yields under semi-arid conditions. This is also why a small amount of irrigation water, particularly if added at a critical stage of growth, can have such a dramatic effect on increasing grain production (high water use efficiency).

26.5 Sustain or Increase Soil Organic Matter Content

Soil organic matter is an important component of soil because if affects all soil processes, chemical, biological, and physical. Perhaps most important in semiarid regions is the effect that soil organic matter has on the infiltration of precipitation and on the water-holding capacity of the soil. It is well known that as the content of soil organic matter declines the infiltration rate decreases, and the amount of plant-available water that the soil can hold becomes smaller. These effects are negative in all areas, but they are particularly bad in semiarid regions because they make the lack of water more severe. Loss of soil organic matter also leads to increased wind and water erosion, and this only exacerbates poor soil water conditions. Soil organic matter serves as a "glue" to hold soil particles together and its loss greatly accelerates the erosion hazard. Organic matter loss also destroys soil structure, which is so important in soil water properties.

Soil organic matter decline is common in all climatic regions when agro ecosystems are formed. Tillage is the main cause of the organic matter decline and the more intensive the tillage, the higher the rate of loss. The negative effect generally increases with increasing aridity because the initial soil organic matter content is usually low and the amount of biomass produced annually to replenish the supply is limited because of the lack of water.

26.6 The Proposal

The principles of this project proposal recall the "Conservative Agriculture" and "Dryland Agriculture" approaches: that is, no-till and minimum-till practices are endorsed because they allow cropping system intensification beyond the long-standing system of wheat fallow, thereby maximizing the interception, infiltration, and storage of rainfall water in the soil for enhanced crop production. Intensive soil tillage operations, on the other hand, increase the drying of surface layers and the risk of erosion in hilly areas.

The overall objective is the identification of dryland crop and soil management systems that maximize plant water use efficiency of the total annual precipitation (both snow and rain).

Our sub-objectives are:
1. Determine the biological feasibility and economic viability of cropping systems with and without fallow periods;
2. Quantify the interrelationships of climate (precipitation and evaporative demand), soil type, and cropping sequences, which involve fewer and/or shorter fallow periods;
3. Quantify the effects of long-term use of no-till management systems and mulching on: (a) Soil structural stability; (b) soil microorganisms and faunal populations; (c) soil organic C, N, and P content, soil infiltrability, all in conjunction with the new intensive cropping systems that have fewer and/or shorter fallow periods; and 4) on pest and beneficial insect population dynamics;

4. Identify crop sequences that minimize soil erosion by crop residue maintenance;
5. Develop a crop sequence database across soil and climate gradients that will allow economic assessment of entire management systems.

The project proposal requires an interdisciplinary approach. The partners involved should be: agronomist, soil physicist, soil chemist, biologist, meteorologist, engineer, economist, and crop modeller.

This project will be implemented in a series of stepwise phases.

26.6.1 Phase No. 1 – Historical

1. **Bibliography.** To review the information on cropping systems, tillage and mulch effect on water balance in semiarid environments
2. **Dataset of concluded researches.** To collect the data of researches with main reference to long-term experiments
3. **Experiments in progress**

26.6.2 Phase No. 2 – Experimental

1. **Application of CA and DA Agriculture.** To define examples of new agriculture practices following CA and DA guidelines
2. **New field experiments.** Selection of pilot sites based on criteria such as: locally important or widespread soil type, typical climate for local crop growing, and social and economic conditions
3. **Use of simulation models for new scenarios.** To use cropping systems simulation models in order to evaluate new DA and CA scenarios in different climates and soil types

26.6.3 Phase No. 3 – Transfer of Knowledge

It is important in the phase of exploitation of the results to expose farmers and stakeholders to alternate CA practices and commence training them in the practical use of new technologies:
1. describe and explain equipment, alternate species (cover crops), field layouts (bed farming, permanent wheel tracks);
2. support implementation of the new practices by encouraging flexible funding mechanisms and incentives, especially in the phase of transition;
3. establish close cooperation and dialogue between scientists, suppliers and farmers, and between government and educational institutes to ensure a dynamic cycle of research and development support and feedback.

26.7 References

Allen RG, Pereira LS, Raes, D, Smith M (1998) Crop evapotranspiration. Guidelines for computing crop water requirements. Irrigation and drainage paper no 56. FAO, Rome, 300 pp

Ball BC, Tebrügge F, Satori L, Giráldez JV, González P (1998) Influence of no-tillage on physical, chemical and biological soil properties. In: Tebrügge F, Böhrnsen A (eds) Experience with the application of no-tillage crop production in the West-European countries. Final report of concerted action N. Air3-CT93-1464. Fachverlag Köhler, Giessen, Germany, pp 7-27

Blum WEH (1990) The challenge of soil protection in Europe. Environ Conserv 17:72-74

Ehlers W (1997) Optimizing the components of the soil water Balance by reduced and no-tillage. In: Tebrügge F, Böhrnsen A (eds) Experience with the application of no-tillage crop production in the West-European countries. Proceedings of the EC-Workshop III, Evora, Portugal. Wissenschaftlicher Fachverlag Dr. Fleck, Giessen, Germany, pp 107-118

El-Swaify ES, Pathak P, Rego TJ, Singh S (1985) Soil management for optimized productivity under rainfed conditions in the semi-arid tropics. In: Stewart BA (ed) Advances in soil science, no 1. Springer, Berlin Heidelberg New York, pp 1-64

FAO (1981) Arid zone hydrology. Irrigation and drainage paper 3. FAO, Rome

FAO (1976) Conservation in arid and semi-arid zones. Conservation guide 3. FAO, Rome

Francis CA, King JW, Nelson DW, Lucas LE (1998) Research and extension agenda for sustainable agriculture. Am J Alternative Agric 3:123-126

García-Torres L, González-Fernández P (eds) (1997) Agricultura de Conservación: Fundamentos Agronómicos, Medioambientales y Económicos, Asociación Española Agricultura de Conservación/Suelos Vivos (AEAC/ SV), Córdoba, España, 372 pp

Hall AE, Cannell GH, Lawton HW (eds) (1985) Agriculture in semi-arid environments. Springer, Berlin Heidelberg New York

Hoare J (1992) Sustainable dryland cropping in southern Australia: a review. Agricult Ecosyst Environ 38:193-204

Kinsella J (1995) The effect of various tillage systems in soil compaction. In: Farming for a better environment. A white paper. Soil and Water Conservation Society, Ankeny, Iowa, USA, pp 15-17

Lal R (1997) Residue management, conservation tillage and soil restoration for mitigating greenhouse effect by CO2-enrichment. Soil Tillage Res 43:81-107

Maiorana M, Castrignanò A, Fornaro F (2001) Crop residue management effects on soil mechanical impedance. J Agric Eng Res 79:231-237

Mastrorilli M, Katerji N, Rana G (1999) Productivity and water use efficiency of sweet sorghum as affected by soil water use deficit occurring at different vegetative growth stages. Eur J Agron 11:206-216

Narayana VV, Dhruva, Ram Babu (1985) Soil and water conservation in semi-arid regions of India. Central Soil and Water Conservation Research and Training Institute, Dehradun, India

Rabbinge R, van Diepen CA (2000) Changes in agriculture and land use in Europe. Eur J Agron 13:85-99

Rinaldi M, Rana G, Introna M (2000) Effects of durum wheat straw on soil evaporation in a semi-arid region. In: M.I. Ferreira, H.G. Jones (eds) Proceedings of the 3rd international symposium on irrigation of horticultural crops, 28 June - 2 July 1999, Lisbon, Portugal. Acta Horticolt 537:159-165.

Rinaldi M, Maiorana M (2000) Sunflower irrigation treatments in interaction with a previous mulch crop in Southern Italy. Proceedings of 15th intern. sunflower conference, Toulouse (F), C 97.

Singh RP, Reddy GS (1988) Identifying crops and cropping systems with greater production stability in water-deficit environments. In: Bidinger FR, Johansen C (eds) Drought research priorities for the dryland tropics. ICRISAT, Patancheru, pp 77-85

Stewart BA, Unger PW, Jones OR (1985) Soil and water conservation in semi-arid regions. In: Ei-Swaify SA, Moldenhauer WC, Lo A (eds) Soil erosion and conservation. Soil Cons Soc Am, Ankeny, Iowa

Stewart BA (1985) Water conservation technology in rainfed and dryland agriculture. Proceedings of the conference 'Water and water policy in world food supplies'. Texas A&M University, College Station, Texas, pp 335-359

Swindale LD (1988) Agricultural development and the environment: a point of view. Sustainable agricultural production: implications for international agricultural research. TAC Secretariat, FAO, Rome, pp 7-14

Troeh FR, Thompson LM (1993) Soils and soil fertility. Oxford Univ Press, New York

Unger PW, Jordan WR, Sneed TV, Jensen RW (eds) (1988) Challenges in dryland agriculture - a global perspective. Texas Agric Exp Sta, College Station, Texas

Van Lynden GWJ (1995) European soil resources. Current status of soil degradation, causes, impact and need for action. Council of Europe Press. Nature and environment, no 71, Strasbourg, France

Ventrella D, Rinaldi M, Rizzo V Fornaro F (1996) Water use efficiency of nine cropping systems in a water limited environment. Proceedings of 4th ESA congress, Veldhoven-Wageningen (The Netherlands), II, 506-507.

27 Soil Degradation and Land Use

Marcello Pagliai

Istituto Sperimentale per lo Studio e la Difesa del Suolo, MiPAF, Firenze; Italy;
E-mail pagliai@issds.it

27.1 Abstract

The main aspects of environmental degradation that involve soil (erosion, soil compaction, soil crusting, deterioration of soil structure, flooding, losses of organic matter, salinization, onsite and offsite damages, etc.) result from human activities. Since conventional agricultural production systems have resulted in excessive erosion and soil degradation, change is needed that will control such ruin.

Scientific results have clearly shown that agricultural management systems can play an important role in preventing soil degradation provided that appropriate management practices are adopted. Long-term field experiments in different types of soils have shown that alternative tillage systems, such as minimum tillage, ripper subsoiling, etc., improve the soil structural quality. Continuous conventional tillage causes a decrease in soil organic matter content associated with a decrease in aggregate stability, leading, as a consequence, to the formation of surface crusts, with an increase in runoff and erosion risk.

In hilly environments, land leveling and scraping are dangers to soil, causing its erosion.. After leveling, slopes being prepared for planting (particularly vineyards) are almost always characterized by the presence of large amounts of jumbled earth materials accumulated by the scraper. In this vulnerable condition, a few summer storms can easily cause soil losses exceeding 500 Mg ha^{-1}year^{-1}. Moreover, land leveling and the resulting soil loss cause drastic alteration of the landscape and loss of the cultural value of the soil.

Subsoil compaction is strongly underevaluated, even though the presence of a ploughpan at the lower limit of cultivation is largely widespread in the alluvial soils of the plains cultivated by monoculture. It is responsible for the frequent flooding of such plains when heavy rains concentrate in a short time (rainstorm), since the presence of this ploughpan strongly reduces drainage. Alternative tillage practices, like ripper subsoiling, are able to avoid the formation of this compact layer.

27.2 Introduction

The main aspects of environmental degradation that involve soil (erosion, soil compaction, soil crusting, deterioration of soil structure, flooding, losses of organic matter, salinization, onsite and offsite damages, etc.) result from human activities. Since conventional agricultural production systems have resulted in excessive erosion and soil degradation, change is needed that will control such ruin. There is a need for crop production in harmony with soil conservation and environmental protection. Soil management practices are sustainable only if compatible with the expectations of farmers and if their influence on the environment is such that they can be practiced indefinitely without undesirable consequences. Since soil degradation in Europe now has an increasing profile on political agendas, there is a real opportunity for the soil science community to influence the way soils are used and how they might best be protected in the future. This chapter reports the most important aspects of soil degradation and emphasizes the importance of adopting an adequate soil management policy for sustainable land use.

27.3 Soil Degradation

Soil degradation is a major environmental problem worldwide, and there is strong evidence that the soil degradation processes present an immediate threat to both biomass and economic yields, as well as posing a long-term threat to future crop yields. Therefore, it is absolutely necessary that such soil degradation practices be put under control.

Many soil degradation processes reflect the result of short-lived extreme events as well of gradual long-term trends. Both these processes can lead to irreversible changes. The main aspects of soil degradation, of concern in Europe today, are:

1. Soil erosion by water and wind
2. Compaction of surface and subsoils
3. Organic matter impoverishment
4. Structural deterioration (soil crusting, soil sealing, etc.)
5. Salinization (and sodication)
6. Desertification
7. Contamination and pollution
8. Acidification

Desertification in reality combines a number of degradation processes including water erosion and sedimentation, salinization, and sodication.

The vulnerability of the Italian soils to degradation processes is certainly higher than in other European countries due to the high variability of the environment.

It has been estimated that in Europe 72,000 km^2 (25% of all agricultural land), 54,000 km^2 (35% of all pasture land), and 26,000 km^2 (92% of all forest and wood land) is affected by some kind of soil degradation (Van Lynden 1995).

Soil is being lost by erosion at about 30 times the sustainability rate (Pimentel et al. 1993). There are very few studies at the European level on the estimation of the economic damage due to soil erosion. The loss of soil degrades the arable land and eventually renders it unproductive. It has been estimated that worldwide about 12 million ha of arable land are destroyed and abandoned each year because of non-sustainable farming practices (Lal and Stewart 1990). In many regions, this loss of land is a major cause of food shortage and undernutrition. In addition to decreasing food production, soil erosion creates serious environmental and economic problems. The use of large amounts of fertilizers, pesticides, and irrigation helps offset deleterious effects of erosion, but at the same time has the potential to create pollution, health problems, destroy natural habitats, and contributes to high energy consumption and unsustainable systems.

Soil degradation is always due to incorrect use of soils. Not only erosion is responsible, but also soil compaction, soil crusting, loss of soil structural stability, etc., strongly contribute to degradation. For example, soil compaction is estimated to be responsible for the degradation of an area of 33 million ha in Europe (Soane and van Ouwerkerk 1995). About 32% of soils in Europe are highly vulnerable to soil compaction and another 18% is moderately vulnerable (Fraters 1996). Due to ever-increasing wheel loads in agriculture, compaction is increasingly found in the subsoil.

The cost of soil degradation in Europe is not precisely known. Results of concerted action on subsoil compaction in the ambit of EC research projects (Van den Akker et al. 1999) have estimated that 38-ton sugar-beet harvesters have caused yield losses of 0.5% per year. Assuming that such harvesters are used on at least 500,000 ha in the EC, this results in an annual loss of sugar beet yield of 100,000 kEURO. It is expected that these heavy harvesters will be increasingly used. In the Netherlands it has been estimated that traffic-induced subsoil compaction has reduced the total production of silage maize by 7%. This results in an annual loss in the Netherlands of 21,000 kEURO. During ploughing, annual fuel consumption is claimed to be 1 million tons higher than necessary because of soil compaction.

One of the impacts of soil erosion and soil compaction is that water and nutrient use efficiency decreases, which means that more irrigation water is needed and the loss of the nutrients in the environment increases.

27.4 Soil Management in a Sustainable Environment

The need to reduce the environmental impact of agricultural activities and to control soil structure degradation are the main aims of land management. In response, farmers are led to consider the possibility of adopting "more simplified" cultivation practices as an alternative to conventional tillage. The abandoning of traditional farming rotations and adoption of intensive monocultures, without application of farmyard manure to the soil, has decreased the organic matter content in soils with evident deterioration of soil structure. In fact, the main consequence of long-term intensive cultivation is the degradation of soil structure, which can reduce the effect of chemical fertilizers. As soil erosion increases, solid soil particles and nutrients can be transported with the consequent risk of surface water pollution.

Long-term field experiments in different types of soils, representative of the typical pedological environments of Italy, have shown that alternative tillage systems, such as minimum tillage, ripper subsoiling, etc., improve the soil pore system by increasing storage pores and elongated transmission pores, as compared with the consequences of continuous conventional ploughing (Pagliai et al. 1998). The resulting soil structure is more open and more homogeneous (Fig. 1), thus allowing better water movement, as confirmed by the greater values of hydraulic conductivity measured in minimally tilled soils. Water content is also generally higher in less tilled soils. Continuous conventional tillage, moreover, causes a decrease in soil organic matter content that is associated with a decrease in aggregate stability, leading, as a consequence, to the formation of surface crusts, with increased runoff/erosion risk, and compacted structure (Vignozzi and Pagliai 1996).

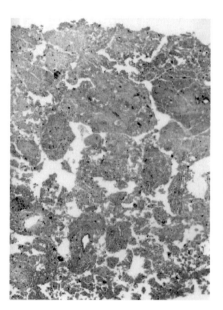 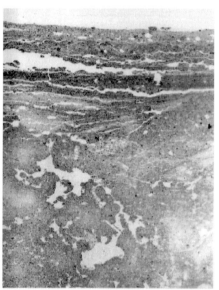

Fig. 1. Macrophotographs of vertically oriented thin sections, prepared from undisturbed soil samples collected in the surface layer (0-5 cm) of a loam soil under minimum tillage (*left*) and conventional tillage (*right*). A subangular blocky structure is evident in minimally tilled soil while the structure appears more compact in conventionally tilled soil, where in the surface a platy structure, due to the formation of a surface crust, is very evident. Plain polarized light; pores appear white; frame length 3 cm

Soil compaction is another important factor responsible for environmental degradation. It causes strong modifications in soil structure and reduces soil porosity. Soil compaction is caused by a combination of natural forces, which generally act internally, and by man-made forces related to soil management practices. The latter forces are mainly those related to vehicle wheel traffic and tillage implements and have a much greater compactive effect than natural forces such as raindrop impact, soil swelling and shrinking, and root enlargement. This is because trends in agricultural engineering over the last few decades have resulted in machines of greater size and weight. Thus, soil compaction has become one of the most significant aspects of soil degradation, and problems of finding tires, inflation pressures, etc., able to reduce soil compaction are far from being solved. It is therefore fundamental to evaluate the impact of wheel traffic on soil structure: porosity measurements can help to quantify the degradation effects of compaction. Results have shown that compaction, both in agricultural and forestry soils, not only reduces total soil porosity but also modifies the pore system. In fact, the proportion of elongated pores, useful for water movement and root growth, is strongly reduced in compacted soil. The modifications to the pore system also change the type of soil structure: a platy structure is a common feature in compacted soil. Results also showed that the reduction of porosity and of elongated pores following compaction is directly related to the increase in penetration resistance, and to the decrease in hydraulic conductivity and root growth (Pagliai et al. 2000). Soil regeneration after compaction depends on the type of soil and on the degree of damage to the soil. Soil compaction, such as the compact layer (ploughpan) formed at the lower limit of cultivation in continuous ploughed soils, is caused not only by wheel traffic but also by the shear strength of tillage implements. In Italy subsoil compaction is strongly underevaluated, even though such a ploughpan is largely widespread in the alluvial soils of the plains cultivated by monoculture. It is responsible of the frequent flooding of such plains when heavy rains concentrate in a short time (rainstorm), since the presence of this ploughpan strongly reduces drainage. Alternative tillage practices, such as ripper subsoiling, help to avoid the formation of this compact layer.

The decrease in organic matter leads, as a consequence, to loss in aggregate stability and to an increase in soil crusting. Surface crusts are one of the dangerous aspects of soil degradation, and are formed by raindrop impact, which causes the mechanical destruction of soil aggregates. After the drying process the dispersed particles that form the crust (the particles of which can also be translocated by runoff) form a compact layer of horizontally oriented plate-like soil particles at the soil surface. This compact layer contains few, if any, large pores. Generally, several thin layers of fine particles form the surface crusts. These are intercalated by thin elongated pores oriented parallel to the soil surface, not continuous in a vertical sense, or by rounded pores (vesicles) formed by air trapped during drying. Such pores in the topsoil formed a vesicular structure that can be regarded as an indicator of an unstable and transitory formation induced by poor stability of soil aggregates. Soil crusting reduces seedling emergence, soil-atmosphere gas exchange, water infiltration, and increases surface runoff. Results have shown that addition of organic materials to the soil and reduced tillage practices help to prevent crust formation (Pagliai and Vignozzi 1998). In intensively cultivated soils, the surface aggregates are less rain stable.

Another aspect of soil degradation: in hilly environments, land leveling and scraping cause soil erosion.. Land leveling is generally applied on undulating land for efficient water application and conservation before terracing. Also, bulldozers are often used to remove natural vegetation or the residues of old plantings, with consequent scalping of the soil. In the Mediterranean basin bulldozing is usually used for clearing and leveling the land to obtain uniform, easy-to-cultivate slopes. Furthermore, this operation is usually performed in summer or autumn, which is the period of the most erosive rainfall. After leveling, slopes being prepared for planting (particularly vineyards) are almost always characterized by the presence of large amounts of incoherent earth materials accumulated by the scraper. In this vulnerable condition, a few summer storms can easily cause soil losses exceeding 500 Mg ha^{-1}year^{-1} (Bazzoffi and Chisci 1999). Moreover, the land leveling and the resulting soil loss cause a drastic alteration of the natural landscape.

In summary, the ultimate responsibility of soil management in a sustainable environment is to provide multiple improvements in human welfare. In this context, the functions of soils for human societies and the environment are of special importance. According to Blum (1998), soils have at least six different roles in the social and economic development of humankind, which can be divided into three ecological functions and three others directly linked to human activities -- these latter defined as technical, industrial and socio-economic functions. However, the necessary harmonization of the uses of the six soil functions is not a scientific problem, but a political one, which means that all people living in a given area or space have to decide which soil functions they may use at a given time and/or a given space (by a top-down or bottom-up approach). Scientists only have the possibility of developing scenarios and explaining the impacts that may occur when different options are exercised. Those scenarios can be condensed into indicators, which may help politicians and decision-makers as well as people living in certain areas to choose the right options.

The results of research can contribute to realizing specific guidelines suitable for specific soil typology in order to develop an economically and environmental friendly agriculture, and to increase the number of the more enlightened and environmentally conscious farmers. New forms of agriculture such as biological farming, ecological farming, and organic farming with low or no input of chemical fertilizers and other agro-chemicals require optimum soil quality and therefore the prevention of soil degradation.

27.5 Conclusions

Prevention of soil degradation is essential for an economically and environmentally sustainable agriculture. For example, knowledge of the susceptibility of soil to degradation and particularly to subsoil compaction, plus a good estimation of the load-bearing capacity of subsoils, would enable manufactures to design subsoil-friendly equipment and would help farmers decide whether, where, and when they should use this kind of equipment. Land evaluation studies generally neglect aspects of soil degradation such as crusting and overall compaction. This neglect is a result of a lack of knowledge of the impact of crusting and subsoil compaction on a soil's physical quality and on seed emergence , rooting possibilities, and crop growth (each diminished). The increase in nutrients and pesticedes in the environment following soil degradation (erosion, compaction) can also be dangerous to human health and food safety. Improved knowledge of these aspects would improve the analysis of the impact of political decisions and agricultural practices on environment, crop production, and the use of natural resources.

As this century unfolds, the role of soil in a sustainable environment will be much more critical than ever before, because we have reached the cross-roads of conflicts between the uses of different soil functions, with severe environmental problems in many areas. Soil use therefore will occur under ecological, technical, and socio-economic conditions quite different from past centuries. This is not only due to increasing competition for space, e.g., through increased urbanization and industrialization with all its socio-economic and environmental impacts, especially in Europe, but also through increasing and severe competition between biomass production on one hand and groundwater preservation on the other. Further problems can also be expected in biodiversity and global change, e.g., through the extinction of species and through the emission of gases from soils into the atmosphere (Blum 2000).

Therefore, a new concept of soil and land management is needed in order to maintain a harmonized use of soil for sustainable development. Soil physics can play an important role in these endeavors because the spatial arrangement of soil materials is decisive for all physico-chemical and biological soil processes.

This holistic approach to the role of soil in a sustainable environment may be helpful in order to define the specific ecological, socio-economic, and technical problems, thus enabling science to develop more comprehensive scenarios for sustainable development in the future. The use of indicators can help in this endeavor because they can be used in a framework that is understandable for those who have to take initiative in order to solve the problem. These are politicians, decision-makers, and administrators. The Driving Force-State- Response (DSR) and the Driving Forces-Pressures-State-Impact-Responses (DPSIR) framework approaches, developed by the European Environment Agency (EEA 1999), seem to be reasonable tools to alleviate soil and land management problems and to create better environmental conditions in the future.

27.6 References

Bazzoffi P, Chisci G (1999) Tecniche di conservazione del suolo in vigneti e pescheti della collina cesenate. Riv Agron 3:177-184

Blum WEH (1998) Agriculture in a sustainable environment – a holistic approach. Int Agrophys 12:13-24

Blum WEH (2000) Challenge for soil science at the dawn of the 21st century. In: Adams JA, Metherell AK (eds) Soil 2000: new horizons for a new century. Australian and

New Zealand Second Joint Soils Conference, vol 1. Plenary papers, 3-8 Dec 2000, Lincoln University. New Zealand Society of Soil Science, Lincoln, NZ, pp 35–42

European Environment Agency (EEA) (1999) Environment in the European Union at the turn of the century. Copenhagen, Denmark

Fraters B (1996) Generalized soil map of Europe. Aggregation of the FAO-Unesco soil units based on the characteristics determining the vulnerability to degradation processes. National Institute of Public Health and the environment (RIVM), Bilthoven, The Netherlands. RIVM Report no 481505006, 60 pp

Lal R, Stewart BA (1990) Soil degradation. Springer, Berlin Heidelberg New York

Pagliai M, Vignozzi N (1998) Use of manure for soil improvement. In: Wallace A, Terry RE (eds) Handbook of soil conditioners. Dekker, New York, pp 119-139

Pagliai M, Rousseva S, Vignozzi N, Piovanelli C, Pellegrini S, Miclaus N (1998) Tillage impact on soil quality I. Soil porosity and related physical properties. Italian J Agron 2:11-20

Pagliai M, Pellegrini S, Vignozzi N, Rousseva S, Grasselli O (2000) The quantification of the effect of subsoil compaction on soil porosity and related physical properties under conventional to reduced management practices. Adv Geoecol 32:305-313

Pimentel D, Allen J, Beers A, Guinand L, Hawkins A, Linder R, McLaughlin P, Meer B, Musonda D, Perdue D, Poisson S, Salazar R, Sieber S, Stoner K (1993) Soil erosion and agricultural productivity. In: Pimentel D (ed) World soil erosion and conservation. Cambridge Univ Press, Cambridge, pp 277-292

Soane BD, van Ouwerkerk C (eds) (1995) Soil compaction in crop production. Developments in agricultural engineering 11. Elsevier, Amsterdam, 662 pp

Van den Akker JJH, Arvidsson J, Horn R (1999) Experiences with the impact and prevention of subsoil compaction in the European Community. Proceedings of the concerted action on subsoil compaction, 28-30 May 1998, Wageningen, The Netherlands, DLO Winand Staring Centre, report 168, 344 pp

Van Lynden GJW (1995) European soil resources. Current status of soil degradation, causes, impacts and need for action. Nature and environment, no 71. Council of Europe Press, Strasbourg, France, 99 pp

Vignozzi N, Pagliai M (1996) La prevenzione della degradazione del suolo attraverso attività agricole a basso impatto ambientale. Bollettino della Società Italiana di Scienza del Suolo 8:207-219

28 Migration Towards the Cities: Measuring the Effects of Urban Expansion in Rural-Urban Interface by GIS and RS Technology

Danijele Brecevic[1], Massimo Dragan[1], Enrico Feoli[1] and Souyong Yan[2]

[1]Department of Biology, University of Trieste, Italy, E-mail feoli@univ.trieste.it,
[2]Institute of Remote Sensing Application, Chinese Academy of Sciences, Beijing, China

28.1 Abstract

The migration of population from rural areas to urban areas is a common phenomenon in developing countries (DC). The recent industrialization and the consequent urban development have already caused a strong deterioration of the rural peri-urban areas where it was usual to practice small-scale agriculture with high water demand. Not only pollution has to be prevented, but plans have to be developed for the optimal siting of new settlements in order to optimize water management and land use and to reduce sanitary risks. Therefore, there is an urgent need to develop new efficient technological tools to support decisions in planning the urban development in order to keep the rural-urban interface reasonably healthy.

This paper presents an application of state-of-the-art geographical information system (GIS) and remote sensing (RS) technology in a pilot study area in the People's Republic of China for the detection, comprehension, and simulation of land cover changes in economically fast-growing areas. China is particularly interesting for this kind of applications since its legislation was modified at the end of August of 2001 allowing free access to cities by the rural population. This may lead to one of the largest migrations in human history – forecasts estimate that more than 600 million people will move from rural areas to urban areas in the next 25 years. The pilot study was conducted on Hainan Island. The data for measuring the urban expansion and the land-use/land-cover changes were obtained from satellite imagery. The results show that the integration of GIS and RS with cellular automata and neural networks can provide neural models capable of evaluating the urban development and its impacts on the rural-urban interfaces.

28.2 Introduction

The increasing flux of population to urban areas is a common phenomenon in all developing countries. This may be a consequence of one or more of the following factors:

1. low capacity of rural areas to provide satisfactory living conditions,
2. insecurity due to conflicts between political and/or ethnic groups,
3. famine due to natural or artificial calamities,
4. demand for unskilled or semi-skilled labor by the industrial and tertiary sectors of the urban economy, and
5. perception that life in the city is better and more exciting.

Urban expansion produces a dynamic landscape system called rural urban interface (RUI). In this system there are conflicts resulting from overlap of different societal interests and demands, and resources constraints. Problems to be faced in the RUI are threefold:

– growing demand for land for urbanization,
– loss of agricultural land and displacement of the agrarian population, and
– demand for services and livelihoods.

The sustainable development of the RUI requires management plans for human settlement that can ensure acceptable quality of environment and life. It is urgent to strengthen an interdisciplinary system research that, by integrating sociology, economics, and ecology, will contribute to supporting management plans for socially acceptable, economically viable, and environmentally sustainable settlements in the RUI. It is important to organize in a participatory approach a set of methodological and technological tools for environmental management. This set of tools may be called the spatial decision support system (SDSS) according to Malczewski (1999). Up to now, there have been very few attempts to study and model the effects of urban expansion in rural areas in a way that integrates sociology, economy, and ecology in line with the suggestions of Folke (1992) and more recently of Slocombe (2001). In spite of all the discussions on integrating sociology, ecology, and economy, we still have to answer the question: "what does integration mean"? We know that there is no single answer to this question, but there are many ways in which "integration" can be achieved under different perspectives. These range from those aiming at establishing monetary costs of natural resources and landscapes (Costanza 1991; Pearce and Moran 1994), to those planning the process of globalization (see Barrett and Odum 2000; di Castri 2000). How to integrate economy with ecology for sustainable development has been discussed in many papers (see Carley and Christie 1993; Smith 1999; Barrett and Farina 2000). The most advanced technologies of the so-called information society have to be integrated among themselves and made available to users (Hogeweg 2001) in order to design and produce an SDSS. There is the need to integrate technology for spatio-temporal analysis and visualization (Liverman et al. 1998; Malczewski 1999; Ott and Swiaczny 2001) in order to offer better tools for decision making. Efforts to quantify the concept of sustainability are prominent today (Chichilnisky et al. 1998; Hall 2000), both from theoretical and practical approaches; however, these need to be applied in different environmental situations. The definition of suitable indicators for sustainability (Moldan et al. 1997; Wacknagel and Yount 2001) is an exercise that requires a process of appropriate successive approximation in each different context.

A major barrier to translating scientific information and judgments into a form meaningful for decision makers is the lack of availability of appropriate tools for an SDSS. Another barrier is the lack of involvement of scientists in decision making. However, the possibility of integrating sociology, economy, and ecology for producing ecological assessment tools (Lessard et al. 1999; Jensen and Burgeon 2001; Kronert et al. 2001) to be used as a platform to measure environmental impact (Treweek 1999) and assist policies (Lester 1989) for sustainable development of the RUI is today realistic and feasible thanks to the spread of technological tools, even at small administrative units in developing countries.

This paper presents an example of an application of an SDSS for a case study in China carried out within an International Cooperative (INCO) project funded by the European Community with the aim of grouping together scientists and administrators (decision makers) to promote actions for sustainable development. The interest to develop such a kind of study in China relies on the fact that Chinese economy is undergoing a fast process of liberalization. As the Chinese economy liberalizes, there are evident increasing pressures on natural resources that make it very difficult to foresee the environmental dynamics without proper scientific and high-level technological tools. It is remarkable for the studies on RUI to consider the fact that in China the gross domestic product (GDP) of about 600 cities is more than 80% of the whole GDP of the country, however, about 70% of the population is still living in rural areas (Feoli 2002). According to the 10th National Five-Year Plan of 2000, new small and middle cities should be developed. Furthermore, at the end of August 2001 local legislation was modified to allow free access to cities by the rural population, starting on 1 October 2001. This may lead to one of the largest migrations in human history – forecasts estimate that more than 600 million people will move from rural areas to urban areas in the next 25 years. Among the many possible areas of interest, the pilot study was developed for the Island of Hainan.

The reasons for selecting Hainan as a case study were the following:
1. Hainan Island is a tropical island rich in biodiversity with a great variety of different environments (mountain, plain, and coast).
2. Since the establishment of Hainan province and the Special Economic Zone in 1988, industries, tourism, and local economy have been developed very quickly. The interaction between human and natural ecosystems is very active and very strong.
3. Abundant and various data have been accumulated since early 1970s.
4. The results of a study carried out in Hainan would also help to introduce improved procedures of environmental management based on new technologies in peri-urban and rural areas under fast industrial development in China and other tropical DC.

The general objectives of the research in which this chapter is framed are:
- to reduce the environmental impact of urban development on the rural areas;
- to avoid the human life degradation in the RUI;
- to fully recognise the role of the RUI with respect to the urban settlements as provider of natural and agricultural resources (vegetation, water, agricultural products), and open space for recreation.

The specific objectives are:
- to obtain a scientific and conceptual framework of interaction mechanisms between the newly urban and the rural surrounding areas;
- to quantify the impact of urban development on natural resources (water and agricultural lands);
- to develop tools for decision makers in order to support the planning of urban settlements.

The focus of this chapter is on the last specific objective, namely, to show how GIS and RS technology can be useful for landscape analysis and for land-use change simulation. Both GIS and RS technology can be considered essential components of spatial decision support systems (Malczewski 1999), which, thanks to the rapid spread of information technology, can be established at low cost in even small administrative units of DC. The problem today is no longer related to fund availability, but to the capacity of the decision makers to deal with scientific technological tools and to organize the work in interdisciplinary groups involving the scientific and technological know-how. In this chapter, the capacity and usefulness of GIS and RS in analyzing the effects of migration toward the urban areas are presented by considering Haikou County, one of the 19 counties on Hainan Island. In the past 10 years this county, in which the capital city Haikou is found, had the island's highest rate of population growth (mainly by migration) and of economic growth. It was therefore chosen as the study area.

28.3 Data and Methods

In order to monitor, understand, and foresee the development of Haikou County, our first objective was the creation of a GIS database by linking cartographic and statistical data. To achieve this aim the following data were used:
- Two Landsat TM 5 images (22 December 1987 and 23 December 1999)
- Reference land-use maps for the considered period
- Statistical Year Book of Hainan (Anonymous 1998)

The image processing analysis was carried out using Erdas Imagine 8.4 (Erdas 1999) and the following classification scheme comprising six classes of area has been adopted: paddy field, agriculture, vegetation, water, urban, and sandy.

Maximum likelihood supervised classification has been performed on the remotely sensed images in order to obtain the land-use/cover maps for the considered years. Location, amount, and direction of land use changes have been identified and quantified through GIS analysis.

The data produced were the bases for the development of a model able to predict the land cover changes as a function of urban development, according to the model suggested by Li and Yeh (2002). The model operates in a GIS environment and is based on the integration between cellular automata and neural networks. Its conceptual framework is briefly illustrated below.

28.3.1 Cellular Automata and Simulation Models

From a theoretical point of view, the cellular automata (CA) were introduced in the late 1940s John von Neumann (1966). John Horton Conway's *Game of Life* brought them to the attention of the scientific community (Gardner 1970) in the 1960s.

Cellular automata are discrete dynamic systems that are able to describe continuous dynamic systems. By saying discrete we mean that both time and space have a finite number of states. Their basic purpose is the attempt to describe a complex system not by complex equations but through the interaction of simple entities that follow simple rules, the so-called *bottom-up approach*.

The basic properties of the cellular automata are:
- an n-dimensional matrix in which every cell has a discrete state.
- a dynamic behavior established by rules. These rules set the state of the given cell in the next time interval as a function of the surrounding cells states.

The basic element of the CA is the cell. We may consider a cell as an elementary memory unit. In the simplest case (Boolean) only two states are considered. In more complex CA, a cell may assume many states during its life, and even have more than one property (attribute), each with many states. This latter is the approach adopted for urban cellular automata utilized in land-use and land-cover-change simulation models for economically fast-growing areas, as in our case. In such automata, cells represent portions of land of a given extension and their states are characterized by variables related to the considered context (e.g., land use, land value, altitude, etc.).

Rules are used to introduce dynamics in the system. Their function is to determine the state of a cell at a time $(t+1)$ by considering the surrounding cells' states at a preceding time (t). The Moore (3×3) neighborhood is commonly used to determine which cells have to be considered in the evaluation (Fig. 1).

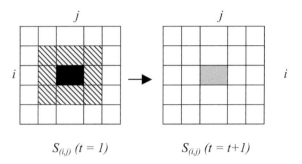

$S_{(i,j)}$ (t = 1) $S_{(i,j)}$ (t = t+1)

Fig. 1. Cell state change and Moore neighborhood

The quoted Game of Life (Gardner 1970) was one of the first applications to demonstrate the capability of the cellular automata to produce dynamic structures and patterns. The Game is carried out in a bi-dimensional matrix, with binary cell states and using the Moore neighborhood. In practice, value 1 represents a cell that is "alive" and value 0 represents a "dead" cell. It was John H. Conway who introduced the following set of rules:

– A cell that is dead at time t becomes alive at a time $t+1$ if exactly three of the eight cells in his neighborhood were alive at time t.
– A cell alive at time t dies in a time $t+1$ if his neighborhood at time t consists of less than two or more than three live cells.

In simulation models based on urban cellular automata (e.g., Sanders 1996), the urban settlements and the surrounding territories are considered a self-organizing system in which the base elements, cells representing portions of land, experience changes of state, i.e., of land-use and land-cover changes (Sui and Zeng 2001). Land-use conversions are governed by transition rules that define the behavior of the single cells. In this way, a global pattern can evolve from the definition of local behavior according to the interactions between a cell and its neighborhood (local interactions lead to global dynamics).

This local, bottom-up approach that characterizes cellular automata models well fits the actual political trends of the P.R. China and to our case study. China is going through a deep transition from a centralized to an open market economy (Wang and Zhou 1999). From the perspective of urban development, this implies a liberalization of the land market and the devolution of the decisional prerogatives from the central government to local administrations.

Acknowledging the simulation capability of urban models (visualization, bottom-up dynamics, self-organizing processes), their calibration and validation are still matters of discussion. Many different approaches have been undertaken in recent years: from basically heuristic models (Besussi et al. 1998) to models based on logistic regression (Wu 2002). Nevertheless, it appears convenient to derive or calibrate land-use development from the observation of occurred changes in land use. These land-use change data can be effectively supplied by remote-sensing imagery through GIS analysis.

We are thus facing the need to express the relationships between land use and the factors that influence the probability of land-use conversions. The number of required variables and their parameters rise considerably taking into account the fact that the development of an area is not based solely on physical attributes (soil types, altitude, etc.), but also on the development sequence and on the neighborhood effect. The growth of system complexity has brought about a new method based on the integration of cellular automata and neural networks that is able to handle the increased data complexity. In this innovative approach (Li and Yeh 2002), the parameters are calculated automatically during the learning process of a neural network, and the land-use change probabilities are represented by the network output values. Furthermore, the site attributes are automatically updated at the end of each cycle making the network operate in an iterative way.

28.3.2 Neural Networks

Neural networks are composed of simple elements that operate in parallel and are inspired by biological nervous systems (Demuth and Beale 2002). As in the biological case, the function of a network is largely determined by the connections between the elements. It is hence possible to train a neural network to perform the desired function by adjusting the values of those connections (weights) between the elements.

We can represent the various inputs to the network by the mathematical symbol, $x(n)$. Each of these inputs is multiplied by a connection weight. These weights can be represented by $w(n)$. In the simplest case, these products are simply summed, fed through a transfer function to generate an output result. Considering that weights are adjustable parameters, we can train the network to show the desired behavior by modifying them until the network functions properly. Typically, a training set containing numerous couples (input – desired value) is used to perform this kind of supervised training.

Sigmoid transfer functions are frequently adopted. These functions receive an argument between $\pm\infty$ and successively restrict their output in a $[0,1]$ range making them suitable to be used in combination with the back-propagation algorithm (Rumelhart et al. 1986) in feed-forward neural networks.

Using the quoted algorithm the learning phase begins by picking up the weight values randomly, and then proceeding to minimize the error function by adjusting the connection values (weights) backwardly from the output layer to the inside layers.

A single neuron certainly cannot do much. However, a considerable power is distinctive of neural networks obtained by combining many neurons in a layer or in many layers (Fig. 2). The different layers play different roles and an input layer, an output layer, and one or more hidden layers between them can be distinguished. It is notable that the output of a layer represents the input for the layer that follows.

There is no universal optimal structure for all applications. The principle is to use as few layers and neurons as possible, and three-layer networks have been commonly used because of their simplicity and effectiveness. Kolmogorov's theorem indicates that any continuous function $\Phi: X^n \rightarrow R^c$ can be implemented by a three-layer neural network that has n neurons in the input layer, $(2n+1)$ neurons in the single hidden layer, and c nodes in the output layer. However, experiments indicate that $2n/3$ hidden neurons can generate results of comparable accuracy requiring considerable less time to train (Wang 1994; Li and Yeh 2002).

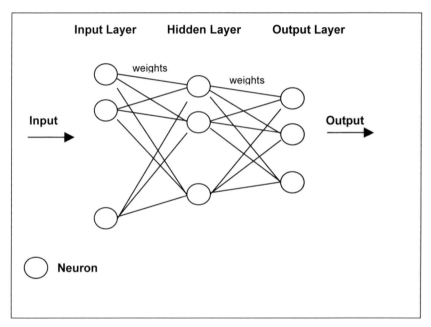

Fig. 2. Artificial neural network scheme

Properly trained back-propagated neural networks provide reasonable outputs when supplied with input data they never saw before. Such a characteristic makes them appropriate for urban simulation models and is the reason we chose them as the tool for our application.

28.3.3 The Simulation Model

To properly simulate land-use and land-cover change in the considered period, a three-layer neural network has been implemented. The neurons of the input layer receive the attributes describing the considered site (cell) as input. By associating a class of land use to each neuron of the output layer the outputs of those neurons will represent the land-use conversion probabilities for the corresponding class.

In order to perform the simulation two different stages have to be completed:

– Training stage – calculation of the network weights using the back-propagation algorithm and an appropriate training set (input – desired output couples)
– Simulation stage – calculation of the output probabilities and consequent land-use conversions.

The number of input neurons is established by the number of the considered spatial variables, e.g., by the number of attributes used to describe individual sites. These kinds of attributes can be conveniently computed by GIS analysis from the two classified satellite images. According to Li and Yeh (2002), the following variables are closely related to urban development and consequent land-use change:

- Distance variables:
  - distance from the cell to the nearest city center
  - distance from the cell to the nearest peripheral urban or rural area
  - distance from the cell to the nearest road
- Neighborhood variables
  - quantity of each land-use/cover type within a 7×7 window
- Existing land-use/cover.

Attributes for a proportional stratified sample of 3,000 sites have been collected. Then, the observed land-use changes that occurred in the 1987-1999 period have been encoded as follows:

- 1 – land-use change
- 0 – no land-use change

and successively embedded into the sample as the desired output values. Such a training set has been used in the training stage with a back-propagation algorithm to calculate the network weight values for the simulation stage. The calculation of the training stage has been accomplished using Matlab Neural Network ToolBox (The Mathworks 2002).

In the simulation stage the neural network has been implemented using the ArcInfo (Esri 2001) GIS and Arc Macro Language (AML) incorporating the previously calculated weights. The choice to work in the GIS environment guarantees a flexible environment and effective tools for map analysis, avoiding at the same time data import/export inconveniences.

The simulation model was executed by using the 1987 land use map as the initial grid. Each cell was characterized by ten attributes which have determined a land-use conversion when a threshold conversion probability was exceeded. In our case, the threshold value has been set to 0.9. Once the whole map has been processed, the new, simulated map can be used itself as an input to the neural network allowing the system to work in a cyclic manner until the defined constraints are satisfied.

The validation of cellular automata models is still matter of discussion. The plausibility of the results, the exact cell-to-cell correspondence, and the structural similarity between the observed and the simulated development have to be taken into account.

In our analysis three validation criteria were adopted:
1. Overall accuracy - the number of correctly predicted cells divided by the total number of cells
2. Comparison matrix between the simulated and the actual land uses in 1999
3. Change matrix by categories

28.4 Results

Land-use/land-cover maps for the years 1987 and 1999 were produced from the remotely sensed images using a maximum likelihood supervised classification (Fig. 3a, b). After that a conversion matrix was created trough GIS analysis capability (Table 1). The interpretation of the obtained data reveals that two separate dynamics took place in the Haikou County.
1. Massive urbanization both in proximity of urban agglomerates and along the communication network: Most of the urban development occurred on agricultural land and reclaimed coastal areas or lagoons. If the need for the conservation of those delicate areas appears clear from an environmental point of view, the preservation of agricultural potentials, in a highly populated country such as China, seems unavoidable.
2. The comprehensive conversion of exposed sandy soils in vegetated or agricultural areas in the eastern part of the county

A contradictory picture emerges from the first part of this study. On the one hand a vigorous and diffused urban development has taken place, but on the other the awareness of landscape management importance has partially balanced the loss of productive land and natural environment.

Table 1. Land-use and land-cover comparison matrix for Haikou County, 1987-1999 (in ha)

		1999						
		Paddy	Agriculture	Vegetation	Water	Urban	Sand	Total 1987
1987	Paddy	1080.72				1022.85		2103.57
	Agriculture		6004.71			2082.06		8086.77
	Vegetation			6111.63		663.66		6775.29
	Water	154.17	423.27	85.59	5214.33	455.85		6333.21
	Urban					2348.82		2348.82
	Sand		773.19	377.82	29.16	298.17		1478.34
	Total 1999	1234.89	7201.17	6575.04	5243.49	6871.41	0.00	27126.00
	Change (ha)	-868.68	-885.60	-200.25	-1089.72	4522.59	-1478.34	6365.79
	Change (%)	-41.30	-10.95	-2.96	-17.21	192.55	-100.00	23.47

Figure 3c shows the results of the land use change simulations for the period 1987-1999 by using the neural network based cellular automata model.

The percentages of the correct predictions for each class are shown in Table 2. Note that the value for sandy and water class are equal to zero. This is because the sandy area was completely converted to vegetation according to a landscape management plan and therefore there are no cells belonging to that class in 1999. Moreover, the majority of the humid environments, such as part of the lagoons, have been lost due to reclamation.

For the remaining classes, the best results were obtained for vegetation with 95.70% and for agriculture with 83.99% followed by paddy field with 63.67%.

Table 2. Confrontation matrix between the observed and the simulated land uses in 1999 (in percentage)

		Simulated						
		Paddy	Agriculture	Vegetation	Water	Urban	Sand	Total
	Paddy	**63.67**	4.20	0.00	8.43	23.70	0.00	**100.00**
	Agriculture	0.00	**83.99**	3.82	2.76	5.02	4.41	**100.00**
Obser-	Vegetation	0.00	1.09	**95.70**	0.75	1.25	1.22	**100.00**
ved								
1999	Water	0.00	0.00	0.00	0.00	0.00	0.00	**0.00**
	Urban	4.86	32.02	9.58	3.72	**48.08**	1.74	**100.00**
	Sand	0.00	0.00	0.00	0.00	0.00	0.00	**0.00**

The achieved 48% of correct prediction for the urban class is a relatively modest outcome. However, there are strong reasons for justifying such a result:
1. There was a lack of slope and soil maps that if available would improve simulation accuracy.
2. It was impossible to introduce the notable changes in the transportation network in the simulation and therefore their effects on the developing areas.
3. In this study, we were concentrated strictly on the Haikou province because of its socio-economical "vivacity." By doing so we omitted consideration of a relevant urban area very near but out of the county borders. This neglected presence has probably prevented the conversion of the adjacent large agricultural area to the urban class and thus negatively influenced the simulation performance.

Even considering that the main objective for simulation models is to capture global change dynamics rather than to obtain a cell-to-cell agreement, the achievement of a 73% overall accuracy has to be judged satisfactory.

Finally the change matrix in Table 3 confirms that the general patterns of the land use changes were captured by the simulation.

Table 3. Land-use changes by categories (in ha)

Land use	1987	1999	
		Observed	Simulated
Paddy field	2103.57	1234.89	1119.96
Agriculture	8086.77	7201.17	8371.71
Vegetation	6775.29	6575.04	7225.83
Water	1118.88	0	607.77
Urban	2348.82	6871.41	4040.1
Sand	1449.18	0	517.14
Total	21882.51	21882.51	21882.51

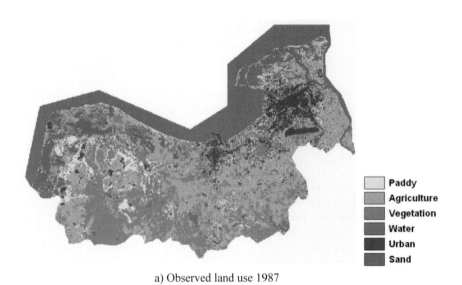

Paddy
Agriculture
Vegetation
Water
Urban
Sand

a) Observed land use 1987

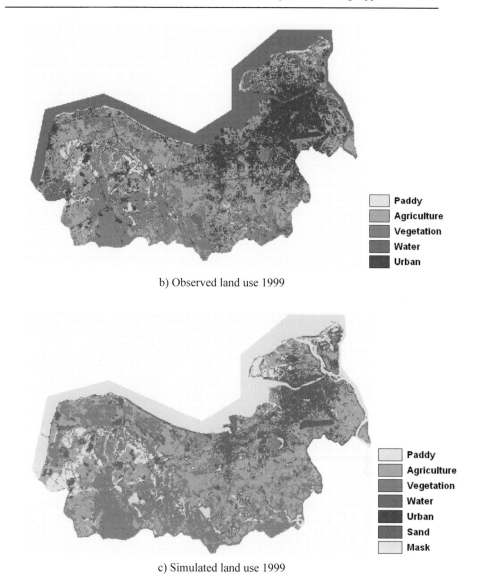

b) Observed land use 1999

c) Simulated land use 1999

Fig. 3. Haikou County, observed and simulated land-use/land-cover maps

28.5 Discussion and Conclusions

On the basis of the scores in Table 3, one can say that the model did not work properly, however, the results are very useful. Actually, the model can be used as a tool to simulate scenarios according the concept of business-as-usual (BAU)

since it produces simulated maps according to the acquired knowledge about land cover characteristics at time 1. Changes due to new events, such as the construction of a highway or policy incentives for building construction that can occur between the two periods of time are neglected by the model. As a consequence, the scores simulated by the model can be considered as describing a "neutral" model. The deviation from the simulated scenario is an indication that the events do not conform to the BAU scenario and therefore the user is forced to look for the reasons that have caused such deviation.

On Hainan Island the migration toward the urban area was very strong. The population of Haikou County increased about 70% in the last 10 years, but it is possible that the expectations of economic development were too high in respect to the real potentialities. The main effect of such expectations was an urban development out of the rules of the "self-organizing system" that represents the basis of the implemented model. According to the 1999 map, the urban area increased 192%, which is far more than the percentage of population growth of the county, while according to the simulated map the growth of urban areas would be of about 70%, a percentage similar to the county's growth. The difference between the simulated and actual scores is due to the fact that the model learned the rules of landscape pattern (derived by the spatial variables analysis) according to the map of 1987. The fact that the model foresaw correctly only 48% of the 1999 urban area suggests that the urban structures have been built following different rules from those learned by the neural network.

Whoever visits Haikou County can see that many buildings are incomplete and empty! Probably the policy incentives, and other factors such as new road construction, have produced an urbanization stronger than the one expected in a BAU scenario. We can conclude that the urban development is not fitting the actual socio-economic development. In fact, the socio-economic trends (Feoli 2002) prove that the economy of Hainan is still based on agro-industry. This being true, economic development should mainly be relate to the carrying capacity of the island in terms of local agricultural resources.

Acknowledgments. This study was funded by INCO-DC (International Cooperation with Developing Countries) Project, title: "Development of Landscape Management Strategies for Sustainability of Natural Resources (Biodiversity, Soil and Water). A Case Study for the Island of Hainan (China) Based on Remote Sensing and GIS Technology." Contract number: ERBIC18CT980283. This work was carried out with the work package "Integrated spatial tools for natural resources assessment and management" of the Center for Excellence in TeleGeomatics, GeoNetLab, University of Trieste, Italy.

28.6 References

Anonymous (1998) Statistical year book of Hainan. China Statistical Publishing House
Barrett GW, Farina A (2000) Integrating ecology and economics. Bioscience 50:311-312
Barrett GW, Odum EP (2000) The twenty-first century: the world at carrying capacity. Bioscience 50:363-368

Besussi E, Cecchini A, Rinaldi E (1998) The diffused city of the Italian north-east: identification of urban dynamics using cellular automata urban models. Comput Environ Urban Syst 22:497-523

Carley M, Christie I (1993) Managing sustainable development. University of Minnesota Press, Minneapolis

Chichilnisky G, Heal GM, Vercelli A (eds) (1998) Sustainability: dynamics and uncertainty. Kluwer, Dordrecht

Costanza R (1991) Ecological economics: the science and management of sustainability. University Press, New York

Di Castri F (2000) Ecology in a context of economic globalization. Biosciences 50:321-332

Demuth H, Beale M (2002) Neural network tollbox for use with MATLAB. The MathWorks Inc, Natic, MA

Erdas Inc (1999) Erdas imagine 8.4. Atlanta

Esri Inc (2001) ARC/INFO. Redlands, California

Feoli E (ed) (2002) Development of landscape management strategies for sustainability of natural resources (biodiversity, soil and water). A case study for the Hainan Island (China) based on remote sensing and GIS technology. Final report, Univ Trieste (available on CD)

Folke C (1992) Socio-economic dependence on the life supporting environment. In: Folke C, Kaberger T (eds) Linking the natural environment and economy: essays from the eco-eco group. Kluwer, Dordrecht, pp 77-94

Gardner M (1970) The fantastic combinations of John Conway's new solitaire game of "life". Sci Am 223:120-123

Hall CAS (2000) Quantifying sustainable development. The future of tropical economy. Academic Press, San Diego

Hogeweg M (2001) Spatio-temporal visualization and analysis: the need for integration. Geoinformatics 4:32-35

Jensen ME, Burgeon PS (eds) (2001) A guidebook for integrated ecological assessments. Springer, Berlin Heidelberg New York

Kronert R, Steinhardt U, Volk M (eds) (2001) Landscape balance and landscape assessment. Springer, Berlin Heidelberg New York

Lessard G, Jensen ME, Crespi M, Bourgeon PS (1999) A general framework for integrated ecological assessments. In: Cordell HK, Bergstrom JC (eds) Integrating social sciences with ecosystem management: human dimensions in assessment policy and management. Sagamore publishing, Champaign, IL, pp 35-60

Lester JP (1989) Environmental politics and policy. Duke University Press. Durham and London

Li X, Yeh AGO (2002) Neural-network-based cellular automata for simulating multiple land use changes using GIS. Int J Geograph Inform Sci 16:323-343

Liverman D, Moran EF, Rindfuss RR and Stern PC (1998) People and Pixels. Linking Remote Sensing and social science. National Research Council, Washington DC

Malczewski J (1999) GIS and multicriteria decision analysis. Wiley, New York

Moldan B, Billharz S, Matravers R (1997) Sustainability indicators: a report on the project on indicators of sustainable development. SCOPE 58. Wiley, Chichester

Ott T, Swiaczny F (2001) Time-integrative geographic information systems. Management and analysis of spatio-temporal data. Springer, Berlin Heidelberg New York

Pearce D, Moran D (1994) The economic value of biodiversity. IUCN, Earthscan Publ, London

Rumelhart DE, Hinton GE, Williams RJ (1986) Learning internal representations by error propagation. In: Rumelhart DE, McClelland JL (eds) Parallel data processing, vol 1. MIT Press, Cambridge, MA

Sanders L (1996) Dynamic modeling of urban systems. Spatial analytical perspectives on GIS. Taylor and Franci, London

Slocombe DS (2001) Integration of physical, biological and socio-economic information. In: Jensen ME, Burgeon PS (eds) A guidebook for integrated ecological assessments. Springer, Berlin Heidelberg New York, pp 119-132

Smith HM (1999) Understanding economics. Sharpe, London

Sui DZ, Zeng H (2001) Modelling the dynamics of landscape structure in Asia's emerging desakota regions: a case study in Shenzhen. Landscape Urban Planning 53:37-52

The Mathworks Inc (2002) Neural network toolbox for use with Matlab. Natic, Maryland, USA

Treweek J (1999) Ecological impact assessment. Blackwell Science Ltd, Oxford

Von Neumann J (1966) Theory of self-reproducing automata. University of Illinois Press, Champaign, Illinois

Wacknagel M, Yount JD (2001) Footprints for sustainability: the next steps. Environ Dev Sustain 2:21-42

Wang F (1994) The use of artificial neural networks in a geographical information system for agricultural land suitability assessment. Environ Planning 26:265-284

Wang FH, Zhou YX (1999) Modelling urban population densities in Beijing1982-90: sub-urbanisation and its causes. Urban Stud 36:271-287

Wu F (2002) Calibration of stochastic cellular automata: the application to rural-urban land conversions. Int J Geogr Inform Sci 16:795-818

29 Population Shifts and Migration in the Sahel and Sub-Sudan Zone of West Africa Reviewed from the Aspect of Human Carrying Capacity

Werner Fricke[1]

Ruprecht-Karls-University Heidelberg, Department of Geography, Im Neuenheimer Feld 348, 69120 Heidelberg, Germany, E-mail W.Fricke@urz.uni-hd.de

29.1 Abstract

The hardship triggered by the climatic change in the Sahel zone is an important factor in the dynamics of the Sahara syndrome, which led to the migration of its rural population to the more humid Sudan-savanna zone. Unfortunately, this migration was to the same sort of physically and socioeconomically fragile system and caused problems in the areas of immigration, and the dangerous dynamics of the biological reproduction process as well as the prevalent economic production process have not yet changed. The statement of Bogue hits the point : "Thus, migration is a process for preserving an existing system" (1969 p. 487). Any attempt to search for the apparent carrying capacity can be seen as a wasted effort, as modeling the carrying capacity by quantitative information does not seem to be possible yet. It is more promising to fight against the vulnerability of the poor population in countries where the average income is less than US$ 1 per person per day. There are some dominant physical and human factors of which only the socioeconomic constellation may possibly be altered directly and, through this, the

---

[1]I would like to express my gratitude for the acceptance of this topic in a biological conference as it reflects a modern concept of science, which overcomes traditional boundaries between disciplines. This has also happened in the interdisciplinary Joint Project 268 between the Universities of Maiduguri, Nigeria, and Frankfurt am Main, Germany: "The Development of Culture and Nature in the Savanna of West Africa" sponsored by the German Research Council from 1988 to 2002. The researchers belong to many disciplines including archeology, archeobotany, anthropology, African languages and history, botany, and physical and human geography, and covering the period from the Pleistocene to the present. I was able to enlist the following colleagues for the final report: the physical geographers, Günter Nagel, Frankfurt, Jürgen Heinrich, Jena, and the African historian, Rudolph Leger, Frankfurt, to deal with this question of the difficult, unbalanced path of society in Gombe State, Northeastern Nigeria. Furthermore, I would like to thank Margaret Friederich for checking the English version and my wife, Dr. Christa Mahn-Fricke, for critical comments.

climate indirectly. In the main they can only be identified on the spot as the geographic situation differs from village to village and region to region. The salient point is to look at the social roots of the vulnerability of the small farmers and their entitlement to the benefits of their local society. Governments, and not only in West Africa, should be alarmed by this vicious circle: the climatic change of increased dryness in the Sahel zone has led to out-migration to the neighboring more humid zones to the south; but that decrease in rainfall in the north has been caused in large part by the deforestation in the south as a result of the population increase there. The problem is embedded in two global processes: global climate developments and the socioeconomic makeup of the so-called *least developed countries*.

29.2 General Aspects of Biology and Human Migration in the Savanna of West Africa

Population geographers have to be aware of two environmental influences on man: the biological and the cultural. Migrations are similarly induced either by natural or by man-made factors, in many cases stress from both may force or stimulate people to migrate. There are numerous factors influencing the mobility of people, as Gould and Prothero (1975) have shown for Africa. Most reasons leading to migration definitely have an economic background. However, my task is to concentrate on the biological aspects, and in this particular case climatic change, especially drought, becomes the most important point of interest.

Migration reflects the challenge of regional differences, either physical or cultural, or both[2]. The physical differences in West Africa are – because of the high temperatures of the tropics in general – governed by the rainfall regime. The annual precipitation decreases from more than 3,000 mm at the coast of Liberia to a few millimetres gained by erratic rainfalls in the Sahara, as is reflected in the zonal arrangement of the isohyets. These have been used to delineate vegetation belts, indicating decreasing humidity in rain forests, deciduous rain forests, Guinea savanna, Sudan savanna and the Sahel zone, which are nowadays regarded as ecological zones.

It is well known that climatically caused seasonal differences in the natural carrying capacity of the vegetation are reasons for the traditional seasonal movements of nomadic pastoralists from the Sahel and the arid north to the more humid southern savanna of West Africa. Exceptional droughts in 1969 and 1970 under-

---

[2] I would like to point out that the vegetation formations correlating to isohyets are just the result of the attempt to make a geographic continuum feasible by the metric system, but experienced botanists in the field as well as climatologists doubt the eco-climate value of such zonation (Tuley 1972, p. 121; Nicholson et al. 2000, p. 2629). It becomes even more difficult if one takes the actual decreasing rainfall into consideration as did Mortimore and Adams (1999, p. 9), based on Koechlin (1997).

line the impact of climate on migration by inducing their permanent settlement in the south (Fricke 1993, maps 25a , 25b, 28, 29).

The recent and quite active rural-urban migration in Africa has often been induced by biogeographical factors -- e.g., drought, the decreased fertility of land caused by overfarming, or diminishing availability of farmland -- as well as by the poor living conditions in the villages, which demand hard work, in contrast to the expectation of an easier life in town. However, in most cases, recent population shifts in Africa are of the dominantly *rural-rural* kind – e.g., the move of agriculturalists to places with better farming conditions. In this situation one has to differentiate between movement of an already prosperous, very active agriculturalist or pastoralist from a good farming/grazing area to a more attractive place because of greater fertility or better marketing possibilities, etc., as opposed to flight of poor people with not enough land for farming or grazing from their crowded village. In the first case, the voluntary movement is governed by the *pull factor* and in the second case the *push factor* is dominant. However, in either case, the migration results from differences in living conditions between the present location and the favored area selected by the most active elements of the population.

The very high cohesiveness of the ethnically biased traditional societies in Africa may often cause tension between migrants and local people. Quite often a high population density in the place of origin correlates with the readiness for migration and a low density with attractiveness of places of destination. As land in Africa cannot usually be bought because it is inherited from the ancestors of the local tribe, bloody feuds between autochthons and immigrants are quite common and very often serve as barriers against those compensatory population shifts.

Especially after a natural disaster in an area, e.g., the drought in the Sahel zone 30 years ago, the question arises whether the number of inhabitants there has exceeded its assumed carrying capacity. This model of a given carrying capacity for human beings considers, on one hand, the impact of pure physical conditions, e.g., climate, relief, fertility of the soil, etc. In such an area they form, in combination with the biological resources, e.g., the vegetation, the physical geographical environment. However, on the other hand, the non-biological aspects of the human existence, based on economy and culture, e.g., social and historical structures, are of no less importance. "The temporal factor, inertia, may well be the most significant of any factor in shaping the size and location of a population," wrote Zelinsky (1970, p. 38). In the case of the Sahel, years of decreasing precipitation, caused by the change in the atmospheric circulation, climaxed in the drought and threatened the people with starvation and drove a great many out of their home areas. In the meantime "the agricultural and pastoral systems have proved surprisingly resilient in the Sahel, and their productivity has been sustained to a remarkable degree" (Mortimore and Adams 1999, p. 5). However, this may be a result of out-migration as this chapter will indicate, because quite a number of migrants followed the lure of land in more humid zones. Unfortunately, the adjacent savanna is just as vulnerable.

29.3 The Fragile Savanna Ecosystem

29.3.1 General Features

Since the end of the 1960s, newspapers and schoolbooks have informed us of the fragile ecosystem of the Sahel zone in West Africa, where many negative physical and cultural factors are dependent on and entwined with each other. However, unfortunately, the same can be said for parts of the adjacent Sudan-savanna zone where the actual problems are not yet as familiar, even to the scientific community, as those of the Sahel. The reason for this may be the very large regional differences in the dynamics of the population. Our area of study in Gombe State in northeastern Nigeria has more than 100 inhabitants/km² (Fricke 2001, p. 140). This is in contrast to other regions of West Africa, e.g., in Burkina Faso, where, under comparable physical and socioeconomic conditions, a population density of more than 40 persons/km² is rare and densities of less than 15 persons in rural areas are common (Braun 1996, p. 20). No wonder that under such conditions soil erosion is seen as only a minor problem (Mazzocato and Niemeijer 2000); I suggest that the high percentage of unfarmed surface prevents accelerated runoff from larger catchment areas that cause soil deterioration. However, in the regions in the Sudan savanna that have a low population density today, one still has to be vigilant as, unfortunately, this zone is interlinked, not only by the climatic dynamic of the tropical circulation, but also by actual migratory processes with the Sahel. Because of a strong out-migration to the Guinea coastal zone, especially to the Ivory Coast, the pressure of the population in the western part of the Sudan savanna zone in West Africa has been low.

29.3.2 The Importance of the Rainfall Regime

The sensitivity of the ecosystem of the West African savannas is a result of its location in the semi-arid periphery of the tropics. Because of year-long high temperatures the availability of water is the decisive factor for plant growth and human life. The precipitation is ruled by the seasonal shift of the Inter-Tropical Convergence Zone (ITCZ) in West Africa. This explains the brief advance north of the equatorial westerlies, maritime (moist tropical air mass) during the northern summer, and the dominance of a dry tropical air mass from the desert during the rest of the year which is dry and hot. In the case of Nigeria – the country of our special concern – only a fringe from the northeastern edge west of Lake Chad to the border zone of the Sokoto Statein the northwest has been classified as belonging to the Sahel, which has less than 20 in. (500 mm) of annual rainfall. The next zone, the Sudan savanna, with 20 to 40 in. (1,000 mm) has been seen until now as a transition zone from dry or thorn savanna to wooded savanna. Recently, Koechlin (1997) has added the zone with less than 550 mm and 8-10 months dry season to the Sahel, and the zone with less than 750 mm and 7-8 months dryness into the sub–Sahelian sector. Such an extension of the classification of the Sahel climate needs more discussion from the agricultural point of view, because agro-botanists had already carved out the study area around and south of Gombe which has more than 750 mm rain, as part of the sub-Sudanic zone or wooded savanna. Koechlin classified this area as the 'northern Sudanian sector.' Do these divergences in classifications take the recent climatic change into account?

Southward, the isohyets of 40 in. (1,000 mm) undisputedly demarcate the northern Guinea zone which has more than six humid months but is less humid than the southern Guinea zone with eight or more humid months. This seasonal differentiation of decreasing precipitation in relation to increasing latitude, which correlates in Nigeria to the distance from the coast (r=0.85) originates from the southwesterly equatorial, the so-called monsoonal, moist and cool air masses following the northern track of the ITCZ in summer. It is opposed in winter by the northeasterly dry Harmattan air mass from the Sahara, therefore, a reverse gradient of the correlation coefficient between the latitude and the annual amount of evaporation can be distinguished (Kowal and Knabe 1972, p. 27)

29.4 The Situation in Gombe State (NE Nigeria)

29.4.1 The Climatic Situation

The diagram in Fig. 1 shows representative data for the sub-Sudanic zone from the meteorological station Dadin Kowa east of Gombe on the Gongola River. It is a comprehensive representation of a number of relevant meteorological data after Walter and Lieth (Brunk 1998, p. 164). The station is situated 230 m above sea level, at a little more than 10° N lat., 850 km north of the coast. A complete sdataset only covers 16 years, from 1977 to 1993. The temperature scale is on the left side. The *line* in the middle shows the mean monthly temperatures with an average of 28.3°C (max. 31.9, min. 24.0°C). The upper and the lower *broken curves* represent the monthly maximum and minimum temperatures. The amount of precipitation is registered on the right and the *columns* in the middle represent the rainfall distribution from April to October. The annual average rainfall is 756 mm with a maximum of 956 and a minimum of 507 mm. The pan-evaporation of 2,626 mm is more than three times higher than the average precipitation and is only exceeded by rainfall in July, August, and September. The *dotted parts* symbolize the dry period which is in contrast to the *vertically hatched* 5.5 humid months. Some irregular rainfall events at the beginning and at the end of this latter period may give the impression of extending it to nearly 8 months.

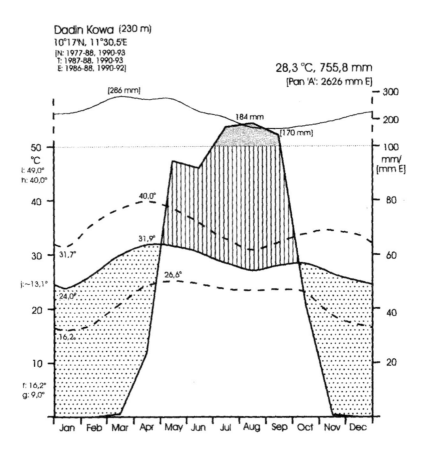

Fig. 1. Characteristics of the Sudanic climate (Brunk 1998)

29.4.2 Vulnerability Accelerated by Specific Climatic Elements

The vulnerability of the farmers in the savanna climate originates from a number of hazards. The variability of the rainfall has to be mentioned first. Neither the beginning nor the duration of the rainy periods is predictable for them, which likewise applies to the start and length of dry spells. Farmers have to be aware that they need to replant their fields after a long dry period, perhaps weeks after normal planting time. Another menace of the sub-savanna climate, the heaviness of the downpours is shown by the diagrams from Moldenhauer and Nagel (1998) in Balzerek et al. (2003, p. 103). The special erosive dynamics of the early rains were observed over a 5-year period on ten test plots in the Gombe hinterland and revealed that average values of precipitation do not explain the strength of the erosion process sufficiently. The erosion by splash (when the raindrop strikes the ground and sends soil particles into the air, making them ready for transport) is important. The initial transport of sandy sediment is correlated to a total precipitation of more than 10 mm in connection with rainfall intensities of more than 0.2 mm/min (Nyanganji 1997). About 70 % of the rainfall events met these threshold values, which occur especially at the beginning of the rainy season in April-May and reach a climax in August. "Hence, precipitation characteristics related to erosion contrasts with precipitation frequencies and amount" (Balzerek et al. 2003, p. 102). Diagrams of results from five stations in the region of Gombe show how often heavy rainfalls are to be expected: each second year 60 mm and each 5th year 100 mm/day (which means 100 l/m²!). Unfortunately, those peaks of erosion coincide with the period of planting when fields are uncovered. After harvesting, the remains of crops are transported to the farmyard, and stubble grazing by cattle, goats, and sheep follows, reducing the coverage of the soil. In preparation for the planting process, even the surface of the soil is usually loosened by hoe (or nowadays by plough), which means that the soil surface is fully open to erosive forces. Heavy rains can wash the young plants away easily. This affects the farmers in the same way as do the dry spells, because the peasants have to replant their fields, and this of course means that more seed and more labor are required, and thus their return per surface area and per labor unit is diminished.

29.5 Population Development and Climatic Change in NE Nigeria

29.5.1 The Pre-colonial Structure

Both the Sahel and Sudan zones in Nigeria, as seen from the point of view of human geography, have since medieval times had a feudal territorial structure. This has been based on grain farming (by men) in the so-called Hausa economy with a well-developed communication network between urban centres, and across the Sahara with the Mediterranean and Arabic cultures. The urban population has relied on a surplus from the rural areas produced by the bondsmen of the aristocracy. As land has been available, this system has required manpower, which was once recruited by slave raids of the mounted warriors in the neighboring fringes (with non-Moslem populations). In the South a great number of different acephalic tribes, without a traditional territorial organization, were able to defend themselves against the mounted slave hunters. Their defense was backed in many cases by a hilly relief and by the semi-humid ecosystems as the horsemen feared for their horses: the wooded savanna outside the settlement areas was infested by the tsetse fly (*Glossina* spp.), transmitter of the deadly sleeping sickness (trypanosomiasis).

The members of these tribes practised intensive hoe cultivation of tubers (only by the women) plus grain cultivation by men. Because of the difference to the territorially organized "Hausa economy", it has been named "Middle Belt Economy".

29.5.2 The Rapid Population Development

The population of northern Nigeria increased between 1952 and 1991, from 16.8 to 47.3 million. These 2 years not only mark the only two fairly reliable census figures in the country, but in 1952, at the dawn of independence, which finally took place in 1960, the British Labor government started to modernize their West African colonies. This included not only investing capital and attracting European trade firms and banks, but also sending young and highly motivated agricultural officers for the development of modern production techniques, e.g., introduction of the ox plough, fertilizer, and commercial long staple cotton seeds for export.

Concurrently, as a result of hygienic measurements, the gap between birth rate and death rate increased greatly, resulting in a rapid population increase from 30.3 to 88.5 million inhabitants in Nigeria, 2.7% per annum. In the NE, the number of inhabitants increased at the same rate as on the national level. In arid Borno State, bordering Lake Chad (Fig. 2), growth tripled from 2.94 to 8.3 million (i.e., +182.3%), which is even higher than the rate of 2.46% p.a., i.e., less than the whole of the north where it was 2.61%. In the Middle Belt the rate was up to 3.12%, and that of the adjacent Bauchi State reached 2.8% p.a. which is also above the national average (Fricke and Malchau 1994).

29.5.3 Uneven Population Development Indicating the Climatic Change

As documented by historical records, the last three generations of people in the northern part of NE Nigeria have suffered a dramatic change, as can be demonstrated by the meteorological records of Maiduguri (near Lake Chad). They show a 20% decrease in precipitation between 1905 and 1994 (Tschirske 1998, p. 63). The water table of Lake Chad sank 4 m between 1870 and 1972. Because of its extreme shallowness, the surface area of the lake decreased from 25,000 to 6,750 km² causing an expansive irrigation project to lie idle. However, this may have also been influenced by the damming of the lake's tributary rivers.

Decrease in rainfall can also be seen by comparing the isohyets for 400 and 600 mm between 1970 and 1991 and between 1949 and 1961. The isohyets shifted more than 100 km south (Fig. 2; Thiemeyer 1997, p. 15). This fits, according to the actual findings of Olanjiran (1999) and Nicholson et al. (2000), into the general picture of the climatic change in West Africa.

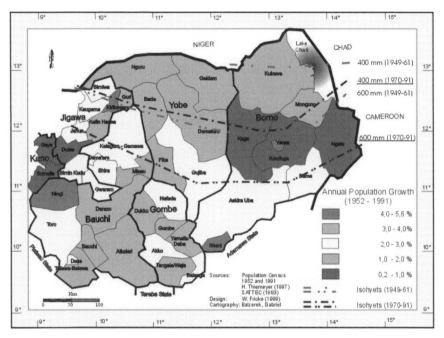

Fig. 2. Annual population growth 1952-1991 (Fricke 2001)

Unfortunately, this calamity to agriculture seems to be man-made. About 60% of the water content in the clouds in middle and northern Nigeria originates from evaporation of the humid rainforest and the adjacent semi-humid rainforest. However, during the last decade, 70% of these forest areas have been cleared.

By grouping local government areas (LGAs) according to annual population increase, it became evident that the increases lie much below the average rate of 2.61% for northern Nigeria. The rate of increase in the most northern LGA of Nguru was only 0.34% p.a. These figures include the births and deaths and in- and out-migration, of which the latter two are not registered. However, according to the findings through the fieldwork of my Nigerian colleagues Ijere and Ngadzama (1993) and the Britons Mortimore and Adams (1999), the figures have to be interpreted as population losses from out-migration to the south. During our investigations in the Gombe districts, we learned about a heavy influx of those northern immigrants; some migrated stepwise as laborers via urban centres in that area, some directly from the northern border areas as well as from the crowded central region of Kano. Evidently, others became attracted by the chance to farm land irrigated by the receding flood on the banks of the rivers near Hadjia and Katagum, now free of tsetse flies(Fricke 2001, p. 140).

By grouping the LGAs according to their density into four classes -- the highest density being above 200 persons/km² -- it becomes evident that the LGAs in the SW have reached this highest level. The LGAs in the Gombe State have an average above 100 inhabitants/km². In contrast, the districts in the NE have less than 30 people/km².

29.6 The Process of Environmental Destruction in the Sudan Savanna in Gombe

29.6.1 First Signs of Man-Made Destruction of the Environment

Geomorphologists and archeologists have been able to time the landscape history of this area by analyzing the landforms, the pediments, and the artefacts. According to their findings, a minor change of vegetation had already taken place 2,000 years ago, presumably caused by the beginning of shifting cultivation. Accelerated erosive activity became evident in the 14th century, coinciding with the rise of the Jukun Empire, which was able to conquer this region with its mounted warriors. These processes are evident through increased sheet erosion, which leads to the denudation of the slopes and the trapping of the sediments in the depressions.

A large change in land use has become visible as a result of accelerated erosion processes since the early 1950s, caused by the rapid expansion of farm land. At that time the British freed the plains around Gombe from the infestation of tsetse flies by application of DDT and introduced the production of American cotton and ox-plough farming. Gombe obtained access to the railway in 1962 and grew as an agricultural centre. This was the beginning of a population boom in that area.

Mimicking the attraction of an irrigable environment in Hadejia and Katagum to settlers from the north, Gombe became similarly attractive because of its fertile land (after the clearance from tsetse flies) and the introduction of ox traction onthe heavy fertile clay soils, which are difficult to cultivate by hoe.

29.6.2 The Process of Rapid Environmental Destruction in the Sudan Savanna of Gombe

Our team in the field registered many signs of destruction of the environment caused by its overuse through farming, grazing, firewood-cutting, etc. The reason for the deterioration lies in its indiscriminate use without any ecologically oriented considerations of its vulnerability.

As a result of the more humid climate, the increased runoff causes sheet as well as gully erosion. The introduction of ox-plough cultivation enlarged the length of the furrows as well as the erosive activity. During my first visit to Gombe division in 1961, signs of gully erosion were already visible. Detailed studies by Patrick (1987) and Brunk (1995), using aerial photography and satellite imagery since 1991, of the area around Gombe and Kaltungo, revealed the large impact of land deterioration by plough farming and firewood collection.

Mapping the land use change in the Tumu area, Fricke and Wolfbeiss (1997) demonstrated the difference between 1964 and 1991 in this area south of Gombe.

During this short span of time, the village of Tulmi grew from 13 taxpayers (approximately 60 persons) to 1,156 inhabitants. Large parts of the area were changed by bush clearance from savanna-bush land and even forest reserves to farmland. Only a small part of the forest reserve has remained and bush land occupies only the outcrops of the buttes of a sandstone plateau. The fertile plain areas with vertisols – the so-called black cotton soils – are under permanent cultivation. The farmers are now, despite the prejudice of the public, market oriented and instead of the original cotton, now produce maize and beans for the national market. They buy, besides cheaper foods such as cassava for their own supply, fertilizer, too. Because of the ecologically unadapted land use, soil erosion (sheet and gullies) are visible. However, after a fallow period, grasses, herbs, and a regrowth of trees appear. The new growth defines this process as different from the final stages of desertification. These observations lead us to a discussion of aridification (Brunk 1995; Heinrich 1995).

29.6.3 The Process of Aridification Fought by Terrace Farming

The term aridification was introduced by soil scientists and originates from the destruction of the vegetation cover. "…[T]he surface of fine-textured soils (with a high portion of silt, very fine sand, or clay) is extremely sensitive to the formation of crust or warp. The effect of such crust or warp is a decrease in infiltration and consequently an increase in runoff. Due to the runoff, there is less water available on high spots, and more in depressions" (De Ridder et al. 1982, p. 199). In the Sahel zone, this process has to be seen as an active contributor to desertification as plants and seeds do not get water even in the rainy season. According to our observation in the southern Sudan savanna zone, the aridification process does not produce an irreversible situation in all places. Only at sites with denuded rocks without any soil cover will there be no chance of rehabilitation. Because of a more humid climate than in the Sahel zone, we do not – together with other authors (Heinrich 1998; Brunk 2000, p. 9) – consider this an irreversible process like the desertification in the Sahel zone.

Indeed, where soil surface is still present careful conservation practice can heal the wounds. An excellent example is given by a number of tribes of acephalic hill dwellers in the "middle belt economy" who practise ecologically adapted terrace farming. The high degree of ecological adaptability is indicated by the use of fruits, leaves, and bark of quite a number of savanna trees.

29.7 Reducing the Danger to the Ecology in the Semi-Arid Zone of West Africa

29.7.1 Technical Measurements and Their Constraints

In West Africa the pressure of the fast-growing agricultural population on the natural resources became visible in Gombe State through the devastation of the natural vegetation. The accelerated denudation induced by farming, indiscriminate grazing, and firewood cutting led to aridification. The attempt to intensify farming by the extension of ploughed land in Gombe Statecaused many cases of most severe sheet and gully erosion. Ecological countermeasures such as reforestation of the watersheds, contour ploughing on slopes, and the application of manure as well as fertilizer would likely stop the deterioration process of the natural environment and stabilize the economy of the farmers. However, who will teach the agriculturalists, and how long will it be until this additional input to the farmers will be repaid by a higher output?

In this context, we have to remember that the British agricultural officers were not able to convince the farmers to integrate the newly introduced plough oxen into the farming enterprise, and, by keeping the oxen in a stable, to utilize their manure (Tiffen 1975). True, none of this is easy to manage in the savanna climate: cattle are generally left to graze on the commons and the farmer cannot therefore collect the manure. As a result of the strong seasonality of the vegetation period, natural grassland grows at the same time as sown *Gramineae,* and therefore the very important weeding of cereals, as well as the cutting of fodder plants for making hay , compete for the manpower of the peasants.

Furthermore, the conflict of choice of whether to plant foodstuff, cash crops, or fodder on the available plots results in the first two being chosen. The collection of fodder is considered to be profitable only by a few wealthy townspeople and settled agro-pastoralists for the fattening of bulls after the harvest until Christmas (Fricke 1993; Demirag 2002, p. 131).

Artificial fertilizer is expensive, and because of corruption, difficult to obtain for ordinary people.

The specific character of the tropical grasses creates additional problems because, as it grows in isolated tussocks, it does not prevent erosion of the surface soil in the same way as a sward grass in a temperate climate would (Fricke 1993, p. 273).

In summary, even the farmers of the "Hausa economy" know by tradition about the necessity of maintaining the soil fertility either by fallow periods or by the use of manure. In case of fallow intervals, additional farm land would be necessary, but this, because of the population pressure, is not available. The application of chemical fertilizer calls for additional money, which the small holder does not have. In the comparable historical case of German smallholders, population pressure was relieved by the fact that many could commute or migrate to factories or urban centres, or even emigrate. Unfortunately, because of the absence of industrialization in Africa, this release does not come into play. Formal education of females and better hygiene (public water supply) would reduce the birth rate. Electrification and eradication of corruption are other key words to be mentioned in this context.

29.7.2 The Search for Sustainability

Mortimore and Adams postulated four resource endowments of a Sahel household in Nigeria which can be extended, at least, to all the other zones with smallholder households:

"The natural resource endowment of a household is the productive capacity of the soil and other living resources to which members of the household enjoy the right of access both to work and enjoy the benefits."

"The economic resource endowment...includes the labour and capital available to it either from the family or through social or commercial means of approaching it from outside...."

"The technical resource endowment which means availability or access to technologies which have practical importance to the households in a wide range from soil management, weed management, harvest management, to irrigation etc."

"The social resource endowment includes the age, gender, health and skills of household members, their availability for economic productive work, ...and the place of the households in the social community, which in turn determines its access to sharing institutions...." (Mortimore and Adams 1999, p. 17)

Their painstaking study of four Sahel villages showed a great variety of activities in the range of endowments mentioned: intensification, agricultural diversification, additional craft, and seasonal long-distance commuting. Perennial out-migration has to be added, which gives not only relief to the local resources but normally a steady cash flow from migrant farmers abroad. However, it has to be added that the different endowments have to be reflected through the mirror of social stratification, which differs according the historical-geographic formation from place to place (Fricke 1993, p. 51) -- a diversity verified by Demirag (2002) for the agro-pastoralists in Adamawa and Gombe States.

Even with the seeming wealth of resources, carrying capacity is not easy to assess, either for a village area or a region such as Gombe or the Sahel or Sudan savanna zone. This shall be demonstrated by reviewing some earlier attempts. Ironically (and not only in traditional agricultural systems), the limit of the carrying capacity (CC) of an area is not registered until the CC has already been reached, and erosion is evident (Street 1969).

29.8 Review of the Carrying Capacity Paradigm

29.8.1 General Description of the Model

Obviously, in the case of the Sahel zone, the out-migration was, among other things, a presupposition of the recovery of the traditional farming system there. From the viewpoint of a carrying capacity model (CCM), migration is a typical element or indicator of an individual adjustment to the equilibrium between a farming population and the endowments available. In contrast, the migrational overflow from the Sahel has caused the maximum ecological carrying capacity of the area of destination (Gombe) to be reached.

Most recently, G. Richard (2002), in a manner very similar to Allan (1965), has defined the carrying capacity of an ecosystem as the "maximum population size of a species that an area can support without reducing its ability to support the same species in future". In fact, such an idea concerning the use of a CCM is, according to Bähr et al. (1992, p. 117), some thousands of years old and still going on as the popularity of Meadows (1972) "Limits of Growth" and the "2. Report of the Club of Rome" (Mesarovíc and Pestel 1974), already a generation ago, have demonstrated.

Even today in the most populous state – China – with more than 1,260 billion inhabitants on 9,6 million km² an official calculation of CC has taken place. This is based on a projected demand of grain and grain yield per km² for each province in 2025. "This capacity will depend on a nation's land resources as well as on technological and socio-economic factors" (Institute of Geography 2000, p. 239). By taking into account the socioeconomic factors and the technology, which may increase in the case of China, the demand on supply, as well as the productivity of the resources, is clearly shown. The calculated carrying capacity is not postulated as absolute but as a relative threshold only. This idea is based on the formula that Ackerman published in 1959 in which he tried to reconnoitre the factors among the divisors and numerators in more detail than anybody before.

$$P = \frac{RQ(TAS_t) + E_s + T_r \pm F - W}{S}$$

The symbols have the following meanings:

P —number of people
S —standard of living
R —amount of resources
Q —factor for natural quality of resources
T —physical technology factor
A —administrative techniques factor
S_t—resource stability factor
W —frugality element (wastage or intensity of use)
F —institutional advantage and "friction" loss element consequent upon institutional characteristics of society
E_s—scale economies element (size of territory, etc.)
T_r—resources added in trade

Fig. 3. Ackermann formula of carrying capacity

It is significant that Ackerman put the standard of living as the decisive factor into the denominator, and not only the resources 'R' but also their stability 'St', and 'F' for institutional friction, into the numerators. As early as 1970 Zelinsky voiced criticism in one of the first modern books published on population geography: "It would be extremely difficult, though not impossible, to quantify some of the factors of this equation, and equally difficult to gather all the basic information for a given area and period" (1969, p. 38). However, Ackerman (1969, p. 622) had already commented that: "Nonetheless, the general composition of the relation involved in the resource-adequacy question would seem to be described in the equation." He also mentioned the difficulty in assessing the amount of resources demanded by the standard of living in a stratified society. Certain cultural features affect the supply side as well as the dynamic or "feedback" elements.

29.8.2 A Few Special Examples

Another simplified model was used by Allan (1965), who found a high correlation between an increasing population density and a decreasing proportion of fallow in different traditional land cultivation systems in Africa. This observation fits into the theory of Boserup (1965), who saw in the curtailment of fallow in Asia and Africa not only a change of the agricultural system but a reflection of the society. She regarded population growth as the independent variable and agricultural change as the dependent variable. Her argument is an antithesis to Malthus, contending that adequate food production depends on technical input. However, in our example of the labor-intensive hill farming in Tula, there is a high population density but not the accompanying social progress. The high input of labor allows an exceptional population density because of a very high food subsistence. but unfortunately the intensive land use limits the market participation as there is neither space nor labor force available for additional products. Furthermore, the small terraces as well as the lack of capital prohibit the employment of innovative technology (Bergdolt and Demirag 1996). This observation confirms the critical arguments of Datoo (1978) as opposed to Boserup's hypothesis that an intensive agricultural system leads to a higher carrying capacity (i.e.., population density) as this will only be the case in non-stratified societies where no ruling class takes its tribute.

In 1973, Borcherdt and Mahnke attempted to solve this problem, as did Mortimore and Adams in 1999, by focusing on the individual farming enterprise. The advantage of their approach is that the focal point of agricultural production is the individual farm enterprise. The influence of manifold social factors, from its size and ownership to capital and innovations, have to be taken into account. Furthermore, one has to consider the influence of the government as well as the organization of the market. The resource-determined factors start with physical environment and end with the proportion of arable land.

Of course, all these manifold factors have to be considered in the context of estimating the carrying capacity of an given area, but it is illusory if one considers the grass roots work necessary. The result of Borcherdt and Mahnke's analysis is a purely agricultural carrying capacity, which may be applied to very different scales from a local subsistence to a world market-integrated economy. However, the connection with the latter can be very dangerous as in the case of Mali, where the farmers lost a main source of their income through the drop in the world market prices for groundnut and cotton. As the national market for foodstuff in Mali is small because of its small population, a possible change in agriculture to produce for the local market had no chance at all. In contrast, the small-scale farmers in a large national market, i.e., Nigeria, used their chance to change production of cash crops (e.g., cotton and groundnuts) for the world market to foodstuffs for the national market, especially for the fast-growing urban sector (Malchau 2001). In addition to this regionally maintained carrying capacity, Nigeria's international trade balance is kept up by exporting oil instead of having to import industrial materials in order to establish a national manufacturing sector sufficiently large to make the exportation of its products feasible.

The support from out-migrants who contribute to their families income has already been mentioned. Malchau (1998, p. 5) called this an economic carrying capacity.

Another example of an economically extended carrying capacity in a desert climate is demonstrated by the existence of the metropolitan area of Las Vegas in the USA, where 1.5 million people generate a profitable existence in a desert area by means of a national gambling and entertainment industry. They also physically tap the huge water catchment area of the Colorado River, bundled by the Hoover Dam of Lake Mead. However, the overuse of the groundwater resource became obvious when the town center sank by 2 m (Glaser and Schenk 1995). This is an example of overstepping an *artificially extended physical carrying capacity*.

29.9 Dwindling Resources: Part of a Syndrome of Local, Regional, and Global Environmental Deterioration

The specific local problems of desertification and aridification in the Sahel zone and Sudan savanna through soil erosion and crusts have already been traced as a result of the colonial impact on the African society: e.g., simple hygienic measures caused the exponential population growth which induced a rapid demand for more farmland *and* for the production of cash crops for the world market. Farmers had to change from the traditional fallow system to permanent use of the land, which led to its desiccation. However, the decrease in precipitation in the Sahel and Sudan savanna is a result of a disturbance of the atmospheric circulation south of them, to wit, by the devastation the zonal natural vegetation in the Guinea and coastal zone of West Africa. The resulting increase in atmospheric dust prevents the rain-producing monsoonal winds from moving north.

As more than 80% of the rainfall is caused by these monsoonal air masses from the Atlantic, which pass on their way from the coast northward in a number of circuits of downpours and evapo-transpiration processes, the fast runoff on the denuded soil surface diminishes the amount of water available to atmospheric circulation (Zheng and Eltahir 1998).

The bush fires of the local farmers, hunters, and pastoralists may additionally prevent rainfall by the chemical reaction of water vapor with carbon particles in the clouds.

A number of scientists have favored a model in which the expanding drought belt in West Africa is caused by effects of the huge energy and chemical output of the industrial world, especially Europe. Others try to find a correlation with El Niño. Schellnhuber (1999) called the climatic change the "Sahel syndrome," which can only be analyzed by interdisciplinary research, and which must include the whole of West Africa. Schellenhuber and his team tackled not only the interdependence of the drought with the biosphere, atmosphere, hydrosphere, and pedosphere, but also the vicious circle of the economy in its concerns with national debt, globalization, emancipation of women, transfer of technologies, etc.

As most of the biogeographic deficits under the label of fading carrying capacity are caused by unfavorable socioeconomic processes, international policy has changed from fruitless academic discussion concerning overpopulation, to the strategic policy of fighting the vulnerability of people, as seen, e.g., in "Water and Sanitation in the World's Cities," UN-HABITAT (2003) and "IFAD: In partnerships for rural poverty reduction" (2001).

Practitioners who do fieldwork in the Sahel prefer a simple model to demarcate the main obstacles to be solved, such as that of Mortimore and Adams (1999), quoted above. Constraints are not only natural factors such as rainfall and bioproductivity, but also human factors such as the availability of labor and capital to implement the actions necessary. Still, in the center stands the single farmer and his individual property -- and his ability to react flexibly to the different challenges confronting him. It is at least a system that enables him to earn a livelihood. Nowadays, it is recommended that vulnerability be overcome by mobilizing local resources, e.g., the social capital (i.e., the position of the single farmer or husbandmen in the social hierarchy of his area). This enables him, in a variety of ways, to gain access to natural capital, e.g., land, water, etc.

29.10 References

Ackerman EA (1969) Population and natural resources. In: Hauser PM, Duncan OD (eds) The study of population. An inventory and appraisal. University of Chicago Press, Chicago, pp 621–648

Akobundu IA (1991) Integrated weed management for striga control in cropping systems in Africa. In: Kim S (ed) Combating striga in Africa. IITA, Ibadan, pp 122–125

Allan W (1965) The African husbandman. Oliver and Boyd, Edinburgh, London, 503 pp

Bähr J, Jentsch C, Kuls W (1992) Bevölkerungsgeographie. De Gruyter, Berlin XIV, 1158 pp

Balzerek H, Fricke W, Heinrich J, Nagel G, Rosenberger M (2003) Man-made flood disaster in the savanna town of Gombe (NE-Nigeria). Erdkunde 57:94–109

Bergdolt A, Demirag U (1996) Changes in settlement pattern and culture – the process of down-hill migration in Tula, Bauchi State. Berichte des Sonderforschungsbereiches

268.8 „Kulturentwicklung und Sprachgeschichte im Naturraum Westafrikanischen Savanne", Frankfurt aM, pp 129–135

Bogue DJ (1969) Internal migration. In: Hauser P, Duncan OD (eds) The study of population. University of Chicago Press, Chicago, pp 486–543

Borcherdt C, Mahnke HP (1973) Die "Tragfähigkeit der Erde" als wissenschaftliches Problem. In: Borcherdt C (ed) Geographische Untersuchungen in Venezuela. Stuttgarter Geographische Studien, Bd 85, pp 7-27

Boserup E (1965) The conditions of agricultural growth. The economics of agrarian change under population pressure. Allan and Unwin, London

Braun M (1996) Subsistenzsicherung und Marktproduktion. Eine agrargeographische Untersuchung der kleinbäuerlichen Produktionsstrategien in der Provinz de la Comoé, Burkina Faso. In: Barsch D, Gebhardt H, Meusburger P, Heidelberger (eds), Geographische Arbeiten, Heft 108, Heidelberg

Brunk K (1995) Zum Landnutzungswandel in der südlichen Sudanzone am Beispiel des Bauchi State (NE-Nigeria). Berichte des Sonderforschungsbereichs 268/5 „Kulturentwicklung und Sprachgeschichte im Naturraum Westafrikanischen Savanne", Frankfurt a.M., pp 51–57

Brunk K (1998) Klima im südlichen Gongola-Becken (NE-Nigeria) – Agrarmeteorologische Daten der Station Dadin Kowa. Berichte des Sonderforschungsbereichs 268/10 „Kulturentwicklung und Sprachgeschichte im Naturraum Westafrikanischen Savanne", Frankfurt a.M., pp 141–183

Brunk K (2000) Formen und Ursachen der geoökologischen Degradation in der südlichen Trockensavanne Nordost-Nigerias. Frankfurter Geowiss Arb 26:9-32

Datoo RA (1978) Toward a reformulation on Boserup´s theory of agricultural change. Econ Geogr 54:135–144

Demirag U (2002) Soziale Räume – Handlungsräume. Eine vergleichende Studie über Fulbe in ländlichen Regionen der Savanne Nordostnigerias (Adamawa und Gombe State). PhD Thesis, Naturwissenschaftlich-Mathematische Gesamtfakultät, University of Heidelberg

De Ridder N, Stroosnijder L, Cisse AM, van Keulen H (1982) Productivity of sahelian rangelands. A study of the soils, the vegetations and the exploitation of that natural resource. PPS course book I. Agricultural University Wageningen, Department of Soil Science and Plant Nutrition

Fricke W (1993) Cattle husbandry in northern Nigeria. A study of its ecological conditions and social geographical differentiations, 2nd edn. In: Barsch D, Fricke W (eds) Heidelberger Geographische Arbeiten, Bd 52, Heidelberg

Fricke W (2001) Factors governing the regional population development in NE-Nigeria. The effects of physical and economic processes since 1952. Berichte des Sonderforschungbereichs 268/14 „Kulturentwicklung und Sprachgeschichte im Naturraum Westafrikanischen Savanne", Frankfurt a.M., pp 133–151

Fricke W, Malchau G (1994)Die Volkszählung in Nigeria. Geographische Aspekte eines politischen Pokers. Z Wirtschaftsgeogr 38:163-178

Fricke W, Wolfbeiss A (1997) Monitoring land use change caused by population development in the Nigerian Savanna from the 1960s till the 1990s. In: Daura MM (ed) Issues in environmental monitoring in Nigeria. Nigerian Geographical Association, Maiduguri, pp 47-54

Fricke W, Heinrich J, Nagel G, Leger R (2003) Natur und Gesellschaft in der Savanne Westafrikas in der Entwicklung zu einem dynamischen Ungleichgewicht (am Beispiel des südlichen Gombe Staates/NE-Nigeria. Berichte des Sonderforschungbereichs 268 „Kulturentwicklung und Sprachgeschichte im Naturraum Westafrikanischen Savanne", Frankfurt a.M. (in press)

Glaser R, Schenck W (1995) Glücksspiel als Entwicklungsfaktor in der Stadtregion Las Vegas. Geogr Rundsch 47:457–463

Gould WTS, Prothero RM (1975) Space and time in African population mobility. In: Kosinski LA, Prothero RM (eds) People on the move. Methuen, London, pp 23–49

Heinrich J (1995) Bodengeographische und geomorphologische Untersuchungen zur Landschaftsgenese und aktuellen Geomorphodynamik in der Trockensavanne Nordost-Nigerias. Habilitationsschrift Geowissenschaften, Universität Frankfurt am Main

Heinrich J (1998) Formen und Folgen der jungholozänen Bodenzerstörung in Trockenlandschaften Nordostnigerias. Petermanns Geogr Mitt (Gotha) 142:355-366

IFAD (2001) Federal Republic of Nigeria country strategic opportunities paper (COSOP). A framework for partnership for rural poverty – reduction between Nigeria and IFAD. Executive Board, 22. Session, Agenda Item 12, 16 pp

Ijere JA, Ngadzama NM (1993) Migration as response to environmental push and pull factors. A case study of northern Borno and Yobe States. Berichte des Sonderforschungsbereiches 268/2 Berichte des Sonderforschungbereichs 268/14. „Kulturentwicklung und Sprachgeschichte im Naturraum Westafrikanischen Savanne", Frankfurt a.M., pp 145-160

Institute of Geography (2000) The atlas of population and sustainable development of China. Chinese academy of sciences etc. Science Press, Beijing

Koechlin J (1997) Ecological conditions and degradation in the Sahel. In: Raynaut C (ed) Societies and nature in the Sahel. Routledge, London, pp 12-36

Kowal JM, Knabe DT (1972) An agroclimatological atlas of the Northern States of Nigeria. Samaru, Zaria, Ahmadu Bello University

Lagoke STO, Parkinson V, Agunbiade RM (1991) Parasitic weeds and control methods in Africa. In: Kim S (ed) Combating striga in Africa. IITA, Ibadan, pp 3–14

Malchau G (1998) Einkommensstrukturen kleinbäuerlicher Haushalte und gesamtwirtschaftlicher Strukturwandel in Südost-Nigeria. Untersuchungen im Rahmen eines erweiterten Tragfähigkeitskonzeptes im dichtbesiedelten Hinterland von Uyo. Arbeiten aus dem Institut für Afrika-Kunde 98. Institut für Afrikakunde, Hamburg

Malchau G (2001) Cultivation and marketing of farm products in the hinterland of Gombe. Berichte des Sonderforschungsbereiches 268/14 Berichte des Sonderforschungbereichs 268/14. „Kulturentwicklung und Sprachgeschichte im Naturraum Westafrikanischen Savanne", Frankfurt a.M., pp 153-159

Mazzocato V, Niemeijer D (2000) Rethinking soil and water conservation in a changing society. A case study in eastern Burkina Faso. Wageningen University, Tropical Research Management Papers 32

Meadows D (1972) Die Grenzen des Wachstums. Bericht des Club of Rome zur Lage der Menschheit.. Deutsche Verlagsanstalt, Stuttgart

Mesarovíc M, Pestel E (1974) Menschheit am Wendepunkt, 2. Bericht an den Club of Rome zur Weltlage. Deutsche Verlagsanstalt, Stuttgart

Moldenhauer KM, Nagel G (1998) Untersuchungen zur Niederschlagscharakteristik in der Sudanzone NE-Nigerias und ihre Bedeutung für die Bodenerosion. Berichte des Sonderforschungsbereiches 268/10 Berichte des Sonderforschungbereichs 268/14. „Kulturentwicklung und Sprachgeschichte im Naturraum Westafrikanischen Savanne", Frankfurt a.M., pp 269–285

Mortimore M, Adams WA (1999) Working the Sahel. Environment and Society in northern Nigeria. Routledge, London

Nicholson SE, Some B, Kone B (2000) An analysis of recent rainfall conditions in West Africa, including the rainy season 1997 El Niño and the La Niña years. J Climate 13:2628-2640

Nyanganji JK (1997) Some basic information on rainfall and runoff related factors on an artificial catchment in Maiduguri (NE-Nigeria). In: Daura MM (ed) Issues in environmental monitoring in Nigeria. Nigerian Geographical Association, Maiduguri, pp 152–165

Olanjiran OJ (1999): Evidence of climatic change in Nigeria based on annual series of rainfall of different daily amounts, 1919-1985. Climatic Change 19:319-341

Patrick S (1987) Gully erosion in Gongola and Bauchi States, Nigeria. PhD Thesis, University of London

Richard G (2002) Human carrying capacity of earth. ILES Leaf, Seattle

Schellnhuber HJ (1998) Syndromes of global change: an integrated analysis of environment and development issues. In: Kochendörfer-Lucius G, Pleskovic B (ed) Development issues in the 21 century. Villa Borsig Workshop Series 1998. Deutsche Stiftung für internationale Entwicklung, Berlin, pp 66-73

Street JM (1969) An evaluation of the concept carrying capacity. Prof Geogr 21:104-107

Thiemeyer H (1997) Untersuchungen zur spätpleistozänen Landschaftsentwicklung im südwestlichen Tschadbecken (NE Nigeria). In: Flügel WA, Mäusbacher R, Menz G, Schramke W, Sedlack P (eds) Jenaer Geographische Schriften 5, Jena

Tiffen M (1975) The enterprising peasant. Economic Development in the Gombe Emirate, North Eastern State, Nigeria, 1900-1968. Overseas research publications no 21. Ministry of Oversea Development, London

Tuley P (ed) (1972) The land resource of NE–Nigeria 5. Land Resources Division, Tolworth Tower, Surbiton, Surrey, England, pp 1–283

Tschirschke K (1998) Statistische Analyse und Interpretation langjähriger Niederschlags- und Temperaturdaten von Klimastationen im Tschadseegebiet. Berichte des Sonderforschungsbereiches 268/10 Berichte des Sonderforschungbereichs 268/14. „Kulturentwicklung und Sprachgeschichte im Naturraum Westafrikanischen Savanne", Frankfurt a.M., pp 11–140

Zelinsky W (1970) A prologue to population geography. University of Chicago Press, Chicago

Zheng X, Eltahir EAB (1998) The role of vegetation in the dynamics of West African monsoons. J Climate 11:2078-2095

30 Desertification and Human Migration

Béatrice Knerr

Dept. of Development Economics and Agricultural Policy, Faculty 11, University of Kassel, Steinstrasse 19, 37213 Witzenhausen, Germany, E-mail knerr@.uni-kassel.de

30.1 Abstract

This paper approaches the question how far out-migration from regions affected by desertification contributes to reducing the pressures on these lands, and hence to reducing desertification. For that purpose different forms of migration are analyzed and exemplified by case studies. The investigation concentrates on case studies representing often-observed conditions and migration patterns, and highlights the desertification-migration dynamics.

Migration of people in reaction to the changing environment is decisively shaped by social and political conditions in the places of origin as well as in the potential places of in-migration. Hence, such movements may generally be less predictable than that of other species. Desertification is not often sufficient reason for out-migration, and where it takes place, it is not necessarily a relief for the region of out-migration. As adaptations depend on a wide spectrum of social, economic, technical, and ecological factors, adaptations to changes in the natural environment may be even self-defeating. Increasingly, temporary forms of migration dominate where the family/household remains at the original place of living and is supported by remittances from the migrant(s). Farming, which otherwise would not be economically sustainable, is usually carried on in such agriculturally marginal areas, subsidized by remittances. This leads to further resource mining. Farm land is usually only abandoned when it is damaged beyond recovery and livelihood no longer secure, or when strategies of migration are so profitable that farming is no longer considered worthwhile. Policies to cope with emerging downward spirals should concentrate on the formation of and accessibility to human, physical, social, and financial capital to make labor more productive, provide income alternatives, and reduce the extensive use of free-access natural resources.

30.2 Introduction

Desertification[1] is one of the most serious environmental and economic problems the world has to face at the beginning of the 21st century. Implying an irreversible decline of the soil's biological potential to produce food for human consumption, it affects a total of around 3.6 billion ha of land, and about 70% of the world's dry land. Most of this land is in regions where the poorest live. More than 250 million people in over 110 countries are directly affected, and about 1 billion are at risk (UNCCD 2000). Although desertification is a global phenomenon which hits the poor as well as the rich countries, in the southern as well as in the northern hemisphere, it is the most severe in the poorest countries. Desertification is particularly destructive in the dry lands of South America, Asia, and Africa. For these three areas combined, 18.5% of productive lands are severely desertified (Canadian International Development Agency 1994). Among these areas, desertification has the most devastating effects in Africa, of which two thirds are deserts and dry lands, mainly in the Sudano-Sahelian zones, and to a lesser extent in some countries south of these. About 73% of Africa's dry lands are degraded (Canadian International Development Agency 1994; UNCCD 2000). The region is frequently affected by droughts, the population suffers from poverty, and for its survival is extremely dependent on natural resources.

Desertification is both a result of and a threat to human activities. Although climatic variations play an important role in the process, human activities are the main factors triggering and promoting erosion of vulnerable land by poor water management, soil mining, deforestation, unsuitable agricultural practices, and overstocking (United Nations Food and Agricultural Organization 2003).

Arid and semi-arid regions[2] hence have become the most endangered living spaces, and worldwide they are subject to growing population pressure. They are mostly in countries with population growth rates of around 1.5% p.a. (see World Bank 2002). Within North Africa (Sahara, Sahel), e.g., the population has doubled over the last three decades. Some particularly affected countries such as Jordan, Iraq, Morocco, and Iran display population growth rates of 2.2% and more, which are among the highest in the world (UNDP 2003).

Under such conditions, people living from agriculture and confronted with declining productivity of their natural environment look for ways to sustain their livelihood. When adaptations such as introducing new farming technologies, gaining off-farm income or simply living from external aid fail, they move to other places[3]. Experts predict that by the year 2020 about 100 million people will have

---

[1]Desertification is defined as "land degradation in arid, semi-arid, and sub-humid areas resulting from various factors, including climatic variations and human activities" (UNCCD 1995).

[2]In keeping with the World Resources Institute's definition, arid regions are defined here as those with an average precipitation of less than 200 mm p.a. Semi-arid regions are defined as those with a precipitation of 200 to 400 mm p.a. (World Resources Institute 1994).

[3]Often the word "environmental flight" is used for the movement of people out of a region where environmental conditions have deteriorated. The author rejects using that word be-

left their homes as, due to soil destruction, they will not be able to survive in their customary places of living (Forum Umwelt und Entwicklung 1999). Many others, although not yet on the brink of survival, decide to look for alternative places of living because they see their labor productivity increasingly decline.

Since the 1980s, this development has been raising increasing concern in the international community. After a period in which the natural environment and the ecological aspects of the phenomenon had been the major focus of interest, since the 1990s the relationship between desertification and population growth has aroused increasing attention and has alarmed politicians and scientists alike. The human consequences of desertification have become of concern to the international community as they do not remain the problem of only the people immediately hit by it, but through social unrest and interregional movements affect other places, too.

Out-migration from desertifying regions is often considered as a relief, as the pressure on the degraded land is thus supposed to decline. This idea is based on the conventional neoclassical theory of migration that movements of people, as they follow an (actual or perceived) income gap, are directed to places where their standard of living, as determined by their labor productivity, is highest. A more recent approach to human migration takes into account that the free movement of people is increasingly restricted by the unwillingness of other regions to integrate additional people, and that modern means of communication and infrastructure allow the establishment of trans-regional and trans-national households. As a result, migration is increasingly temporary, implying family and household strategies which include remittances to those left at home, as well as regular out-and-return migration as a lifestyle. High rates of population growth in arid regions, where the limits of ecological carrying capacity are obvious, seem to promote those strategies. Hence, negative consequences of out-migration are regarded more as a problem of the receiving than of the sending regions.

People leaving their homes to migrate over long distances because conditions are better elsewhere is not a new phenomenon in human history. New is the extent of the movements, the speed of the processes; the unprecedented population growth behind them; the barriers against in-migrants that have never been higher on a global level; the scope of mobility and communication; and the associated flows of capital that go largely in the direction opposite to that of the migration movements. Economic, social, and ecological capacities to integrate additional in-migrants are already strained in many parts of the world, however, and for this reason, conflicts resulting from migration tend to escalate within desertifying regions as well as far from them (Bächler 1995). These developments have led to an urge for political action. As the number of migrants is growing on a global scale, the policy focus is increasingly directed toward the group of international migrants.

In order to shape preventive measures against undesired consequences, it is essential for policy makers to gain deeper insights into the logic and consequences

---

cause migration always requires a region which attracts, i.e., where the conditions are considered better. Otherwise, the migrant would not move.

of migratory movements and into their interrelationship with the process of desertification. To collect, compare, and draw conclusions from experiences in this area
is of vital and growing importance, implying questions of food security, of social
peace, and of international conflicts.

Desertification is not an isolated cause of migration. People (households) have
a portfolio of resources; and if one of them degrades, they usually adapt. Faced
with desertification, and trying to maintain their livelihood, they might: (1) introduce new technologies, (2) gain off-farm income, (3) reduce their standard of living, (4) try to receive resources from outside, 5) change their social organization
and institutions, (6) migrate, or implement a combination of these elements.
Which option is chosen depends on the results of a personal cost-benefit analysis.
The exit option is only chosen if it is more profitable than staying at the same
place. Hence, not all regions affected by desertification display significant out-
migration. Desertification leads to migration only if the other alternatives that
might be pursued are less profitable. Yet, within the spectrum of possible reactions, demography has far-reaching impact on the social, political, economic, and
ecological conditions on a worldwide scale and comes more than ever into the
foreground of public attention.

In addition to the push initiated by worsening local conditions, a migratory
process, once started, usually develops a self-sustaining dynamic as, due to social
networking, transaction costs of migration typically decline with the number of
out-migrated friends and relatives. In addition, migration experiences are a form
of social capital to draw on when conditions get worse. So, due to extensive experience, transitory agricultural systems (such as nomadic pastoralism and shifting
cultivation) can, when population density increases, and living space becomes nar-
rower, pave the way for and turn into systems of large-scale out-migration, (Mannion 1995).

The above explanations should make clear that the desertification-migration
link is first of all relevant in agricultural areas and for the farming population. It
likewise should be clear that the most severe threat – and the greatest challenge to
the international community – is the degradation of agricultural lands in the poorer
countries where many depend upon farming for their livelihood[4].

Against this background, the present contribution analyzes the interactions between human migration strategies and desertification. In particular, it challenges
the hypothesis that human out-migration implies a relief for the desertifying environment. The study concentrates on the ways, directions, and consequences of migration. For this purpose, it draws on theoretical reflections and experiences made
in different low-income countries, relying essentially on case studies from Asia,
Africa, and Latin America. It considers only the "south", i.e., the poorer countries,
because desertification in the "north" hardly leads to spectacular human migration.
The cases analyzed exemplify typical patterns of response that emerge under different interactions of ecological, economic, political, and social conditions. The
analysis starts from the hypothesis that out-migration contributes to improving the

---

[4]One of the countries in the world with the highest rate of desertification is the United
States; nevertheless, there is no significant desertification-induced human migration.

situation in the region of out-migration due to reduced population stress, and that migration in that sense leads to a more balanced distribution of the population between more and less strained regions. Further, it considers the role migration plays in precipitating conflicts;and if, by way of human migrations, desertification problems are exported. The exposition is organized along a gradient which leads the reader from the analysis of the economically worst situation for those involved to the economically best situation. In conclusion, some policy recommendations are presented as a basis for discussion.

30.3 Migration at the Interfaces

When reviewing global migration patterns, four major types stand out: (1) out-migration resulting in a far-reaching dissolution of the social structures for those who moved away as well as for those staying behind (if any are left at all); (2) out-migration in a weak social fabric, where essentially through out-migration of the most productive and negligence of the source region by the migrants, the region of out-migration is drawn into a self-perpetuating downward process; (3) household strategies to bridge temporary shortages of revenues, income or capital, often born of a long tradition in risky environments; (4) highly profitable international migration which leads to significant income growth in the migrants' households and constitutes a major source of foreign exchange for many poorer countries. Transitions from one category into another, as well as co-existence of different systems within the same region, are observed.

These types are exemplified by case studies from different regions affected by desertification. In addition, an example is presented of a region which due to changing external conditions turned from one of out-migration to one of in-migration.

30.3.1 Erosion of Social Structures: Sub-Saharan African Aspects

As "life in Africa revolves around the land, and when the land is degraded the quality of people's lives deteriorates accordingly" (Suliman 1994, p. 116), land degradation is a major push factor for migration in the region. As a result, Africa has the largest number of migrants in the world with refugee status[5] and of displaced persons who have moved due to environmental reasons.

In many parts of the continent, migration has traditionally been an integrated element of life. According to Suliman (1994), migration has been one of the most important mechanisms for adjusting to ecological changes, and at the same time the most persuasive reasons to move may have been ecological. With the formation of nation states, this option for adaptation has been drastically reduced. More-

---

[5]These are persons offically recognized as refugees by the United Nations High Commissioner on Refugees (UNHCR).

over, mounting population densities and the spread of large-scale mechanized farms restrict resettling possibilities in the region.

As restrictions on across-border movements have been increasing, intranational displacement has become more likely to be the final outcome of desertification. Accompanied by an increasing collapse of traditional subsistence economies, internal migration has indeed grown dramatically. The exodus from rural to urban areas aggravates already serious urban problems. At the same time, it makes it more difficult to rehabilitate and develop rural areas as manpower is lacking and the land is neglected. Moreover, desertification entails an erosion of the rural infrastructure as the market ignores those who cannot sell and buy, rendering conditions even more difficult.

In the mid-1990s intranational migrants in Africa were estimated to be four times the number of international migrants. Nevertheless, the number of people moving across international borders has also increased. Many of them are staying in refugee camps. When considering the relationship between migration and environmental degradation, Suliman (1994) emphasizes the following Africa-specific aspects.

There is a difference between the socioeconomic situation of those who are internally displaced and those who are refugees in foreign countries. Internal displacement is predominantly rural-urban, while external refugees are usually confined to rural border regions. For security reasons, displaced people prefer to move to urban centers, where they are visible to the international community, their safety is maintained, and there are more resources available to live from. Hence, in all arid and semi-arid countries, cities are spreading and the growth rates of urban populations are significantly higher than the average of the total population. The internally displaced join the competing millions in the informal urban sector, while those with a refugee status are entitled to the protection and assistance of the UNHCR. As a result, the internally displaced suffer greater food security problems than those recognized as refugees. Moreover, many African governments hold back information about the situation of the displaced in their country, and deny international access and assistance to them.

Women and children constitute the majority of the displaced people, with a relation of about one man to every five women and/or children. A growing number of the internally displaced men are being drawn into armed conflicts; a significant share join guerrilla movements or the forces of regional warlords, or become mercenary soldiers against governments of neighboring states.

Foreign refugee communities as a rule remain indefinitely alien over two or three generations, and later generations often become refugees themselves. Groups under pressure increasingly attempt to displace other people living in richer areas. But the latter also often suffer from drought and dispossession. In that context any fight for gain is often politically misused by third parties.

As an example of the African context, Suliman (1994) describes the case of Sudan, as "broadly representative of the entire African continent" (Suliman 1994, p. 120). Since the 1970s, due to low rainfall, drought and unsustainable farming methods, the vital equilibrium in the vast arid and semi-arid areas in the north of the country has been upset, and millions of people have abandoned their homeland. As a result, Sudan might have been the country with the highest proportion of internally displaced people in the world. The movement of people and herds to areas already occupied by other ethnic groups has created tensions and hostilities. One consequence is widespread violence within the country.[6] Under the circumstances, people have disposed of their animals and moved to the expected security of the towns. Various sources estimate the number of displaced persons between 3.5 and 4 million with the vast majority being women and children[7]. In the urban regions they live in camps or slums, not well tolerated and not integrated in the host communities. "Destitution often drives them to begging, virtual slavery as domestic servants, prostitution or crime" (Suliman 1994, p. 125). As families disintegrate and group cohesion collapses, children often end up on the streets. Those on rural roads become easy prey for warlords as child soldiers.

Strategies to cope with adverse environmental conditions have existed in arid regions since ancient times, and often have proved essential for the population's survival. In many places they continue according to established patterns, although assuming different dimensions. Adaptations to desertification take place within this given framework, built on long-term adaptations to arid environments, such as pastoralism or patterns of regular seasonal and circular labor migration (see, e.g., Scoones 1995; Prothero 1998).

The case of Sudan is paralleled in a number of other Sub-Saharan African countries, e.g., Mali. Referring to the Malian Gourma, which has been hit by severe desertification, Randall has demonstrated how traditionally based strategies to temporarily escape from drought can end up in permanent displacement, loss of the traditional socioeconomic life style, and acceleration of desertification (Randall 1998). When too much physical and social capital is lost, both people who are maintaining a nomadic life style and those who have migrated to towns modify their use of natural resources in ways which lead to over-exploitation. The livelihood in the arid rural region can be neither maintained nor resumed because the resources which would provide the basis of existence are no longer available. The loss of herds and herewith common property rights are the key to this development. "Unfortunately, human exploitation is probably less controlled than animal numbers and poverty is leading both rural and peri-urban populations to depend more and more on the only 'free' resources around" (Randall 1998, p. 172).

In contrast to the described scenario, case studies from Senegal, Zimbabwe, Mauritania and Cameroon demonstrate functioning household migration systems, but also stages of transition to social dissolution.

Examples of the transition from systems with occasional migration to bridge temporary shortages to regular and permanent out-migration to secure the survival of the family household are found among the people of Senegal. At the same time, their migratory patterns display typical inter-ethnic differences which are predetermined by historically established traditional survival strategies. This is dem-

---

[6]A classical example is the war of Arab groups under pressure of drought and desertification against the non-Arab Fur and Nuba in West-Sudan.

[7]Suliman reports that of 3,527,000 displaced persons in Sudan in the mid-1990s, 447,000 were men, 1,200,000 women and 1,880,000 children (Suliman 1994, p. 115).

onstrated by Dia's detailed study of the migration strategies of the Kaskas, the Soninké, the Seres, and the Haal Pular (Dia 1992).

The Kaskas live under climatic conditions which make the demand for farm labor peak over a short period. Because of insecure rainfall, irrigation is essential for sustaining and increasing agricultural productivity. On the irrigated lands, external labor is particularly important due to the narrow calendar of cultivation. Farm/household development strategies include specific forms of temporary migration which mainly comprise the younger age groups. On average, each household has 1.5 out-migrants. Their remittances are the most important source of off-farm income among the Kaskas.[8]

This average setting is characterized by important differences between household groups. Three typical groups can be identified: (1) households where about 75% of the men are migrants who subsidize their farms by remittances that pay for inputs and external labor; (2) large production units with few migrants. In spite of large areas of land per household (74 ha) only an average of 2.6 ha is cultivated due to lack of external income to hire labor, buy inputs, and finance irrigation. Here, land productivity is low, farm households are indebted and suffer from food deficits; (3) small production units without migrants; they are the worst off. Although they employ innovative technologies, such as direct seeding, etc., they are not able to compensate for the lack of labor and inputs.

Among the Kaskas, temporary migration has lead to reduced permanent out-migration. The number of migrants within a family is positively correlated with the extent of its land, with its irrigated area, and with its scope of farm mechanization. However, these farm activities are not sustainable from their own resources, as the net return on their investment is negative. The migration-remittance strategies applied permit the families to continue a lifestyle which otherwise would not be sustainable.

Similar strategies are common in other ethnic groups of Senegal's population. Dia reports that more than 90% of the male Haal Pular between 30 and 60 years of age have migrated at least once in their life; 58% of them to towns within Senegal, 35% to neighboring Mauritania, and 6% further away. Households in the home villages on average consist of 1.4 men present, 2.2 women present and two absent persons, not including the seasonal migrants. Close ties maintained within the clan support highly organised migration patterns. So, for example, households established in Dakar take over the responsibility for young migrants arriving there.

Migration rates of 48% are observed among the Seres, whose movements have intensified significantly with the increasing desertification in the region. They have dispersed over Dakar, the Terres Neuves, and their home region in Central Senegal. Among these regions there are intense movements, supported by strong social networks.

The Soninké are specialized in long-distance migration, in particular, since in the 1960s they took part in a labor-force agreement between Senegal and France. When in 1975 the French government decided to stop immigration from Africa, 83% of the out-migrated Soninké were already in France. On that basis, illegal

---

[8] In 1988, e.g., each household on the average received 65,800 F CFA p.a., which is equivalent to a salary for 188 to 268 working days.

movements to France continued, and in addition new international paths were established, in particular to central and western Africa.

Similar strategies for maintaining the reproductive unit in the rural area by subsidizing farm activities with migrants' remittances are common in other arid regions of Africa. In the communal areas of Zimbabwe, where the subsistence needs of the smallholder families can only be met by migrants' remittances, with which to buy the necessary inputs (Hedden-Dunkhorst 1993). This is complemented by long-term migration-remittance strategies which aim at providing children with a school education that later on will put them in a position to earn a higher income, allowing higher remittances.

All of the described strategies of coping with adverse ecological situations can only be successful as long as the migrants are able to find employment that provides them with a surplus to transfer home. With increasing desertification, accompanied by mounting population pressure, these strategies are increasingly threatened to fail. Mauritania provides an example where they definitely have collapsed, but where, at the same time, international aid organizations have played a decisive role in the process (Fahem 1998). While the country's nomadic areas emptied, urbanization increased from 8 to 47% between 1965 and 1988. This development has been promoted by the availability of water and food supplied by international aid organizations to urban centres. Out-migration from rural areas is highly selective in favor of the younger and most productive males, but hardly any resources flow back. Women outnumber men in rural areas while the opposite holds true for the urban areas. One third of all households in the country are headed by women. As desertification increases and population growth continues to be high, most of Mauritania's population is threatened by hunger and thirst. For many, only international aid secures survival.

A case study done by Schrieder and Knerr in rural regions of Cameroon suggests that migration-remittance strategies support first of all those in the home region who possess productive resources (Schrieder and Knerr 2000). Not all of those in the rural regions who have migrant family members are supported in case of need. The amount of remittances received is positively correlated with the number of animals and amount of land the remittee holds, and which the remitting person might inherit, There is no negative correlation with the income of the migrant's family at home, as would be expected under an assumption of altruism. The example of Cameroon demonstrates that migration strategies might be more a way to preserve productive capital in a strained region, than a strategy to maintain the living standard of those left behind. An earlier study by Lucas in the mid-1980s in Botswana points a similar direction (Lucas 1985; Lucas and Stark 1985).

30.3.2 Out-Migration in a Weak Social Fabric: For Example the Valle Grande

While in the Sub-Saharan African context large-scale out-migration from desertifying areas usually starts from household-family integrated systems, which nevertheless may erode over time, in other places conditions exist in which social coherence is comparatively weak from the beginning, so that desertification-induced migratory processes are characterized by behaviour of "leaving the sinking ship," which implies leaving the weakest members of the society behind.

The case of out-migration in a weak social fabric where sending remittances to secure the survival of the family/household left behind is uncommon. It has been analyzed by Müller (1993) for the Valle Grande, a smallholder region in Bolivia characterized by long-term net population loss and selective out-migration. With the applied farming techniques, 10% of the province area can be cultivated by field crops. In the 1980s and 1990s, the region was hit by repeated periods of drought. Desertification has been promoted by deforestation and soil erosion, which is mainly due to cattle grazing in an unregulated pasture economy. Between 1950 and 1992, the Valle Grande lost 20% of its population. Out-migration has been promoted by the fact that the Valle Grandinos dispose over attractive non-agricultural income alternatives in a macro-economic environment that offers chances to the skilled and dynamic. As traditionally they are engaged in trade and transport, agricultural innovations to improve their farms have not been in the center of their economic strategies.

About 76% of the population of Valle Grande live in rural areas where non-agricultural sources of income are largely lacking. In that setting, instead of reducing the pressure on the land, out-migration has promoted land degradation, , as it has led to more concentrated cattle holding without reducing the number of animals, and to accelerated deforestation due to lack of labor. As the fields weed up rapidly, and herbicides are expensive, it has been more profitable for the farmers to burn down forest land to gain new fields rather than weed old ones. Many households are female-headed, and women and children, for lack of other income alternatives, make a living by selling fuel wood which they collect in the free-access forest.

In addition, out-migration has brought a permanent erosion of the region's productive human capital. Migrants tend to be the younger and more productive, while the weaker sections of the population stay behind. As the better qualified leave, more demanding jobs cannot be filled adequately, either in the private sector or in the provincial administration. Due to the erosion of both physical and human resources, the region suffers from steep economic and social decline. Declining productivity has been particularly pronounced in agriculture. For almost all crops for which Valle Grande once had an almost monopolistic standing, such as maize, land productivity has fallen far beyond the average of the department and the country. Life expectancy is significantly below and child mortality above the Bolivian average. As the economic potential of the region has dramatically declined, the remaining population finds it increasingly difficult to maintain itself. As this socioeconomic pattern has dominated for decades, social problems, such as over-aging of the population, high dependency rates, alcoholism, and high suicide rates prevail. Since unfavorable areas are increasingly emptying and isolated, critical numbers of inhabitants for maintaining the public infrastructure in many places no longer exist, all adding up to further incentive for out-migration.

30.3.3 Highly Profitable International Migration: For Example Mexico

Strategies of labor migration in the direction of high-income countries to escape an ever-more desertifying environment are pursued on a large scale at the international level, often supported by governments. Arid countries are among the world's major labor-exporting countries[9] (Knerr 1998). In these countries, remittances are so high that they have a significant, and in many cases a dominant, influence on the macro-economic development of the whole country[10].

Since the monetary gain from migration typically is so large, reactions of households involved differ significantly from those of intranational migration or migration to poor neighboring states. Therefore, this form of migration requires separate consideration. Most often, remittances exceed all other sources of household income. While a large part of the most productive labor force is absent for a longer span of time, their remittances are rarely spent on productive farm investment[11]. Against such a background, it is not unusual that international outmigration of this variety is accompanied by declining agricultural production

Mexico is a striking example of this development. Around 80% of its land area is affected by or vulnerable to desertification, mainly in the north and northwest. Each year about 2,250 km^2 are taken out of cultivation, mainly due to land degradation. At the same time, annual population growth in the poor rural areas has been found to reach 2.5 to 3% (Natural Heritage Institute 2003). Around 900,000 people each year leave the country's arid and semi-arid areas. As a consequence of internal migration, desertification becomes ever more widespread as smallholders who abandon their degraded lands move to marginal lands unsuited for agriculture, while higher quality land is occupied by large farmers (Schwartz and Notini 1994). The major reasons for land degradation are unsuitable farming practices, and as the migrants continue to pursue these practises, desertification is carried on. In addition, as access to forest land is largely free in rural poverty-stricken areas, slash-and-burn systems are used. Analyzing the situation in Yucatan, Pascual and Barbier concluded that "poverty makes it optimal to increase allocation of labour to extensive shifting cultivation" (Pascual and Barbier 2003).

Mexico is situated close to the United States, a country offering comparatively highly paid employment to Mexicans. The US census of 2000 reported 9.3 million persons living in the US who were born in Mexico (Durand 2003). In 2002, the

---

[9]Labor-exporting countries are defined as those that receive more than 50% of their foreign exchange through migrants' remittances (Knerr 1998). It is a striking fact that those arid countries which do not earn a significant amount of their foreign exchange by oil exports are labor exporters.

[10]For details see Knerr (1998).

[11]Typical investment categories are houses, furniture, and vehicles, and in some regions, e.g., Pakistan, the marriage of the migrant himself, his brothers and sisters (Batzlen 2000). This last investment category might provide the family with a social safety net helpful in adverse situations in a hostile environment. A similar pattern has been observed by Reichert in his case study on six Egyptian villages, where the major investment goals of international migrants were house construction and marriage (Reichert 1993).

amount of remittances sent by US migrants to Mexico reached almost US$ 10 billion (Bank of Mexico 2003). Mexican-US migration is characterized by the establishment of transnational communities, and high remittances to rural areas where most of the migrants come from (Moctezuma and Rodriguez 1999; Delgado Wise and Knerr 2003). US-bound migration concentrates in arid and semi-arid states, as for example Zacatecas.

Being aware of the relationship between desertification in Mexico and migration to the US, the US government supports research on sustainable farming in Mexico with a view to curbing migration (see US Commission on Immigration Reform). Desertification is as yet not the only cause for migration. Both rural-to-urban, as well as international migration have accelerated with the institution of the North American Free Trade Agreement (NAFTA) in 1994, which by reducing trade barriers between the two countries has led to a deterioration of the terms of trade for Mexico's major agricultural products, in particular maize.

The case of Zacatecas demonstrates how international migration combined with remittances may contribute to local prosperity, but at the same time to long-term de-industrialization and a decline in economic activities, especially farming, in the region of out-migration. As a result, Zacatecas possesses a high per capita social product, but one of the lowest per capita domestic products of Mexico. This might have harsh consequences if US migration is curbed and the population has to turn back to local resources for gaining a livelihood. Many of the "migration pockets" display negative population growth.

In a second step, out-migration from the Southern states of Mexico, namely Chiapas and Yucatan, promotes out-migration from degrading regions of neighboring states. Guatemalan farm laborers in southern Mexico work for lower wages than Mexicans are prepared to, implying a tendency for wage rates to adapt to declining soil productivity (Schroth 2003).

International out-migration as a strategy to cope with desertification coupled with increasing population density is also practiced in Egypt. The case of Egypt exemplifies the conditions which prevail in the Middle East/North Africa region.

Almost 100% of Egypt is arid land, characterized by widespread desertification (WRI 2000). At the same time, it has, with almost 3% p.a., one of the highest population growth rates in the world (World Bank 2002). Under these conditions, the move from rural regions to urban centers has been steadily increasing. The rate of urbanization has reached more than 45%.

Population movement that continues to take place in reaction to increasing desertification in Egypt can hardly be overestimated. Nevertheless, there is a striking lack of research in this area. Hence, a paper commissioned by the World Bank with the initial intention of providing an overview over Egypt's demographic development, its causes and consequences, starts with the diagnosis: "Data on migration are so unreliable and the prospects for migration to relieve population pressure are so uncertain that this paper concentrates on changes in mortality and fertility" (Cochrane and Massiah 1999 p. 1). Only few studies have been done on the impact of international labor migration on the rural home regions of the migrants, in spite of the fact that it is a dominant factor in the country's economic development. This applies even more to studies focusing on the impact of internal migration.[12]

In 2000, 2.7 million Egyptians were officially registered as working abroad, while the official outflow was given as 1.9 million, which implies a pronounced fluctuation (International Labour Organization 2003). The overwhelming portion of the migration has been to Saudi Arabia, Libya, and Jordan. In 1998, Egypt officially received US\$ 4.36 billion in workers' remittances, which is equal to \$75.14 per capita of Egypt's population[13]. It amounts to more than 10% of the country's GDP and about 25% of its export earnings. These figures underestimate the total amount of remittances, as much of the money earned abroad - according to estimates by Adams, almost one-third of the total remittances (Adams 1991) - is transferred through informal channels.

Egypt is at a critical stage where the gap between population growth and growth of food production becomes increasingly wider, desertification proceeds, and population movement out of rural regions accelerates. Yet, in spite of its close relationship with core political issues of the country, the consequences of migration are not a priority field of research in Egypt (Toth 1999). In her study about the consequences of migration on Egypt's labor market, el-Hawari states that although it is a widespread phenomenon with far-reaching economic and political consequences, internal migration in Egypt is not reflected in the current literature (El-Hawari 1998 p. 127)[14].

A long-term strategy to cope with ever more eroding natural resources in an arid environment is pursued in Jordan. The country suffers from droughts and widespread desertification, and for decades a large share of its labor force has been employed abroad, remitting billions of dollars each year, by far the country's largest source of foreign exchange. In 2000, an estimated 400,000 Jordanians were earning their living in foreign countries, which is one third of the country's labor force (Directorale of Inner Security 2002). Their remittances reached more than two billion US\$ (Central Bank of Jordan 2002) which is almost US\$ 400 per capita of the population.

---

[12]Studies from the 1980s (Arman 1983; Serageldin et al. 1983) and early 1990s (Adams 1991) indicate that international out-migration implies a skill drain for rural regions and improves food supply for the migrants' families.

[13]Calculated with data from the World Bank (2000). Remittances are calculated from the category "net current transfers from abroad".

[14]More attention is paid to international migration. It is important for Egypt's economic development, and due to its contribution to the country's inflow of foreign exchange and its relief for the labor market, has received much attention on the political level. At the same time, it has been investigated far more from the scientific side than from the internal migration side. One reason for this might also be that data about international migration are more easily available as they imply across-border transactions (out-migration, in-migration, bank transfers). Hence, a number of sophisticated studies about the impact of international migration exists, often concentrating on the macro-level impact of remittances (see, e.g., Farrag 1995). Investigations of the economic and demographic determinants and social implications of international migration allow some rough and indirect conclusions about the impact on the source regions (see, e.g., Adams 1991; Nasrat and Mohiey 1999).

Jordan's population as well as its government have realized that the country's major asset for earning external income is human capital, and hence have intensified private and public investment in education (Knerr and Zaqqa 2002). Jordan has specialized in the formation of a highly skilled labor force, large numbers of which eventually earn income abroad, a strategy which appears highly profitable for the country. In addition, due to its high-level education system, foreign students are attracted to Jordan's universities, making them an important earner of foreign exchange.

30.3.4 From Labor Export to Labor Import: Almería

Almería, an arid province in the south of Spain, represents an example of how a changing economic framework and technical innovations can turn a former region of extensive out-migration of its agricultural population into one of agricultural expansion and large-scale in-migration.

For more than a century, until the 1970s, Almería displayed a negative migration balance, and up to the 1960s a negative rate of population growth (Instituto de Eastadistica de Andalucia 2002). As with high population growth and on-going soil degradation, the conditions for agricultural production eroded, thousands of farmers left their land to work in other regions of Spain, in other European countries, and overseas.

When Spain joined the EU in 1986, the country's chances of export to European markets improved. From the 1980s onward, investment in Almería in modern greenhouse technologies, which made production almost independent of the local soil, increased sharply. Combined with the cheap foreign labor force from African countries, mostly from Morocco, they proved highly profitable[15]. Intense vegetable production in greenhouses (on about 35,000 ha), mainly for export, dominates[16]. It strongly depends on imported inputs, from soil and bumblebees from the Netherlands to fertilizer from Germany.

With the spread of greenhouse production, the population growth rates in the region accelerated. While over the decade 1950 to 1960 population declined at an average rate of 3.2% p.a., the growth was 15% in 1998 (Instituto de Eastadistica de Andalucia 2000)[17]. In the communities where greenhouses are concentrated it was significantly higher, and as a result the population in the province officially reached 505,448. More than 10,000 were labour force from non-EU countries holding an official work permit, 95% of them came from Africa (IEA 2002). 7,685 of the Africans were Moroccans. Including those staying without being registered, the number of Moroccans is much higher. So, e.g., in 1998, the district govern-

---

[15]The employment of foreigners in Almería's farm sector partly follows a pattern found in other former regions of out-migration: the money earned abroad is invested in a home enterprise where a cheap foreign labor force is employed.

[16]Labor costs in tomato production in Almería, for example, are ‚only one-third of those in the Netherlands (Hartkemeyer 2002).

[17]The overall population growth rate in Spain was 0.2% in 2002 (World Bank 2003).

ment of Almería estimated the de facto share of Moroccans in the region's foreign population at 64%, while the official figure was less than 40% (Hartkemeyer 2002).

Migration from Morocco to Spain was initiated by a significant and increasing income difference. The GNP per capita in Spain being more than tenfold that of Morocco,[18] Moroccans started to migrate to Spain in large numbers in the early 1990s. At the beginning of 2000, an estimated 1.2 million Moroccan citizens were living throughout the EU[19]. Since the 1970s, workers' remittances to Morocco have steadily increased, reaching more than US$ 2 billion in 2000 (World Bank 2002). Labor export is supported by the Moroccan government in hopes that this will help to cope with unemployment and with the country's balance-of-payment deficit.

Migration to foreign countries, in particular to countries in Europe, is primarily from rural areas due to pronounced rural-urban income and welfare differences (van der Erf and Hering 2002 p. 12).[20] Farmers' incomes are low compared to the rest of the society: the agricultural sector employs 40% of the labor force, but generates only 20% of the GDP. In addition, employment in agriculture is insecure and vulnerable. Drought has been a powerful factor leading to migration from rural to urban areas (Berrada 1993). Deteriorating natural conditions for agricultural production, combined with high population growth, have been accompanied by large-scale rural-to-urban migration flows, and have turned Morocco from a predominantly rural to a predominantly urban country. Over the last four decades, the share of rural population has dropped from 70 to 46% (World Bank 2000). Nevertheless, striking income differences between urban and rural areas have persisted. Step-wise migration first from rural to urban areas and then abroad is common.

The overwhelming share of Morocco's international migrants[21] are young men from rural regions. According to van Erf and Hering, the major reasons for out-migration are population pressure[22], remittances, lack of social security, and the availability of networks to channel them into host countries. A survey indicates that international migration is considered as the best way to improve one's income and that the children of migrants are expected to have a better future.

In spite of increasing formal barriers against in-migration into the EU, international migration from Morocco in the early 2000s is sustained by family reunion, family formation, and social networks (van der Erf and Hering 2002). Increasing

---

[18]Calculated with data from the World Bank.

[19]They were mainly in France (with 47% of the migrants), Spain (14%), Italy (9%), Belgium (9%), Netherlands (9%), and Germany (6%; Eurostat 2001).

[20]An estimated 27% of the rural, but only 12% of the urban population lives below the poverty line. While 98% of the urban population has access to safe water, only 14% does so in rural areas (World Bank 2000). Twenty-five percent of the young men (15 to 24 years old) and 44% of the young women are illiterate (World Bank 2000).

[21]Apart from Spain, their major host countries are France and Italy.

[22]The population of Morocco in the year 2000 was estimated at 28.4. million: It has tripled since 1950 (World Bank, var. issues), and displays a growth rate of 2% p.a., which provides a future migration potential.

remittances from international migration, which constitute a major source of the country's foreign exchange, demonstrate Moroccans' attachment to their home country and their strong family ties. As most migrants come from rural areas, re-mittances are primarily sent there. A major spending category of migrants in Mo-rocco's rural areas is the education of their children. Farming, in contrast, is not considered as a profitable investment category.[23]

Yet, although for a large part legalized, employment of Moroccan workers in southern Spain is not stable, mainly because the presence of large numbers of young men from Morocco has lead to social unrest and political problems[24]. In-creasingly, Moroccans are being replaced by workers from Sub-Saharan African countries, especially from Ghana and Senegal, as well as predominantly female farm workers from Poland who are particularly welcome due to their religious af-finity.

In addition, the region has tremendous water problems, and a controversial dis-cussion is going on about the possibility of transferring water from the north of Spain to the south.

30.4 Conclusions

The chapter has addressed the question of whether human out-migration from de-sertifying regions reduces the pressure on land resources and thus contributes to reduce the extent of desertification.

The case studies presented to exemplify different forms of migration taking place in reaction to desertification leave little hope that out-migration from regions under pressure will bring a halt to desertification processes. In contrast, through different ways, migratory processes seem to accelerate desertification more often than not.

The observed large migration movements out of desertifying regions are no single indicator of the degree of desertification in the region of origin. Availability of other forms of capital – human, social, and manmade physical – play a decisive role as well, as do the aspirations and expectations of the populations.

It is demonstrated that the process is decisively influenced by the social and po-litical conditions which prevail in the region of origin as well as in potential host regions. Hence, the direct relationship between desertification and out-migration is superimposed by complex interactions of the global human community. Paradoxi-cally, migration strategies seem to be successful for the degrading environment primarily at their two extreme ends of profitability: first, where it is distress migra-

---

[23]Yet, the author received oral information in Almería that there are Moroccan migrants thinking about transferring the observed greenhouse technology to their home country.

[24]In 2000, violent clashes between Moroccan migrants and the Spanish population in the small town of El Ejido brought a turning point in the employment strategy pursued in the region of Almería, these fights having demonstrated that social conditions cannot be ig-nored in spite of apparent economic advantages (see Knauf et al. 2000).

tion, i.e., when people completely resettle simply because they are not able to survive in their current environment; and second, where migration is so profitable that it implies an abundance for the migrants' households which makes agricultural activities dispensable.

The analysis has also shown that – due to external restriction or to personal preferences – migrants' families and households increasingly remain in the region of origin. The transfer remittances helps them to keep their home at the traditional location, giving rise to the formation of transregional and transnational households. As a result, farming is often carried on in marginal regions in spite of low productivity, subsidized by remittances. Hence, if there is no transition to environmentally more sustainable technologies, desertification may become even more severe.

In cases where migration proceeds for resettlement, relief for the region of origin seems to occur only if entire families leave. Regularly, however, out-migration is selective, and the less productive part of the population remains behind. If they do not receive sufficient remittances to secure their livelihood, they often resort to resource-mining for lack of capital, income, and a strong labor force. Often, those staying behind are left to the mercy of international aid organizations.

Moreover, out-migration from regions under desertification, even if it constitutes a relief to the region of origin, may not really imply an improvement on a global level, but rather simply entail the export of desertification, as other marginal areas come under pressure, and hitherto unaffected environments and countries are impacted.

However, there are also groups who welcome the migrants as a source of cheap labor force. Where migrants have the exit option, and can reach an economically well-off region, they might find a productive place in that economy, to the material benefit of both sides. If their remittances are high enough, they might even put an end to the migrant's family's destructive farm activities and thus contribute to reduce the desertification process.

Yet, in the regions of in-migration conflicts may arise between the migrants and those with whom they compete for jobs, housing, and other things. As, however, for many countries, regions and households who send migrants remittances have become a stimulating determinant of economic development, there might be a minor incentive to curb migration in the regions of origin.

Individual economic interests often stand against what would be the best solution for the environment. The following policy recommendations result from the above reflections (and, in addition of course, from the general desire to fight desertification).

Lack of capital formation or of access to capital is a key issue in the process of environmentally destructive migration. Promotion of credit programs such as, for example, micro-finance programmes; promotion and support of institutions of social security geared to reduce incentives to survive on free-access natural resources; promotion of education and extension services to make labor more productive, and promotion of technologies which do not foster soil degradation -- all these should be high on the agenda.

In this context, it also appears worthwhile to support household strategies of temporary and/or seasonal migration in order to reduce pressure on the land and soil and promote the inflow of capital.

A central task is the promotion of non-farm employment to reduce the pressure on land, and at the same time reduce incentives to migration and thus reduce the shift of problems to other regions. In contrast to often-expressed expectations, the inflow of remittances does not by itself seem to initiate activities which establish alternative income sources in desertifying regions. Where highly profitable international migration occurs, it has been observed that the long-term bases of economic development have eroded, pushing people back to the land when migration possibilities decline. In such cases, external assistance may incite positive developments.

Special projects for women in the regions of out-migration should be high on the agenda of politicians and aid organizations, as women are often left behind without the support of either a male labor force or significant remittances. Even where remittances are sent, women might be expected to have problems in taking over the role as head of the household, due to personal restrictions and/or external discrimination.

On a global level, (it applies for many regions) that general economic development seems to be just as important as direct measures taken to fight desertification.

Meanwhile, in many parts of the world, especially on the African continent, the process of desertification is apparently irreversible. Nevertheless, the public reaction to this disaster is comparatively moderate. For example, the Havana Conference on Desertification in 2003 did not receive much attendance even within the development community. A major reason may be that technical innovations that allow substitution of soil by capital input and/or cheap labor promise to secure food production and livelihood. Seemingly successful examples, which flourish under certain favorable socioeconomic conditions, however, may give rise to false hopes, since they might not be economically, socially, and/or ecologically sustainable in the long run.

30.5 References

Adams RH (1991) The effects of international remittances on poverty, inequality and development in rural Egypt. Research report no 86. International Food Policy Research Institute (IFPRI), Washington, DC

Arman IMJ (1983) Labour migration and its impact on Egyptian labour market. The Institute of National Planning of the Arab Republic of Egypt. Cairo. Memo no 1385

Bächler G (1995) Umweltflüchtlinge. Das Konflikt-potential von morgen? Agenda global, Münster

Batzlen C (2000) Migration and economic development. Peter Lang, Frankfurt a.M.

Berrada A (1993) Migration, mutation and economic development in Morocco. Paper presented at the conference on migration and international co-operation: challenges for the OECD countries. 29-31 March, Madrid

Canadian International Development Agency International Development Information Centre (1994) Desertification: serious threat or global myth? Development Express 94-07

Central Bank of Jordan (2002) International financial statistical yearbook. Amman

Directorate of Inner Security (2002) as published in the yearbooks of the Department of Statistics (DoS), Amman, Jordan

Cochrane, M (1999) Recent changes in population growth, their causes and consequences. World Bank, Washington, DC, http://www.worldbank.org

Delagado Wise R, Knerr B (eds) (2004) El Impacto de la Migración Internacional en Zacatecas (in press)

Dia I (1992) Les migrations comme strategie des unités de production rurale. Une étude de cas du Senegal poverty and development. Sustainable development in semi-arid Sub-Saharan Africa. Ministry of Foreign Affairs, The Hague, pp 57-64

Durand J (2004) Políticas emigratorias en un contexto de asimetría de poder: El caso mexicano 1884-2003. In: Delagado Wise R, Knerr B (eds) El Impacto de la Migración Internacional en Zacatecas (in press)

El-Hawari H (1998) Die Auswirkung der Migration auf den Arbeitsmarkt in Ägypten. PhD Dissertation. Wirtschaftswissenschaftliche Fakultät der Universität Leipzig

Europäisches Bürgerforum (Hrsg.) (2000) Anatomie eines Pogroms: z.B. El Ejido (Bericht einer Delegation europäischer Bürgerinnen und Bürger über die rassistischen Ausschreitungen vom Februar 2000 in Andalusien). Verlag EBF/CEDRI, Basel.

Fahem AK (1998) Population and desertification in Mauritania. In: Clarke J, Noin D (eds) Population and environment in arid regions. Man and the biosphere series 19. UNESCO, New York

Farrag M (1995) Black markets in foreign exchange and international migration: the case of Egypt. Int Migration 33(2):177-207

Forum Umwelt und Entwicklung, Arbeitsgruppe Desertifikation (1999) Desertifikation. Entwicklung und Umwelt in den Trockenregionen der Erde und die globalen Zusammenhänge. Frankfurt

Hartkemeyer T (2002) Die Rolle der ausländischen Arbeitsmigranten im Agrarsektor der EU-Staaten am Beispiel der Provinz Almería (Spanien). Master Thesis. University of Kassel, Department of Development Economics and Agricultural Policy

Hedden-Dunkhorst B (1993) The contribution of sorghum and millet versus maize to food security in semi-arid Zimbabwe. PhD Thesis, Stuttgart Hohenheim

Instituto de Estadistica de Andalucia (2000) Anuario de Estadisticas de Andalucia www.iae.junta-de-andalusia.es

Instituto de Estadistica de Andalucia (2002) Un siglo de demografia en Andalucia. La poblacion desde 1900 www.iae.junta-de-andalusia.es

International Labour Organization (ILO) International Labour Migration Data Base (ILM). www.cobise.ch (accessed on 11/11/2003)

Knerr B (1998) Impacts of labour migration on the sustainability of agricultural development in arid regions. In: Clarke J, Noin D (eds) Population and environment in arid regions. Man and the biosphere series 19. UNESCO. New York

Knerr B, Zaqqa N (2002) Economic Costs and Benefits of Human Capital Migration from Jordan. Paper presented at the first World Congress of Middle East Studies (WOCMES), Mainz

Lucas REB (1985) Migration amongst the Botswana. Econ J 95:358-382

Lucas REB, Stark O (1987) Motivations to remit: evidence from Botswana J Polit Econ 93(5):901-917

Mannion AM (1995) Agriculture and environmental change. Wiley, Chichester

Moctezuma Longoria JM (1999) Redes Sociales, Comunidades Filiales, Familias Y Clubes De Migrantes. El circuito migrante Sain Alto, Zacatécas-Oakland, Ca. PhD Thesis, Universidad Autonoma de Zacatécas

Moctezuma L, Rodriguez R (1999) Impacto de la Migración y las Remesas en el Crecimiento Económico Regional. Mexico City

Müller PM (1993) Tragfähigkeitsveränderung durch Bevölkerungsverlust. Beispiel Valle Grande/Bolivien. Geogr Rundsch 3:173-179

Nasrat SM, Mohiey El-Din (1999) The relationship between out-migration and some structural and functional changes of the rural families. Bull Fac Agric Cairo Univ 50:575-590

National Heritage Institute (2003) Breaking the cycle: desertification an migration on the U.S.-Mexican Border. www.n-h-i.org/Projects/PeopleGlobalResources/People/Dryland/US_Mexico.html (28.01.2003)

Pascual U, Barbier EB (2003) Modelling land degradation in low-input agriculture: the "population pressure hypothesis" revised. Paper presented at the 25th conference of the International Association of Agricultural Economists (IAAE), Durban, South Africa, 2003

Prothero (1998) Circular mobility in part of the West African dry zone. In: Clarke J, Noin D (eds) Population and environment in arid regions. Man and the biosphere series 19. UNESCO, Paris New York and Casterton Hall, pp 61-76

Randall S (1998) The consequences of drought for populations in the Malian Gourna. In: Clarke J, Noin D (eds) Population and environment in arid regions.Man and the biosphere series 19. UNESCO, Paris, New York and Casterton Hall, pp 211-246

Reichert C (1993) Labour migration and rural development in Egypt. A study on return migration in six village. Sociol Ruralis XXXIII(1):42-60

Schrieder G, Knerr B (2000) Labour migration as a social security mechanism for smallholder households in Sub-Saharan Africa: the case of Cameroon. Oxford Development Studies 28, pp 223-236. Oxford

Schroth G (2003) Die Bracheoptimierung als Beitrag zur Produktionssicherung auf marginalen Standorten der Tropen. Presentation at the University of Kassel, Fac 11, Witzenhausen

Schwartz ML, Notini J (1994) Desertification and migration: Mexico and the United States. Natural Heritage Institute (ed). US Commission in Immigration Reform, San Francisco

Scoones J (1995) Living with uncertainty. London

Serageldin I et al (1983) Manpower and International Migration in the Middle East and North Africa. World Bank. Washington

Suliman M (1994) The predicament of displaced people inside Sudan. Agenda. Münster

Toth J (1999) Rural labor movements in Egypt and their impact on the state, 1961-1992. University Press of Florida, Gainesville

United Nations Development Programme (UNDP) (2003) http://www.undp/hdro/population.htm

United Nations Conference to Combat Desertification (2000) Informationsblätter. Bonn

United Nations Food and Agricultural Organization (FAO) Sustainable development of drylands and combating desertification http://www.fao.org/docrep/v0265e/VO265E01.htm. 20.01.2003

Van der Erf R, Heering L (2002) Moroccan migration dynamics: prospects and future trends. Netherlands Interdisciplinary Demographic Institute (NIDI), The Hague (prepared for IOM)

World   Bank   (2000)   Country   data;   Egypt,   Arab   Rep http://www.worldbank.org/data/countrydata/aag.htm

World Bank (2002) World development report. Washington, DC

World Bank (2003) World development report. Washington, DC

World Resources Institute (WRI) World resources reports. Washington, DC (var issues)

World Resources Institute (WRI) (2000) Egypt at a glance. http://www.wri.org

31 Airborne Migration of Obligate Nomads Demonstrates Gene Flow Across Eurasia

Eckhard Limpert[1*], Klaus Ammann[2], Pavel Bartoš[3], Werner K. Graber[4], Gerhard Kost[5], Jacques G. Fuchs[1,4]

[1]Phytopathology, Institute of Plant Sciences, Swiss Federal Institute of Technology Zurich (ETH), 8092 Zürich, Switzerland
[2]Botanical Garden, Bern, Switzerland
[3]Research Institute of Crop Production, 161 06 Praha Ruzyně, Czech Republic
[4]Paul Scherrer Institute, 5232 Villigen, Switzerland
[5]Philipps-University Marburg, Special Botany and Mycology, Karl-von-Frisch Str., 35032 Marburg, Germany
*Present address: Lecturer Aerobiology, University Zurich, Scheuchzerstr. 210, 8057 Zurich, Switzerland, E-mail eckhard.limpert@bluewin.ch

31.1 Abstract

Understanding migration is important for the adequate use of biological resources. A new level of understanding is demonstrated with cereal pathogens recognized to be obligate nomads of the atmosphere. From basic reasoning, a hypothesis is put forward: virulence complexity, i.e. the number of virulences per pathogen genotype, is expected to increase in the direction of predominant winds. The hypothesis was confirmed by all of a variety of data from own investigations and from the literature and by modelling. For instance, virulence complexity of cereal mildews and rusts increased from western to eastern Europe and as far as Siberia by approximately one to two virulences per 1,000 km. The impact of our findings for general population genetics and gene flow across Europe and Asia and for further geographical areas is supposed to be considerable and worth elucidating further. *Obligate nomad* is a novel term that appears to be advantageous in several respects of population biology and life. Therefore, the consideration of obligate nomadism is extended to a spectrum of cases including plants, fungi and animals, as well as to neonomads and invasive plant species as a consequence of our present traffic and civilization. Most often obligate nomads are forced to migrate as substrates are ephemeral and ecological niches exist for a short period of time only. Chances and risks for health and the use of biological resources are discussed.

31.2 Migration and Obligate Nomads

Migration and biological resources have always been of major importance for human society. Repeatedly, migrating pests and pathogens on plants gave rise to heavy losses and even catastrophes such as the great famine caused by potato blight and its fungus *Phytophthora infestans* (Day 1977; Dixon 1998). Migrating insects can either cause damage by themselves or serve as vectors of other organisms. Examples include Phylloxera caused by the aphid *Viteus vitifolii* attacking vineyards, and the potato beetle *Leptinotarsa decemlineata*, both reaching Europe from North America in the 19th century. More recently, Dutch elm disease caused by a fungus-beetle association has led to devastating elm decline in many European countries, and the pinewood nematode, again associated with a beetle, was disastrous in Japan. A well-known disease of wild and domestic rabbits caused by a virus is myxomatose, whereby, after introduction of a killer strain to Europe, mutants that were less severe pointed out to be of selective advantage and spread. Moreover, a number of invasive plant species as well as fungi have conquered many regions and even whole continents as neophytes since the Middle Ages, and many of the species pose threats to native flora and fauna. Their migration is well documented. For instance, spores of the fungus *Anthurus müllerianus* were introduced with military blankets from Australia during World War I; starting from the Netherlands this fungus spread throughout western Europe as far as Russia.

Quite a number of species are even forced to migrate and change their habitat in order to survive. Their reasons for moving are well known and quite different. Seasonal climate is one of them, often leading to the temporary loss of living conditions. Moreover, a number of heterotrophic organisms are adapted to unique and exceptional substrates exclusively used by them. Often, the substrates are ephemeral and only available for a short period of time. Therefore, these organisms have transitory habitats only and need to find new substrates frequently. In a recent study of plant pathogens, those species that need migration to survive were called *obligate nomads* (Limpert et al. 1999).

A closer look at these pathogens and their cereal hosts appears to be adequate, as well as an analysis of the forces of, e.g. *dispersal* and *virulence* which are important for host-pathogen systems, because such systems may provide a model for so many systems of interacting organisms in general. Finally, we will widen the scope towards a general look at obligate nomads, including neonomads.

Cereals including their wild relatives are major biological resources around the world. The rusts and mildews living on them are important wind-dispersed pathogens (Oerke et al. 1994; Limpert and Bartoš 1997; Hau and de Vallavieille-Pope 1998). The barley mildew pathogen caused by *Blumeria* (*Erysiphe*) *graminis* f.sp. *hordei*, as well as the leaf rust pathogen on wheat, *Puccinia triticina*, have been extensively studied (Bartoš et al. 1996; Park and Felsenstein 1998; Limpert et al. 1999; Mesterházy et al. 2000). In the present context it is worth stressing that they are obligate parasites which need living hosts to survive.

Living on ephemeral tissue only, these cereal pathogens frequently change their habitat. Starting from a number of points of interest concerning dispersal (Andrivon and Limpert 1992; Hau and de Vallavieille-Pope 1998; Aylor and Irvin 1999; Brown and Hovmøller 2002), it is our aim here to focus further on the understanding of population biology and wind dispersal in particular, and its possible relevance for strategies to use the cereal hosts in time and space (Limpert et al. 1999, 2000; Limpert, unpubl. data).

So far, the term wind dispersal is generally used without further specification. However, wind dispersal is worth differentiating as it can affect populations of organisms quite differently. A key to the understanding is, as briefly mentioned above, the longevity of the organisms, depending on their conditions of life. Despite wind dispersal of spores, life of e.g. the basidiomycete *Armillaria* can remain quite sedentary due to perennial mycelium in soil and decaying wood. Ergot on rye, in contrast, is an obligate nomad once a year, when spores have to reach the seed vessel of a rye ear. Wheat leaf rust and barley mildew in northwestern Europe are obligate nomads as well and, moreover, highly mobile as they change their place of living almost 10 and even more than 20 times per year, respectively, (Limpert and Bartoš 1997).

Changing the place of living also means pathogen attack which often creates considerable losses in crop quality and quantity. Losses can be avoided through the use of chemicals or, ecologically safer, through the use of genes conferring resistance in the host. As a genetic consequence, there will be selection of chemical resistance or pathogen virulence that re-allows the pathogen to grow. Thus, as a general consequence, dispersal means selection (Flor 1946; De Wit 1992; Limpert et al. 1996, 1999, 2002). As another established consequence, virulence complexity (as well as fungicide resistance complexity), i.e. the number of virulences per pathogen genotype, increases with time (Munk et al. 1991). Therefore, virulence complexity is worth considering in strategies to improve the use of host resistance and to reduce the selection of super-races able to attack most or all cultivars. Much less established are the consequences for the evolution of virulence complexity in space, with migration favored in one direction.

Cereal pathogens are well suited for a corresponding investigation. Evidently, for migrating humans, language knowledge can be beneficial and of selective advantage. Similarly, from knowledge of meteorology, aerobiology and population genetics, the hypothesis was derived that virulence of nomadic pathogens increases with predominant winds (Limpert et al. 2002). In corresponding studies, virulence complexity was analysed for different pathogens using established procedures of pathotype (or race) analysis. Spores of the barley mildew pathogen were sampled and analysed according to Limpert et al. (1984, 1990), Müller (1993) and Müller et al. (1996). Representative samples were obtained on the motorway by driving across regions of interest for cereal production in Europe, with a jet spore sampler mounted on the car roof. Petri dishes with fresh leaf segments were exchanged after approximately 100 km, taking into account the use of barley varieties and regional agroecological conditions. The sample size was close to 2000 isolates per year in regions north of Geneva. The virulence spectrum of individual genotypes that developed after incubation was analysed on leaf segments on agar under controlled conditions on differential sets consisting of near-isogenic lines (NILs) (Kolster et al. 1986). For leaf rust, data from the literature were re-evaluated. Methods were based on either those for barley mildew, with 20 NILs plus 5 differential cultivars (Park and Felsenstein 1998), or field sampling and 15 NILs (Mesterházy et al. (2000).

Host-pathogen systems on cereals are unique: nowhere else appears to be more biomass produced that is genetically more uniform. Thus, the reverse of the well-known problems of genetic uniformity and vulnerability of the crop is that the corresponding systems can be excellent models to analyse population patterns and gene flow and their effects on biological resources, in order to develop adequate strategies to solve the problems.

Thus, in addition to (1) specifying the effects of wind dispersal, we (2) modelled the dispersal effects on the evolution of virulence complexity in space, (3) analysed comprehensive pathogen samples in time and space throughout Europe, and (4) re-evaluated data from the literature; moreover, we (5) extended the scope to further obligate nomads with reference to fungi, lichens, mosses and plants, and discuss our findings concerning chances and risks for plant, animal and human health and for the use of biological resources.

31.3 Eurasian and Further Paths of Population Genetics

What are the consequences of obligate nomadism in corresponding models and reality? The air over Europe up to 600 m above ground moves on average some 10 km/h eastward, a movement that is thought to continue throughout Asia (Limpert et al. 2000; Graber and Limpert, unpubl.). Therefore, we modelled the effects of predominant winds and obligate nomadism for a selected trait over three successive regions. With one selection cycle of the pathogen per region, virulence complexity increased consistently and considerably by 0.25 virulences per step and selection cycle, from 1.5 virulences originally to 2.25 in the end (Fig. 1).

There are a number of data sets consistently demonstrating the increase of virulence complexity in the direction of major dispersal. Early observations of the barley mildew pathogen across Europe go back to the beginning of the 1980s (Limpert et al. 1999; Limpert, unpubl.). During an initial phase, western and central Europe were thought to be one unit of population genetics in this respect (Limpert et al. 1996). However, we soon learned that it is worth separating Great Britain from continental Europe. Both areas independently exhibit the phenomenon described.

Much to our surprise, the phenomenon was even better expressed both in degree and consistency (R^2) on the island of Great Britain, which can be due to the canalizing effect of the island shape, with predominant dispersal from Scotland to England. This demonstrates the presence of a *Scottish-English path of population genetics*. In comparison, the path considered on the continent is much wider, with conditions for host and pathogen being more diverse (Limpert et al. 2000, and unpubl.). A far-reaching question and hypothesis was put forward for continental Europe and Asia recently: "Does virulence accumulate along the way from Paris to Beijing?" (Limpert et al. 2000, 2002). Impressive data sets for leaf rust are available for this area in the literature (Chester 1946; Park and Felsenstein 1998; Mesterházy et al. 2000).

Historical data (Chester 1946) are, after re-evaluation (Limpert et al. 2000, 2002), in agreement with the hypothesis of a *Eurasian path of population genetics* that is supported by recent data. Virulence complexity of rust isolates in Europe was studied in 1998 with data from five countries (Mesterházy et al. 2000). Based on predominant pathotypes, virulence complexity again increased from France (close to 2) to Poland (approximately 7; Limpert et al. 2002). In addition to the west-east gradient, a north-south gradient of increasing virulence complexity becomes obvious here. This can be due to climatic conditions favoring epidemics and, as a consequence, an increased number of host resistance genes used in the south. Similar and other effects cannot be excluded in the west-east direction either, in some contrast to conditions for mildew attack that are best in the maritime climate of the northwest. Further results from Europe of interest for the concept come from crown rust of oats caused by *Puccinia coronata* and are currently being re-evaluated (Šebesta et al. 1999; Šebesta, unpubl.).

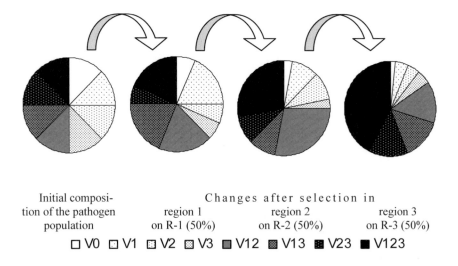

Initial composi-   Changes after selection in
tion of the pathogen   region 1   region 2   region 3
population   on R-1 (50%)   on R-2 (50%)   on R-3 (50%)

□ V0 □ V1 ▨ V2 ▨ V3 ▦ V12 ▨ V13 ▨ V23 ■ V123

Fig. 1. Effects of wind dispersal and selection on virulence complexity, vc, of a nomadic pathogen: a model. Indicated is, from *left* to *right*, the initial composition of the pathogen population consisting of eight pathotypes with a frequency of 12.5% each, and changes after selection in three successive regions; 50% of the host is sensitive, the remainder resistant, with R-gene 1, 2, and 3 (*R-1, R-2, R-3*) present in regions 1, 2, and 3, respectively

Most impressive are comprehensive data on leaf rust of wheat from the area north of Geneva, where virulence complexity increases quite steadily from less than 4 in western France to approximately 7 in Poland, and more than 14 in Siberia close to Omsk (Park and Felsenstein 1998, re-evaluated by Limpert et al. 2002). Although most data are from the beginning of the Eurasian wind path, the effects on population genetics and gene flow for the entire area appear evident. Of course, there is a considerable way left from Omsk to Beijing. However, due to major conditions remaining the same, i.e. predominating winds from the west and presence of the host, the conclusion of virulence complexity increasing along the way from Paris to Beijing appears to be justified and important enough to merit

further consideration. The Scottish-English path of gene flow and population presumably appears to be but one additional case worth knowing and considering.

How do spores migrate? As spores start from plants, concentrations of viable
units are highest close to ground level. Many short distances covered per life cycle
of these polycyclic diseases are thought to sum up to long-distance migration in
the end (Limpert et al. 1999). Presumably the conditions are similar to those of
gene flow via pollen concluded from population genetic studies of Pinaceae
(Ziegenhagen et al., this Vol.). Single-step, long-distance transport of viable
spores may occur occasionally in or below clouds protecting against UV radiation,
as is the case with dust clouds and interhemispheric transport of viable microorganisms (Prospero, this Vol.).

31.4 Obligate Nomads: Extending the Scope

The phylogenetic old kingdoms of organisms, fungi and mosses were very early
land-inhabiting organisms living as obligate nomads. The reasons for moving are
quite different for today's vegetation. One is that the substrates of the heterotrophic organisms are rare and used up so quickly that new ones have to be found
frequently. Plants and fungi have to reach new places with spores or seed. In areas
with seasonal climate, the growing conditions can restrict the organisms to a short
life cycle only. However, also in tropical rain forests with continuous climate,
host-depending organisms have great difficulty in finding the next host. This is
due to the high degree of diversity causing large spaces between individuals of the
same species and substrates in these habitats (Kost 2004).

Some of these very ephemeral substrates are only available for a short period.
As with the cereal rusts and mildews, epiphytes and even more so epiphylls lose
their habitat if the life cycle of the plant or the leaf is terminated. More than half
the tropical plants depend on such habitats. Liverworts, lichens, fungi and algae
living as epiphylls are adapted to this nomadic way of life in tropical rain forests.

Dung of various chemistry and ecology spread by animals is another prominent
example of ephemeral substrate. Dung is very interesting for a number of so-
called coprophilous organisms due to its high content of nutrients and especially
nitrogen (Ellis and Ellis 1988). Dung beetles and many other species from different kingdoms are well known. The main problem of important fungi and mosses,
however, is how to reach the new substrates. This is achieved through an intricate
dispersal biology in harmony with the erratic substrate occurrence (Yu and Wilson
2001). One trick to cope with the problem is to enter the digestion system of the
dung producer. To this end, the spores of the dung fungus have to get from the
place of production to the area of browsing or grass consumption by the animal.
This can be done by active discharge and wind-dispersed spores of the fruiting
bodies of these asco- and basidiomycetes. Indeed, the ejection of the spore balls of
Pilobolus is so effective that even some nematodes use it as a means to reach the
new host (Eysker 1991).

A more efficient method of transport from one "oasis" to the next is to use "camels", like human nomads do. In the case of fungi and mosses, the transfer will be made by insects attracted by conspicuous fruiting bodies. Gluing spores adhere to the chitin surface or water droplets filled with spores are sucked in by flies. Dung mosses like genera of the Splachnaceae have spore capsules with enlarged apophysis which can be remarkably colored. This attraction can serve as a landing place for insects. Thus, interestingly enough many of the coprophilous mosses are entomophilous, enhancing the probability of spore dispersal in appropriate dung habitats by coprophilous insects (Marino 1991).

Cystofilobasidium capitatum is an extremely specialized ~~coprophilous~~-fungus (Oberwinkler et al. 1983). This very rare mycoparasite only lives inside apothecia of dung-inhabiting Ascomycota. Thus, the nomadic way of life of this species is threatened also, as it combines the low availability of the coprobic substrate with the short life span of an ascoma of a dung fungus.

Fruiting bodies of fungi are, again, isolated substrates for many organisms. Insects, imagines or larvae of flies and beetles develop inside them. Specialized mycoparasites use the carpophores as nutrient sources (Helfer 1991) and a high diversity of species is closely adapted to such systems. Mycoparasitism of the ephemeral fruiting bodies is described from many asco- and basidiomycetes. *Spinellus fusiger* (zygomycetes) produces sporangiophores on the surface of the tiny cap of *Mycena* sp. *Tremella* species select as species-specific substrate lichens, *ascomycetes, dacrymycetales, Stereum* species or other wood-inhabiting *aphyllophorales.*

The fungal fruiting bodies can be infected by nomadic species of all classes of fungi such as:

- Basidiomycota (*Asterophora, Atractiella, Chritiansenia, Collybia p. pte.,* Cryptomycocolax, *Cystofilobasidium, Nyctalis, Platygloea, Tremella, Volvariella p.pte.*)
- Ascomycota (Hypocreales, *Cordyceps* p.pt., Orbiliaceae p. pt., *Mycogone rosea*)
- Zygomycota (*Spinellus, Syzygites*)
- Deuteromycota (many species).
- Myxomycota (*Bahamia spp.*)

Fungal parasites of insects survive similarly. After digestion of the larvae or imago the substrates are depleted. In most cases, the fungal life turns into a spore production phase with an additional spore discharge. This behaviour is widespread and most organisms act in this way, if the substrate is used. Only through nomadism can a new basis of nutrition be found. It is very obvious that the sexual stage is produced exactly at that phase of life, before the beginning of the nomadic phase, thus creating additional diversity of advantage for survival in an unknown future.

Many fungal species depend on specific and rare substrates to be digested by them (Alexopoulos et al. 1996, Kirk et al. 2001). So they are forced to find new bases for surviving. This can be observed many times, for example in the wood decomposer which only inhabits selected parts of specific tree species, which have a definite age and which grow under definite climatic conditions.

An unusual growing place of mosses and fungi is charcoal and old fireplaces (Ellis and Ellis 1988). In natural habitats these carbonicolous organisms regularly appear after forest fires. As one of the first plants, *Funaria hydrometrica* regularly covers this heat-sterilized and dead habitat with small gametophytes. At the same time some carbonicolous fungi (*Anthracobia spp., Geopyxis carbonaria, Rhizina undulata, Plicaria spp., Myxomphalina maura*) appear in these places. To spread they utilize the sterile areas with reduced competition from other organisms. These species are able to live there, however, only during the short initial phase of these disturbed areas. As soon as other plants and organisms start recolonizing these places the carbonicolous mosses and fungi disappear.

Of course, deserts are extreme habitats and life needs to be highly adapted. During the dry periods fungi and mosses usually cannot grow. One survival strategy of mosses is to dry up and stop life activities until enough humidity refreshes them. Another strategy of mosses and fungi is to complete the life cycle during the few weeks that the desert is wet enough (Düring 1979). Conversely, both groups of organisms produce drought-resistant spores with thick walls. They will be dispersed by sandstorms over hundreds of kilometers and germinate as soon as conditions for life are better again. Desert Basidiomycota with ephemeral fruiting bodies are members of the genera *Agaricus, Agrocybe, Coprinus, Psathyrella, Psilocybe* and *Stropharia* (Kost, unpubl. data).

A well-known group of nomadic plants in steppe vegetation is adapted for distribution as whole plants by wind (*Eryngium, Falcaria vulgaris, Seseli* div. spec., *Phlomis pungens, Centaurea diffusa, Salsola kali, Gypsophila paniculata*; Walter and Breckle 2002). Due to their distribution dynamics some of these plants also form well-defined geoelements (Turano-Saharian elements). Striking cases of nomadic plants are also adapted to unusual ecological events like fire, land slides, etc. The case of *Funaria hygrometrica* is well known: the basis for the nomadic occurrence is fire, also on rather small spots caused by human activity. Further members of this well-defined community, all based on high demand for nitrogen, are *Ceratodon purpureus, Marchantia polymorpha* and *Lunularia cruciata*.

The semi-saprophyte *Buxbaumia aphylla* occurs nomadically and extremely sporadically in disturbed forest habitats, usually caused by big fires (Jones 1970; Wiklund 2002, 2003). *Geranium bohemicum* is an extremely rare nomadic species in forests, occurring only after big fires (Granstrom 1993). One explanation for its surprising presence is its extraordinary capability to produce seeds that can survive in soils for more than 100 years (Milberg 1994).

A number of neonomads can be regarded as obligate nomads of today's traffic and civilization. Such invasive plant species have conquered many regions and indeed whole continents, and many of the species pose threats to native flora and fauna. Agronomists and ecologists use the terms *weed* and *weediness* in different ways, and this is often a source of misunderstanding, especially in discussions concerning the release of transgenic plants. For agronomists, the problem of weediness is solved if the aggressive weed can be removed from the agrosystem by means of adapted measures. They are not interested in its behavior outside of agrosystems, i.e. in (semi-)natural habitats.

From the ecologist's perspective, invasions of weeds into (semi-)natural plant communities are potentially risky. Highly competitive invaders are able to disturb the pattern of species. As a result, rare indigenous species of tropical islands, of regions and even continents with an old flora, only weakly influenced by migrations of the last few millennia, are often weak competitors, and can be eliminated or markedly reduced in coverage. In worst case scenarios, the invader succeeds in occupying the entire surface of certain areas as a near monoculture, this at least for several years or decades until new and 'natural' enemies fight back. Maybe seemingly less dramatic, but for the eyes of trained field ecologists still worrying, are dynamic situations where ecological niches are invaded, precisely wiping out rare indigenous species. This can happen (and unfortunately has happened in the past) with ecological dynamics which is triggered off by agriculture and horticulture (Ammann 1997, 2000). A typical example of an agricultural weed is the globally distributed *Galeopsis tetrahit*, an allotetraploid, derived from two rather tame species, *G. pubescens* and *G. speciosa* (Müntzing 1930).

A balanced discussion on the fate of exotic (and thus potentially invasive) species in agricultural habitats is given by Meiners et al. (2002). Agricultural practices may also influence the future impacts of exotics. Frequent plowing associated with row-crop agriculture prevents the accumulation of exotic perennial cover. When these sites are abandoned, both natives and exotics start invading at the same time, resulting in a plant community that does not show significant effects of exotic species. In contrast, agricultural practices with repeated biomass removals such as hay fields, meadows and grazing result in perennial exotic communities that resist invasion by other species (Mack and D'Antonio 1998). These effects may persist for 15 years or more. One-time plowing was not sufficient to reduce the impacts of these species on community development at these sites. It is important to note that it is not the invasion of an exotic plant, per se, that reduces species richness but the dominance of a patch by exotic species which may result in reduced species richness. Species richness of natives and exotics is positively associated, showing no effects of exotic invasion on native species richness. However, when exotic plants make up a large proportion of the total cover of a plot, we observed reductions in community richness (Meiners et al. 2001). Therefore, managers should focus control efforts on species that have the potential to dominate local plant communities.

Some lichen species as well as many moss species are extreme niche specialists and can be regarded as nomadic in their dispersal and reproduction biology. Lichens, for instance, can reproduce very effectively by soredia units. Soredia are vegetative reproductive structures that contain hyphae and algal cells. They break off the parent thallus, are dispersed by various agents, and produce new lichens in a new specific niche location. They are tiny structures that form in the medulla and break through the upper cortex to erupt on its surface, which can be easily blown by wind or washed off by water. Another lichen, *Lecanora esculenta*, known in the Bible as Manna, is basically a crustose and saxicolous species, which can break off in dry conditions and eventually assemble in desert depressions as edible Manna used to feed animals (Donkin 1981).

31.5 Conclusions and Outlook

A number of species belong to the group of obligate nomads. For the host-pathogen systems focused on in this chapter, a hypothesis was put forward according to which virulence, the character under selection, and virulence complexity in the pathogen population should increase in the direction of predominant dispersal (Limpert et al. 2002). A prerequisite is diversity for the trait considered in both the host (where single genes are sufficient) and the pathogen (which needs complex genes). Being endemic over vast geographic areas covering several continents contributes to the diversity present even at the local scale. The hypothesis based on fundamental logic is well supported by modelling and comprehensive results both from our own investigations and from the literature after re-evaluation.

Dispersal is selection. As a consequence of predominant dispersal in one direction there appears to be a *Eurasian path of population genetics and gene flow* along which genes and genotypes move, as well as, presumably amongst others, a *Scottish-English path*. Along these paths, pathogen composition and evolution should be predictable to a certain degree. There are important conclusions for disease resistance: where to find the most valuable resistance sources (downwind) and how to make the best use of host resistance (shifting the culture of hosts with a certain resistance gene upwind). Moreover, the hypothesis throws new light on the question of costs of unnecessary virulence. These and further conclusions have been described elsewhere in more detail (Limpert and Bartoš 2002).

Obviously, in addition to wind, further factors such as virulence combinations, climate and temperature or alternate hosts can affect virulence complexity in various ways and would thus benefit from further analyses. Warmer climate, for instance, is more favorable for wheat leaf rust in eastern Europe, whereas the opposite is true for cereal mildews. Moreover, the co-evolutionary processes with epidemic dynamics in natural populations would also be worth studying for the effect on virulence complexity, both from the regional to the pandemic scale.

One very general finding appears worth stressing: presumably, there is no other biomass produced around the globe as big and genetically uniform as that of cereals. As a consequence, this should be true also for the biomass of their pathogens. Host pathogen systems involving cereals are thus supposed to have a pilot function for the recognition of interacting systems in general.

Evidently, dispersal affects populations of species in different ways, depending on the degree of nomadism. What about the other obligate nomads, neonomads, invasive plant species and interacting systems mentioned, such as, e.g., the highly complex coprophilous systems of mosses, fungi and insects? Do they have genes of selective advantage for migration too, and genes affecting gene flow? Can the same or similar rules of interaction be recognized? Presumably, for most of the questions we merely do not yet know the answer.

What are typical traits of the extremes, highly mobile obligate nomads on the one hand, and highly sedentary counterparts on the other? What are typical traits of successful neonomads, and what would be the consequences of knowing them, amongst others, for the use of biological resources, and for measures of quarantine and practice?

Of course, the hypothesis should be of importance not only for plant health, but also for human and animal health, and for further populations affected by wind such as species of insects, amongst others those causing malaria (Limpert and Bartoš 2002). Also spiders are known to travel by wind, and even some crustaceae living in temporary waters like certain *Ostracodae*: their eggs appear to be as easily wind-dispersed as are mildew spores (Vieberg pers. com.).

Fascinating challenges are ahead of us, leading to novel insight into geo-biology and geo-medicine. Evidently, there is the need to quantify nomadism. Perhaps the most striking questions of today, however, relate to viruses (Shortridge and Stuart-Harris 1982; Drosten et al. 2003, Guan et al. 2003; WHO 2004; Becker, this Vol.). Does the Eurasian path of population genetics and gene flow exist for them too?

Foot and mouth disease is known to spread through the air (Donaldsen et al. 2001). How is the viability of coronavirus spreading SARS by air and of the causal agent of human and avian influenza, with a number of further hosts? Even if outdoor conditions were unfavorable in general, it might have been sufficient for the evolution of influenza over the last thousands of years if conditions were favorable for aerial spread once per month, or per year, or even per 5 or 10 years.

China and East Asia are thought to be the origin of influenza pandemics around the world (Shortridge and Stuart-Harris 1982), and they were the origin of both SARS (Drosten et al. 2003, Guan et al. 2003) and the present avian influenza A(H5N1) (WHO 2004). This is in line with our concept predicting that the most dangerous pathogens gather and evolve at the end of this *Eurasian path of population genetics*.

Acknowledgements: We are grateful to COST at Berne and Brussels for financial support.

31.6 References

Alexopoulos CJ, Mims CW, Blackwell M (1996) Introductory mycology. Wiley, New York, 868 pp

Ammann K (1997) Botanists to blame? Plant Talk, 4

Ammann K, Jacot Y, Rufener Al Mazyad P (2000) Weediness in the light of new transgenic crops and their potential hybrids. J Pl Dis Prot, Special Issue 19-29
http://www.botanischergarten.ch/debate/weeds1.pdf

Andrivon D, Limpert E (1992) Origin and proportions of components of populations of *Erysiphe graminis* f.sp. *hordei*. J Phytopath 135:6-19

Aylor DE, Irvin ME (eds) (1999) Aerial dispersal of pests and pathogens. Spec Issue. Agric For Meteorol 97:233-252

Bartoš P, Stuchlíkova E, Hanušová R (1996) Adaptation of wheat rusts to the wheat cultivars in former Czechoslovakia. Euphytica 92:95-103

Brown JKM, Hovmøller MS (2002) Aerial dispersal of pathogens on the global and continental scales and its impact on plant disease. Science 297:537-541

Chester KS (1946) The nature and prevention of the cereal rusts as exemplified in the leaf
rust of wheat. Chronica Botanica, Waltham, Massachusetts

Day PR (ed) (1977) The genetic basis of epidemics in agriculture. New York Academy of
Sciences, New York, 400 pp

De Wit PJGM (1992) Molecular characterization of gene-for-gene systems in plant-fungus
interactions and the application of avirulence genes in control of plant pathogens. Ann
Rev Phytopathol 30:391-418

Dixon B (1998) The fungus that made John F Kennedy president, and other stories from the
world of micro-organisms (in German). Spektrum Akad, Berlin

Donaldsen AI, Alexandersen S, Sørensen JH, Mikkelsen T (2001) Relative risks of the un-
controllable (airborne) spread of FMD by different species. Vet Rec 148:602-604

Donkin RA (1981) The Manna lichen - *Lecanora-Esculenta*. Anthropos 76:562-576

Drosten C, Gunther S, Preiser W, van der Werf S, Brodt HR, Becker S, Rabenau H, Panning M, Kolesnikova L, Fouchier RA, Berger A, Burguiere AM, Cinatl J, Eickmann
M, Escriou N, Grywna K, Kramme S, Manuguerra JC, Muller S, Rickerts V, Sturmer
M, Vieth S, Klenk HD, Osterhaus AD, Schmitz H, Doerr HW (2003) Identification of
a novel coronavirus in patients with severe acute respiratory syndrome. N Engl J Med
348:1967-1976

Düring HJ (1979) Life strategies of Bryophytes: a preliminary review. Lindbergia 5:2-18

Ellis MB, Ellis PJ (1988) Microfungi on miscellaneous substrates. Croom Helm, London,
244 pp

Eysker M (1991) Direct measurement of dispersal of *Dictyocaulus viviparous* in sporangia
of *Pilobolus* species. Res Vet Sci 50:29-32

Flor HH (1946) Genetics of pathogenicity in *Melampsora lini*. J Agric Sci 73:335-337

Granstrom A (1993) Spatial and temporal variation in lightning ignitions in Sweden. J Veg
Sci 4:737-744

Guan Y, Zheng BJ, He YQ, Liu XL, Zhuang ZX, Cheung CL, Luo SW, Li PH, Zhang LJ,
Guan YJ, Butt KM, Wong KL, Chan KW, Lim W, Shortridge KF, Yuen KY, Peiris
JSM, Poon LLM (2003) Isolation and characterization of viruses related to the SARS
coronavirus from animals in Southern China. Science 10:1126/Science 10:87139

Hau B, de Vallavieille-Pope C (1998) Wind-dispersed diseases. In: Jones DG (ed) The epi-
demiology of plant diseases. Kluwer, Dordrecht, pp 323-347

Helfer W (1991) Pilze auf Fruchtkörpern - Untersuchungen zur Ökologie, Systematik, und
Chemie. Libri Bot 1:1-157

Jones EW (1970) Ecology of *Buxbaumia aphylla* Hedw. Trans Br Bryol Soc 6:139

Kirk PM, Cannon PF, David JC, Stalpers JA (2001) Ainsworth and Biby's dictionary of the
fungi. CAB International, Wallingford, 655 pp

Kølster P, Munk L, Stølen O, Løhde J (1986) Near-isogenic barley lines with genes for resistance to powdery mildew, Crop Sci 26:903-907.

Kost G (2004) Ecology and morphology of some tropical fungi (Contributions to tropical
fungi II). In: Agerer, Blanz, Piepenbrinck (eds) Frontiers in Basidiomycote Mycology". IHW, Eching; in press

Limpert E, Bartoš P (1997) Analysis of pathogen virulence as decision support for breeding
and cultivar choice. In: Hartleb H, Heitefuss H, Hoppe H-H (eds) Resistance of crop
plants against fungi. Fischer, Jena, pp 401-424

Limpert E, Bartoš P (2002) Wind-dispersed nomadic pathogens - conclusions for disease
resistance. J Genet Plant Breed 38:150-152

Limpert E, Schwarzbach E, Fischbeck G (1984) Influence of weather and climate on epidemics of barley mildew, *Erysiphe graminis* f.sp. *hordei*. In: Lieth H, Fantechi R, Schnitzler H (eds.) Interaction between climate and biosphere. Progr Biomet **3**, Swets & Zeitlinger B.V. Lisse, 146-157.

Limpert E, Andrivon D, Fischbeck G (1990) Virulence patterns in populations of *Erysiphe graminis* f.sp. *hordei* in Europe in 1986. Pl Pathol 39:402-415.

Limpert E, Finckh MR, Wolfe MS (eds) (1996) Integrated control of cereal mildews and rusts: towards co-ordination of research across Europe. Proc Worksh COST 817, European Commission, Luxembourg, 276 pp

Limpert E, Godet F, Müller K (1999) Dispersal of cereal mildews across Europe. Agric For Met 97:293-308

Limpert E, Bartoš P, Graber WK, Müller K, Fuchs JG (2000) Increase of virulence complexity of nomadic airborne pathogens from west to east across Europe. Acta Phytopath Entomol Hung 35:261-272

Limpert E, Bartoš P, Buchenauer H, Graber WK, Müller K, Šebesta J, Fuchs JG (2002): Airborne nomadic pathogens: do virulence genes accumulate along the way from Paris to Beijing? Plant Prot Sci 38:60-64

Mack MC, D'Antonio C.M. (1998) Impacts of biological invasions on disturbance regimes. Trends in Ecology & Evolution 13:195-198

Marino PC (1991) Dispersal and coexistence of mosses (Splachnaceae) in patchy habitats. J Ecol 79:1047-1060

Meiners SJ, Pickett STA, Cadenasso ML (2001) Effects of plant invasions on the species richness of abandoned agricultural land. Ecography 24:633-644. http://www.botanischergarten.ch/Weeds/Meiners-Ecography-2001-Effects.pdf

Meiners SJ, Pickett STA, Cadenasso ML (2002) Exotic plant invasions over 40 years of old field successions: community patterns and associations. Ecography 25:215-223. http://www.botanischergarten.ch/Weeds/Invasive-40years-Exp-Ecography.pdf

Mesterházy Á, Andersen O, Bartoš P, Casulli F, Csõsz M, Goyeau H, Ittu M, Jones E, Manisterski J, Manninger K, Pasquini M, Rubiales D, Schachermayr G, Strzembicka A, Szunics L, Todorova M, Unger O, Vanco B, Vida G, Walther U (2000) European virulence survey for leaf rust in wheat. Agronomie 20:793-804

Milberg P (1994) Germination of up to 129-year-old, dry-stored seeds of *Geranium-bohemicum* (Geraniaceae). Nord J Bot 14:27-29

Munk L, Jensen HP, Jørgensen JH (1991) Virulence and disease severity of barley powdery mildew in Denmark. In: Jørgensen JH (ed) Integrated control of cereal mildews: virulence patterns and their change. In: Proc 2nd Eur Worksh on Integrated Control of Cereal Mildews, Risø National Laboratory, pp 55-65

Müntzing A (1930) Outlines to a genetic monograph of the genus *Galeopsis*. Hereditas 13:185-341

Oberwinkler F, Bandoni RJ, Blanz P, Kisimova-Horowitz L (1983) *Cystofilobasidium*: a new genus in the Filobasidiaceae. Syst Appl Microbiol 4:114-122

Oerke EC, Dehne HW, Schönbeck F, Weber A (1994) Estimated losses in major food and cash crops. Elsevier, Amsterdam, 830 pp

Park RF, Felsenstein FG (1998) Physiological specialization and pathotype distribution of *Puccinia recondita* in western Europe, 1995. Plant Pathol 47:157-164

Šebesta J, Zwatz B, Roderick HW, Harder DE, Stojanović S, Corazza L (1999) Biological (genetic) control of fungal diseases of oat in Europe. Pflanzenschutz-Ber 58:152

Shortridge KF, Stuart-Harris CH (1982) An influenza epicentre? Lancet 2:812-813

Walter H, Breckle S (2002) Walter's vegetation of the earth: the ecological systems of the geo-biosphere, 4th edn. Springer, Berlin Heidelberg New York

WHO   (2004)   Report   on   avian   influenza   A(H5N1).
    http://www.who.int/csr/don/2004_02_13/en/
Wiklund K (2002) Substratum preference, spore output and temporal variation in sporo-
    phyte production of the epixylic moss *Buxbaumia viridis*. J Bryol 24:87-195
Wiklund K (2003) Phosphorus concentration and pH in decaying wood affect establishment
    of the red-listed moss *Buxbaumia viridis*. Can J Bot 81:41-549
Yu DW, Wilson HB (2001) The competition-colonization trade-off is dead; long live the
    competition-colonization trade-off. Am Nat 158:9-63

32 Discussion Sessions

Dietrich Werner

Philipps-University, Faculty of Biology, Cell Biology and Applied Botany, Karl-von-Frisch-Str., 35032 Marburg, Germany, E-mail werner@staff.uni-marburg.de

Several topics and questions from the discussions of the conference have been integrated in the chapters by the authors. The following summarizing questions were raised and the auditorium agreed that no simple answers are available:
- How much are the major agricultural resources (cereals, legumes, livestock, fishes, trees) threatened by new or migrating parasites?
- Methodologies to monitor migration and dispersal: what techniques can be applied to several ecosystems and resources types, e.g. GIS (geographical information systems) or "BIS" (biological information systems, DNA marker techniques)?
- What can science and technology do to reduce the pressure of human rural-urban and intercontinental migration?
- How much nature reserve and conservation will remain when in the year 2023 more people are starving than today?
- What shall we do if global warming is not "man-made" but nevertheless occurs?
- What limits food production by plant breeding, plant protection and modern techniques such as risk-free genetically modified plants to win the race against continuing population growth in many countries?
- Are there other resources such as water supply and sustainable soil quality that will lead to a reduction of food production instead of a further increase?
- What is the role of the OECD in protecting the intellectual property rights of the breeders or small companies?
- What are the limitations for the fast transfer of knowledge from the OECD member countries to developing countries in sustainable use of biological resources?

The scheme of the final discussion by H. van der Borg (the Netherlands) was as follows:
1. Definitions and terminology on migration and mobility
2. Related aspects:
   - cultural and social
   - production and fishing
   - commodities
   - nutrients
3. Interactive dialogue needed for mobility

- intra-biology
  - ˜ intrinsic elements in bio-systems
  - ˜ pre-condition for dynamics
  - ˜ thread for consolidated systems
  - ˜ reflections about theory-forming
- natural science - social science
- society-science
  - ˜ social
  - ˜ economic
  - ˜ sustainable
  - ˜ political

The various types of impacts and interdependencies of biological, physical, social, financial and human resources are summarized in Fig. 1.

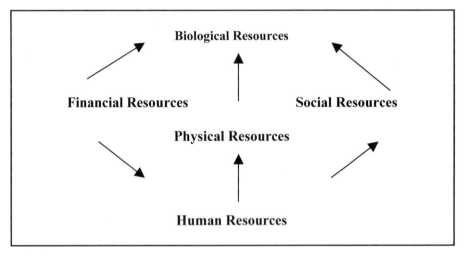

Fig. 1.

Fons Werry (The Netherlands) emphasized that a better understanding of migration is necessary and interactive elements belong to the positive aspects of migration and that we should reduce bad presumptions concerning migration. Jim Schepers (USA) pointed out the fact that the huge amount of transport of grains throughout the world is also a migration of nutrients to the waters and waste dumps inside the countries, but also on an intercontinental scale. Hugh Loxdale (UK) added that farming practices are still on a worldwide scale far away from sustainability. Elisabeth Johann (Austria) continued this idea by reminding us that in the European history of forestry, 200 years ago there was more than 10 million ha of waste land with erosion of soils and reduced fertility, so we still have a huge historical burden to overcome this mismanagement of soils over many centuries, especially in marginal soils covered with trees not suitable for these regions and areas (for example in Slovenia). There is a general agreement that soils are probably the most threatened resource, although this is not so obvious as water pollution or air pollution. S. Uddin from Bangladesh reported that in his country 120 million people are living on 260,000 km², resulting in a population density of 460 people/km², threatened by huge floods and surely beyond the sustainable system. J.J. Nsoh from Cameroon summarized that the conference was very important for many African countries, which have great difficulties in becoming more united, as other continents have managed. The impact of wars on biological resources is very obvious and only three countries are at peace (Cameroon, Nigeria and Tchad). A much better co-operation of scientists from Europe, North America and Asia with scientists from Africa is necessary for production, conservation, control and management of biological resources, particularly the water resources. A problem affecting OECD member countries as well as almost all other countries in the world, as raised by Klaus Riede, is how much we are threatened by new invaders and pathogens (see Chapters by Eckard Limpert, Stephan Becker, Hugh Loxdale and Joseph Prospero). In the CMS Bulletin from November 2003, Klaus Riede briefly reported on the introduction of the GROMS information system (www.groms.de).

Index

Printing and Binding: Strauss GmbH, Mörlenbach